Fernando Barrera-Mora
Linear Algebra

Also of Interest

Abstract Algebra
An Introduction with Applications
Derek J. S. Robinson, 2022
ISBN 978-3-11-068610-4, e-ISBN (PDF) 978-3-11-069116-0

Abstract Algebra
Applications to Galois Theory, Algebraic Geometry, Representation Theory and Cryptography
Celine Carstensen-Opitz, Benjamin Fine, Anja Moldenhauer,
Gerhard Rosenberger, 2019
ISBN 978-3-11-060393-4, e-ISBN (PDF) 978-3-11-060399-6

The Structure of Compact Groups
A Primer for the Student – A Handbook for the Expert
Karl H. Hofmann, Sidney A. Morris, 2020
ISBN 978-3-11-069595-3, e-ISBN (PDF) 978-3-11-069599-1

Commutative Algebra
Aron Simis, 2020
ISBN 978-3-11-061697-2, e-ISBN (PDF) 978-3-11-061698-9

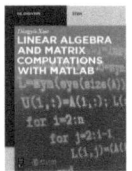

Linear Algebra and Matrix Computations with MATLAB®
Dingyü Xue, 2020
ISBN 978-3-11-066363-1, e-ISBN (PDF) 978-3-11-066699-1

Fernando Barrera-Mora

Linear Algebra

A Minimal Polynomial Approach to Eigen Theory

DE GRUYTER

Mathematics Subject Classification 2010
97H60, 15A04, 15A06, 15A15, 15A18, 15A21, 15A23, 15A63, 15A69

Author
Prof. Fernando Barrera-Mora
Universidad Autónoma del Estado de Hidalgo
Carretera Pachuca-Tulancingo, Km 4.5
42184 Mineral de la Reforma
Hidalgo
Mexico
dsrobins@illinois.edu

ISBN 978-3-11-113589-2
e-ISBN (PDF) 978-3-11-113591-5
e-ISBN (EPUB) 978-3-11-113614-1

Library of Congress Control Number: 2023931376

Bibliographic information published by the Deutsche Nationalbibliothek
The Deutsche Nationalbibliothek lists this publication in the Deutsche Nationalbibliografie;
detailed bibliographic data are available on the Internet at http://dnb.dnb.de.

© 2023 Walter de Gruyter GmbH, Berlin/Boston
Cover image: Fernando Barrera-Mora
Typesetting: VTeX UAB, Lithuania
Printing and binding: CPI books GmbH, Leck

www.degruyter.com

This work is dedicated to the memory of my father, Antonio Barrera Pelcastre (1903–2005), from whom I learned that work makes men respectful and that a deep friendship between them is based on *esprit de corps* work.

Foreword

One of the most popular topics in mathematics, where many textbooks have been written, is linear algebra. This is not a coincidence at all. Linear algebra appears naturally in practically all disciplines, both within mathematics and other sciences, including areas of social sciences and humanities, having significant presence in engineering and especially in physics.

From our first studies, let us say at the level of high school, linear algebra is utilized, even if such a name is not used, for solving linear equations.

Due to the above, when a new textbook related to linear algebra appears in the market, one wonders what this new textbook can contribute with, that has not been presented and studied ad nauseam in the innumerable texts that already exist.

Fernando Barrera Mora's book is dedicated to the linear algebra of a first course of a college major in mathematics. The first thing that I would like to point out is that the book privileges starting with concrete problems that are presented to us in our daily life as in economy, in productive companies, etc. These concrete problems are the starting point to elaborate and combine the ingredients to help students visualize and relate these components, when they are raised to more general situations.

Also, these specific problems that are studied serve to establish both the methods and the necessary theory either to solve them, to study them, or to place them in a more general context.

A valuable point of view to highlight in this work is the approach to what we could call the "eigentheory," that is, the theory that deals with "eigenvalues" (proper values) and "eigenvectors" (proper vectors). This is the heart of the book.

The most common approach to address the solution and study of eigenvalues and eigenvectors is the study of the characteristic matrix, that is, one needs to find the values for which this matrix is singular, which brings us immediately to the calculation of its determinant and, therefore, to the characteristic polynomial.

Although the analysis of the determinants is indispensable in the study of linear algebra, presenting a comprehensive study of its basic properties is a laborious or unclear problem, depending on the approach we select.

In this work, a different approach is taken. The main focus of interest is on the minimal polynomial rather than in the characteristic one. The emphasis is placed on properties inherent to the matrix that gives rise to the problem under study. More precisely, the operator associated to the matrix is studied with respect to a basis appropriately selected. In addition to the natural advantage of having an intrinsic study of the operator, the presentation given here makes it unnecessary to invoke determinants. In any case, the determinants are introduced as areas and volumes of parallelepipeds in two and three dimensions, respectively.

There are several other novelties that differentiate this text from others. For instance, in this work linear algebra is made to interact with analytic geometry; subspaces are introduced and worked with, without even having formally defined what a vector

https://doi.org/10.1515/9783111135915-201

space is; there are several novel or less known proofs, such as the existence of the adjoint operator or that any linearly independent system has cardinality less than or equal to the cardinality of a set of generators; the necessary theoretical base is constructed from a two-dimensional space, it is passed to a three-dimensional space, and finally to any finite-dimensional space; an algorithm is presented for the computation of the minimum polynomial of a matrix.

Other aspects worth mentioning are the way in which the product of matrices is motivated, which is derived from a concrete example about production, and that there are included several exercises, either original or uncommon in other textbooks.

The book may very well be divided into two parts. The first consisting of the first five chapters that have a more basic nature than the other four. The first part introduces the student to the study of linear algebra and the second part is a higher level study.

A final point that needs to be emphasized is, as mentioned at the beginning, that linear algebra is of great importance in all science and engineering curricula and even in other disciplines. This importance is found in the mind of the author throughout this book, which can be perceived by the conception of linear algebra that is presented throughout the treatise.

Gabriel D. Villa Salvador
Departamento de Control Automático
CINVESTAV del I.P.N.
Ciudad de México
May, 2023

Introduction

Some people think that God is a geometer, a hard to prove statement. A more down to earth one, which goes along with Mother Nature, is: Mathematics is the greatest human mind creation that brings with, beautiful aesthetic feelings, as well as a better understanding of Nature.

Linear algebra, together with calculus, is the cornerstone in the mathematics curriculum of science and engineering. This might explain why there are many written books to approach those topics, which are included in a bachelor degree curriculum program. Taking this into account, it is natural to ask: Is it necessary to write another book on linear algebra? A general answer can go along the following lines. All branches of mathematics are in a continuous development, exploration, and review of new results and methods. This might justify why new linear algebra books or textbooks are needed all the time.

When writing this book, we kept in mind a couple of ideas. Firstly, to present our view about linear algebra and its learning; secondly, to provide a reference that might help students in their learning processes of this important branch of mathematics.

Usually, the discussion of a topic starts by clarifying which objects are going to be dealt with. Our case is not the exception, however, it would be very difficult to give a precise description of what linear algebra is all about.

To this respect, we want to point out that linear algebra could be thought as the part of mathematics that studies the equations

$$AX = B \quad \text{and} \quad AX = \lambda X, \tag{1}$$

where A, B, and X are matrices, and λ is a scalar.

Taking this into account, we could say that this book is developed from that point of view, that is, linear algebra is the study of equations in (1), adding three ingredients which we consider important in the mathematical activity: foundations, methods, and applications.

The starting point of the book was the writing of lecture notes that arose from teaching linear algebra courses during several years, particularly at the Autonomous University of Hidalgo State (UAEH), hence the discussion approach is influenced by the content of the curriculum in the bachelor's degree program in Applied Mathematics, offered in UAEH. However, this book can be used in any engineering or science program, where one or two linear algebra courses are required.

Concerning the structure and approach of the topics contained in the book, we think it is appropriate to point out some ideas.

The first five chapters are thought as an introductory linear algebra course. In Chapter 1, we start by presenting hypothetical production models, which are discussed via a system of linear equations. The main idea behind this is to illustrate how linear equations could be used to solve real world problems. Continuing with this idea, in Chapter 2 we use again a hypothetical production model to motivate the definition of a matrix

https://doi.org/10.1515/9783111135915-202

product. This approach to introduce the matrix product has its origin in the observation that for students it seems to be a little bit "artificial," since the sum of matrices is defined by adding the corresponding entries. In Chapters 3, 4, and 5, basic geometric and algebraic properties of a finite-dimensional euclidean vector space are presented. The main topics in these chapters are the concepts and results related to linearly independent sets, dimension, and basis of a vector space, as well as general properties of a linear transformation.

The discussion level in Chapters 6 through 9 is a little bit more advanced. In Chapter 6, starting from a geometric approach, we introduce the determinant function as a means to represent area or volume of parallelograms or parallelepipeds, respectively, determined by vectors. Basic results of the theory of determinants are discussed. Also in this chapter we present a nice proof, credited to Whitford and Klamkin (1953), of the well-known Cramer's rule.

Chapter 7 is the inspiration of the book. We want to mention a few ideas that originated the way the theory of eigenvalues and eigenvectors is discussed. In most linear algebra books, the approach to the theory of eigenvalues and eigenvectors follows the route of the characteristic polynomial and, when needed, the minimal polynomial is invoked. While using the minimal polynomial approach to discussed the theory of eigenvalues and eigenvectors, those and other related problems are resolved once and for ever.

On the other hand, reviewing other branches of mathematics such as algebraic extensions of fields, number fields, the theory of ordinary differential equations, among others, we find that the minimal polynomial of a matrix plays a crucial role in understanding several problems. For instance, in the theory of number fields, the very definition of an algebraic element is given in terms of a minimal polynomial that it satisfies. This polynomial provides such an important information concerning algebraic elements as it is shown in [10, p. 156]. These ideas, concerning the minimal polynomial and other algebraic properties, led us to consider approaching the presentation of the theory of eigenvalues and eigenvectors via the minimal polynomial of a matrix or operator.

Additionally, there are a couple more reasons to justify our approach. The first has to do with the theory itself. That is, using the characteristic polynomial, sooner or later one needs to use properties of the minimal polynomial to characterize when an operator is diagonalizable or triangularizable. Also, one needs to know basic results and concepts from the theory of determinants to discuss eigenvalues. On the other hand, using the minimal polynomial, we need only basic concepts and results from vector spaces, linear transformations, and matrices, which usually have been discussed already. The second approach has to do with the conceptual learning of linear algebra. Taking the minimal polynomial approach, the discussion of basic results in the theory of eigenvalues and eigenvectors is closely intertwined with fundamental concepts such as linear dependence, bases, and dimension of a vector space. Thus, this approach offers students a valuable opportunity to systematically review the cited concepts. We argue that this approach could help students build a robust understanding of the main concepts and results of linear algebra. A couple more points that are gained from the approach that we follow are: first, two algorithms are developed to compute the minimal polynomial of a matrix; second, we can prove the famous Cayley–Hamilton theorem by proving that the minimal polynomial divides the characteristic polynomial.

It should be mentioned that a similar approach to eigenvalues, via the minimal polynomial, appears in [4]. However, in the approach followed there, the author strongly uses properties of a complex vector space, which, according to our view, brings up some difficulties, especially for students that are not familiar with basic concepts and results from the complex numbers.

In the writing of this book, we consulted some basic references; among the most important ones, we mention [3, 8, 11, 16, 12, 20, 26], and [23]. The later is recommended to the reader who wants to explore geometry and linear algebra.

I would like to point out that there is a Spanish version of this book, published in the year 2007. However, I want to mention that the version we are presenting now has several important improvements. First of all, Chapters 3 and 6 of the Spanish version have been split into two, with the aim of making the contents in those chapters clearer to the reader. Concerning Chapter 6, it has been completely rewritten, besides splitting it into two chapters, we have included an algorithm to compute the minimal polynomial of a matrix. With this algorithm at hand, we are able to discuss several examples, in order to offer students the opportunity to practice concepts related to eigenvalues and eigenvectors. An important topic not included in the Spanish edition is the singular value decomposition of a matrix, which is included now in Chapter 9. There are more changes included in this new edition. For instance, more problems have been added to each chapter. This idea has the aim to offer students opportunities to acquire wider insights when discussing specific topics. For example, Problem 5 in Chapter 1 has the purpose of giving a geometric meaning to the elementary operations applied in the process of solving a system of linear equations.

Finally, I would like to express my high recognition to all the people who, directly or otherwise, made important contributions to the success of this work. Special thanks go to Rubén Martínez-Avendaño, who worked at Autonomous University of Hidalgo State from 2003 to 2018, with whom I had very interesting and productive discussions concerning some of the results in Chapter 7; to Fidel Barrera-Cruz, who wrote the code of the algorithms presented in Chapters 7 and 8. Fidel also helped me at any time to overcome technical difficulties when using the wonderful platform, *CoCalc*, developed and supported by Professor William Stein, to whom I am also grateful.

Last but not least, I want to express my gratitude to the Editorial team at De Gruyter, who did a splendid job during the publishing process of this book.

Pachuca Hgo., México, May 2023   Fernando Barrera-Mora

Contents

List of Tables

https://doi.org/10.1515/9783111135915-203

List of Figures

https://doi.org/10.1515/9783111135915-204

1 Systems of linear equations

In a wide range of mathematical problems, applied or not, the approach to their solution is based on the study of a system of linear equations. For instance, in Economics, Wassily W. Leontief,[1] using a system of linear equations, proposed an economic model to describe the economic activity of the USA. Roughly speaking, Leontief's model consists in dividing the USA economical system in 500 consortia. Starting from this, a system of linear equations is formulated to describe the way in which a consortium distributes its production among the others, see Example 1.1.2.

From the mathematical point of view, some problems are formulated as a system of linear equations, while many more can be formulated using one or more functions, which, under good assumptions, are represented linearly, bringing linear algebra into action. These aspects show the importance of starting a systematic and deep study of the solutions of a system of linear equations. In order to achieve this, a conceptual discussion of such ideas is in order. We will need to discuss concepts such as: vector space, basis, dimension, linear transformations, eigenvalues, eigenvectors, determinants, among others.

We will start by presenting some examples that will illustrate the use of systems of linear equations when approaching a problem.

1.1 Examples

In this section we present some examples that illustrate the use of a system of linear equations to approach hypothetical problematic situations. This idea will lead us to discuss terms and results that can help us understand how linear equations are used to model some mathematical problems.

Example 1.1.1. Assume that there are two companies, E_1 and E_2, and that each company manufactures products, B_1 and B_2. Furthermore, assume that for each monetary unit invested, the production is as described in Table 1.1.

Table 1.1: Production relations.

	E_1	E_2
B_1	0.8	0.6
B_2	0.4	0.7

---

[1] Wassily W. Leontief, *Input–Output Economics*, Scientific American, October 1951, pp. 15–21. Leontief won the Nobel prize in Economic Sciences in 1973 in recognition for his contributions to the economic theory by using a system of linear equations to formulate an important economic model to describe the production of the USA economic system based on 500 companies.

https://doi.org/10.1515/9783111135915-001

The second row in the table is read as follows: for each monetary unit, company E_1 produces 0.8 monetary units of product B_1 and company E_2 produces 0.6 monetary units of the same product. From this we have that for each monetary unit, both companies produce 0.8 + 0.6 monetary units of product B_1. In the same way, the third row reads as follows: for each monetary unit, company E_1 produces 0.4 monetary units of product B_1 and company E_2 produces 0.7 monetary units of the same product.

As for product B_1, we have that for each monetary unit, both companies produce 0.4 + 0.7 monetary units of product B_2.

For instance, if the investment in companies E_1 and E_2 is 20 and 18.5 million dollars, respectively, then the value of the production is described by:

$$0.8(20) + 0.6(18.5) = 27.10 \quad \text{million dollars of } B_1,$$
$$0.4(20) + 0.7(18.5) = 20.95 \quad \text{million dollars of } B_2.$$

What is the total profit? We should mention that the profit is the difference between the value of products and the production cost. From this assumption, one can calculate the profit, since the total production cost is what was invested in the two companies, that is, 20 + 18.5 = 38.5 million dollars. While the value of the products is 27.10 + 20.95 = 40.05 million dollars, thus the profit is 48.05 − 38.5 = 9.55 million dollars. It seems like a good business.

Generalizing, if the investment in companies E_1 and E_2 is represented by x and y, respectively, and the total values of products B_1 and B_2 are represented by b_1 and b_2, respectively, then we have:

$$\begin{aligned} 0.8x + 0.6y &= b_1, \\ 0.4x + 0.7y &= b_2. \end{aligned} \tag{1.1}$$

Rewriting decimal numbers as quotients of integers and solving for y in each of the equations, we obtain an equivalent system (explain how system (1.2) was obtained):

$$\begin{aligned} y &= -\frac{4}{3}x + \frac{5b_1}{3}, \\ y &= -\frac{4}{7}x + \frac{10b_2}{7}, \end{aligned} \tag{1.2}$$

which is represented graphically in Figure 1.1.

An interesting question could be: How can the investment be distributed in each company in such a way that the production value of products B_1 and B_2 attains the specified values b_1 and b_2, respectively? Given that the investment is represented by nonnegative quantities, the question can be posed as follows: For which values of b_1 and b_2 the system of equations (1.1) has a solution with nonnegative values of x and y?

Figure 1.1 represents system (1.2) for the case $b_1 = b_2 = 1$. We can observe that the system has a solution in the first quadrant, that is, the values of x and y are positive.

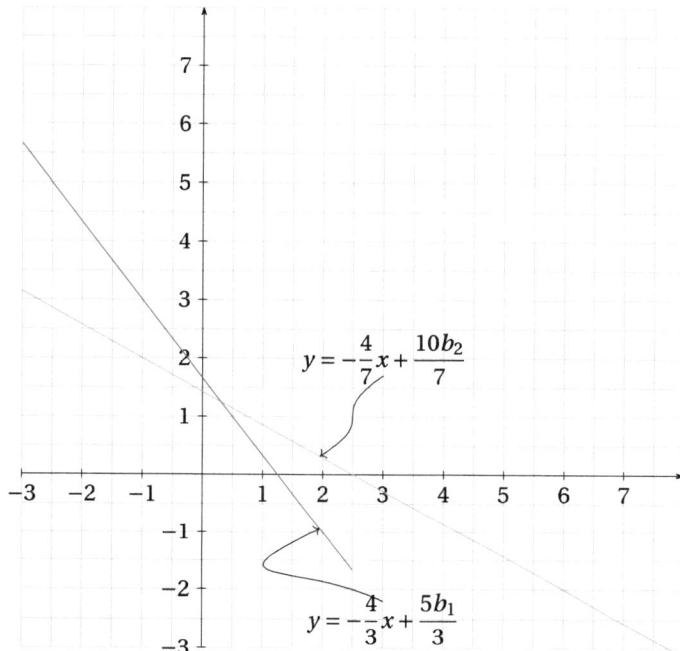

Figure 1.1: Geometric representation of equations (1.2) for $b_1 = b_2 = 1$.

The geometric interpretation of the former question is equivalent to: For which non-negative values of b_1 and b_2 do the lines represented by the system (1.1) intersect in the first quadrant? Given that the systems (1.1) and (1.2) are equivalent and, in the former, y is represented in terms of x, one approach to solve the system is by equating and simplifying. By doing this, we obtain the value for x and, substituting in (1.2), the value for y. Finally,

$$\begin{aligned} x &= \frac{35b_1 - 30b_2}{16}, \\ y &= \frac{10b_2 - 5b_1}{4}. \end{aligned}$$

(1.3)

We notice that the solution is given in terms of b_1 and b_2, which are assumed to be known. The constrains $x, y \geq 0$, lead to

$$\begin{aligned} x &= \frac{35b_1 - 30b_2}{16} \geq 0, \\ y &= \frac{10b_2 - 5b_1}{4} \geq 0. \end{aligned}$$

(1.4)

By performing some calculations, using properties of inequalities, the system (1.4) is equivalent to

$$7b_1 - 6b_2 \geq 0,$$
$$2b_2 - b_1 \geq 0, \tag{1.5}$$

which can be represented as

$$\frac{7b_1}{6} \geq b_2,$$
$$b_2 \geq \frac{b_1}{2}. \tag{1.6}$$

Figure 1.2 represents part of the region in the b_1–b_2 plane where the inequalities (1.6) hold. This region can also be interpreted as the image of the first quadrant under the transformation described below, see Example 5.1.2, p. 120.

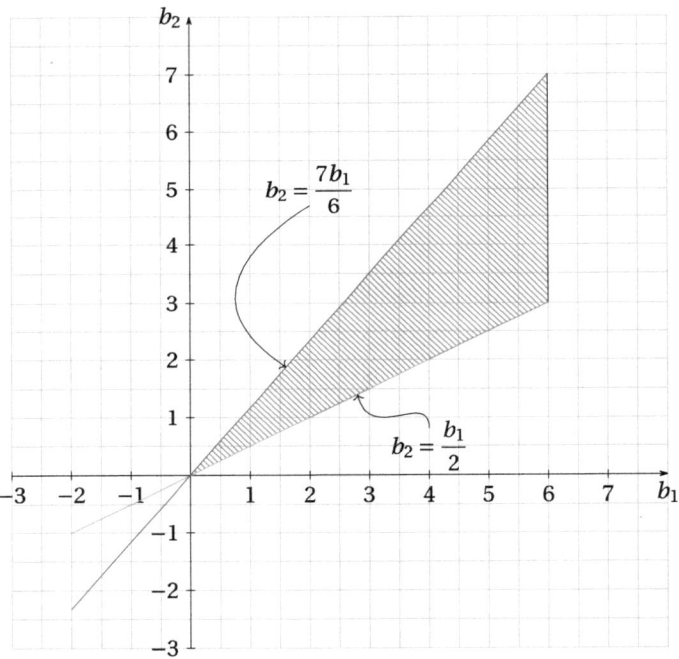

Figure 1.2: Image of the first quadrant under the action of T, from equation (1.7).

From a purely mathematical perspective, the production process can be formulated by means of the function

$$T(x,y) = (0.8x + 0.6y, 0.4x + 0.7y) = (b_1, b_2), \tag{1.7}$$

which transforms each point (x,y) from the first quadrant of the x–y plane into a point of the b_1–b_2 plane represented by the shaded region, see Figure 1.2.

The function T given by (1.7) is an example of a linear transformation. Linear transformations will be discussed widely in the following chapters.

To explore this example further, answer the following questions using different representations (geometric, algebraic, words, etc.):

1. What dollar amount must be invested if one wants to obtain values of products B_1 and B_2 equal to 4 and 3 million dollars, respectively?
2. What is the meaning, in terms of investment, for $b_1 = 4$ and $b_2 = 2$? Does the system (1.1) have a solution for these values of b_1 and b_2?
3. Is it possible to invest in such a way that $b_2 = 2b_1$? Explain.
4. What is the result of investing the same amount of money in both companies?
5. What is the minimum value for the slope of a line that passes through the origin and has one point in the region determined by inequalities (1.6)?
6. Is there a production scheme such as the one discussed in Example 1.1.1 so that $b_2 = 2b_1$? Discuss this case from an algebraic and geometric points of view, and give an interpretation taking into account production assumptions.

Example 1.1.2 (Leontief's model). Assume that an economic system consists of n economic sectors. Furthermore, assume that the economic system is closed, that is, the production is distributed among the n sectors. Additionally, assume that part of the production goes to the public. Leontief's model has a couple of additional assumptions:

1. Each sector has a profit.
2. The economic system is in equilibrium, that is, the total production is distributed among the sectors and the consumers.

Here we present a concrete example of Leontief's model. Assume that there are only three economic sectors: agriculture, transportation, and energy. Let us denote by x, y, and z the amount of money that each sector produces, respectively. The production requirements for each sector are represented in Table 1.2.

Table 1.2: Requirements by sector.

	Transportation	Energy	Agriculture
Transportation	$0.2x$	$0.1y$	$0.2z$
Energy	$0.4x$	$0.3y$	$0.3z$
Agriculture	$0.3x$	$0.5y$	$0.35z$

The second column of Table 1.2 reads as follows: the transportation sector, to produce x dollars, requires $0.2x$ from its own product; $0.4x$ from the energy sector, and $0.3x$ from the agriculture sector. We should notice that the transportation sector requires $0.9x$ dollars to produce x dollars, this means that it will get a 10 % profit out of its investment. The third and fourth columns in the table are interpreted likewise.

If the consumers require 2, 3, and 5 monetary units from transportation, energy, and agriculture sectors, respectively, could these requirements be satisfied?

To discuss the question, we formulate a system of linear equations, which is obtained as follows. The amount of money produced by the agriculture sector must be the sum of what it itself needs and what the transportation and energy sectors need, together with what the public needs. In symbols, this translates to $x = 0.2x + 0.1y + 0.2z + 2$. Likewise, the production of the transportation and energy sectors can be represented by corresponding equations. Collecting all equations we obtain:

$$\begin{aligned}
x &= 0.2x + 0.1y + 0.2z + 2, \\
y &= 0.4x + 0.3y + 0.3z + 3, \\
z &= 0.3x + 0.5y + 0.35z + 5.
\end{aligned} \tag{1.8}$$

One approach to solve the system above could be by trial and error, which is not efficient, since it is difficult to get a good guess. A better method is coming up in the following pages.

Exercise 1.1.1. Assume that in the system of linear equations

$$\begin{aligned}
x &= ax + by + c, \\
y &= dx + ey + f,
\end{aligned} \tag{1.9}$$

all the coefficients are positive real numbers. Furthermore, assume that $a + d < 1$ and $b + e < 1$. Show that the system has a solution (x, y) with $x, y > 0$.

Example 1.1.3. In a hypothetical country, the population is divided in two regions, A and B. It is also assumed that the population growth is zero and[2] a reported data measurement indicates that each year 25 % of population from region A moves to region B and 5 % from region B moves to region A. Assume that at the beginning of the observation there are 8 million people in region A and 2 million in region B. How many people are there in each region after 1, 2, 3, 4, 5 years?

Discussion. To understand a learning mathematical task, it is important to identify and represent the information of the task using different means. In this case, we will use a table to represent how the population is changing each year.

According to data provided in Table 1.3, could you predict how the population would be distributed after 20 years?

For the nth year, if we denote by u_n and r_n the number of people in regions A and B, respectively, then u_{n+1} and r_{n+1} are given by

---

2 The zero population growth, sometimes abbreviated ZPG, is a condition of demographic growth, where the number of people in a specified region neither grows nor declines.

Table 1.3: The change of population in time.

Year	Population of region A	Population of region B
0	8	2
1	$(0.75)8 + (0.05)2 = 6.10$	$(0.25)8 + (0.95)2 = 3.9$
2	$(0.75)(6.10) + (0.05)(3.9) = 4.77$	$(0.25)(6.1) + (0.95)(3.9) = 5.23$
3	$(0.75)(4.77) + (0.05)(5.23) = 3.839$	$(0.25)(4.77) + (0.95)(5.23) = 6.161$
4	$(0.75)(3.839) + (0.05)(6.161) = 3.1873$	$(0.25)(3.839) + (0.95)(6.161) = 6.8127$
5	$(0.75)(3.1873) + (0.05)(6.8127) = 2.73111$	$(0.25)(3.1873) + (0.95)(6.8127) = 7.26889$

$$u_{n+1} = 0.25r_n + 0.95u_n,$$
$$r_{n+1} = 0.75r_n + 0.05u_n. \tag{1.10}$$

From the assumptions, the total initial population is 10 million. Could the population be distributed in such a way that each year the population, in both regions, stays the same? If at the beginning of the observation, population in regions A and B is represented by r_0 and u_0, respectively, then we need that

$$r_1 = r_0 \quad \text{and} \quad u_1 = u_0. \tag{1.11}$$

In general, it is needed that $r_n = r_0$ and $u_n = u_0$.

From these assumptions, (1.10) becomes

$$u_0 = 0.25r_0 + 0.95u_0,$$
$$r_0 = 0.75r_0 + 0.05u_0. \tag{1.12}$$

Simplifying equations (1.12), one has that the system of equations reduces to only one, $0.25r_0 - 0.05u_0 = 0$; and this is equivalent to $u_0 = 5r_0$, whose interpretation is that population in region B must be a fifth of the population in region A in order for it to be stable.

Example 1.1.4. Assume that a company manages three oil refineries and each refines three products: gasoline, diesel, and lubricant oil. Furthermore, assume that for each barrel of oil the production in gallons is as indicated in Table 1.4.

Table 1.4: Production (in gallons) in each refinery.

	Refinery 1	Refinery 2	Refinery 3
Gasoline	20	21	19
Diesel	11	12	13
Lubricant oil	9	8	8

Data from Table 1.4 can be interpreted as follows. From each barrel of oil, Refinery 1 produces 20 gallons of gasoline, 11 gallons of diesel, and 9 gallons of lubricant oil; Refinery 2 produces 21 gallons of gasoline, 12 gallons of diesel, and 8 gallons of lubricant oil; Refinery 3 produces 19 gallons of gasoline, 13 gallons of diesel, and 8 gallons of lubricant oil.

Let us assume that the demand in a community is as follows: 1250 gallons of gasoline, 750 gallons of diesel, and 520 gallons of lubricant oil. How many barrels of oil should be processed in order to satisfy the demand?

Discussion

Identifying important information
1. *Known data.*
   (a) Needed production:
       i.   1250 gallons of gasoline.
       ii. 750 gallons of diesel.
       iii. 520 gallons of lubricant oil.
   (b) Production by refinery. Notice that this information is contained in Table 1.4.
2. *Unknown data.* The number of barrels that each refinery must process is not known. Let us denote by x_1, x_2, and x_3 the number of barrels that refineries 1, 2, and 3 should process, respectively.

Relationship among data and unknowns
We should notice that the total production of gasoline, diesel, and lubricant oil is the sum of the production of each refinery, thus, according to the information from Table 1.4, we have:
- (Total gallons of gasoline) $20x_1 + 21x_2 + 19x_3$;
- (Total gallons of diesel) $11x_1 + 12x_2 + 13x_3$;
- (Total gallons of lubricant oil) $9x_1 + 8x_2 + 8x_3$.

We need to know the values of x_1, x_2, and x_3 in such a way that the equations

$$20x_1 + 21x_2 + 19x_3 = 1250,$$
$$11x_1 + 12x_2 + 13x_3 = 750, \tag{1.13}$$
$$9x_1 + 8x_2 + 8x_3 = 520$$

are satisfied.

Using some strategies to solve the system of linear equations
One approach to solve (1.13) could be by *trial and error*, that is, to propose values for the variables x_1, x_2, and x_3, and evaluate the expressions. For instance, if $x_1 = x_2 = x_3 = 20$, then we have the results

$$20(20) + 21(20) + 19(20) = 1200,$$
$$11(20) + 12(20) + 13(20) = 720, \quad\quad (1.14)$$
$$9(20) + 8(20) + 8(20) = 500.$$

We notice that the proposed quantities need to be increased to satisfy the demand. We try one more case, $x_1 = 20$, $x_2 = 20$, and $x_3 = 21$, and obtain

$$20(20) + 21(20) + 19(21) = 1219,$$
$$11(20) + 12(20) + 13(21) = 733, \quad\quad (1.15)$$
$$9(20) + 8(20) + 8(21) = 508.$$

As the reader can imagine, based on identities (1.14) and (1.15), it can be difficult to obtain a good guess. Hence it is natural to ask: Is there a method to solve the system of equations (1.13)? Generally speaking, can we solve a system of n linear equations in m unknowns? In order to give an answer to this question, it is necessary to introduce some terminology. We will start by stating some properties that we have when performing some "operations" with equations.

Property 1. If both sides of an equation are multiplied by the same number, we obtain another equation. Example, given the equation $4x + 2y = 2$, multiplying by $\frac{1}{2}$ both sides, one obtains $2x + y = 1$.

Property 2. Given two equations, if one is multiplied by a number and the result is added to the other one, we obtain another equation. Example, given the equations $x+y = 3$ and $3x + y = -1$, if we multiply the first one by -3 and add the result to the second one, we have $-3(x+y) + (3x+y) = -3(3) + (-1)$. This latter equation is equivalent to $-2y = -10$.

Property 1 allows us to solve equations of type $ax = b$, with $a \neq 0$, that is, multiplying by $\frac{1}{a}$, we obtain $x = \frac{b}{a}$.

With these properties at hand, we can solve system (1.13) as follows:

$$20x_1 + 21x_2 + 19x_3 = 1250, \quad\quad (1.16)$$
$$11x_1 + 12x_2 + 13x_3 = 750, \quad\quad (1.17)$$
$$9x_1 + 8x_2 + 8x_3 = 520. \qu\quad (1.18)$$

Multiplying equation (1.17) by -20, equation (1.16) by 11, adding the results and simplifying, we obtain

$$-9x_2 - 51x_3 = -1250. \qu\quad (1.19)$$

Multiplying equation (1.18) by -20, equation (1.16) by 9, adding the results and simplifying, we have

$$29x_2 + 11x_3 = 850. \qu\quad (1.20)$$

With equations (1.19) and (1.20), we obtain the system

$$9x_2 + 51x_3 = 1250, \tag{1.21}$$

$$29x_2 + 11x_3 = 850. \tag{1.22}$$

It is important to notice that by the previous process, now we have a new system of two equations in two variables. Applying the same idea to the new system, that is, multiplying equation (1.22) by −9, equation (1.21) by 29, adding the results and simplifying, we have

$$x_3 = \frac{1430}{69}.$$

Substituting the value of x_3 into equation (1.21), one obtains $9x_2 + 51\frac{1430}{69} = 1250$; from this equation we arrive at $9x_2 = 1250 - 51\frac{1430}{69} = \frac{86250 - 72930}{69} = \frac{13320}{69} = \frac{4440}{23}$. From the latter equation, we have that

$$x_2 = \frac{1480}{69}.$$

Substituting the values of x_3 and x_2 into equation (1.16), we have $20x_1 + 21\frac{1480}{69} + 19\frac{1430}{69} = 1250$. This implies

$$20x_1 = 1250 - 21\frac{1480}{69} - 19\frac{1430}{69} = \frac{86250 - 31080 - 27170}{69} = \frac{28000}{69}.$$

Finally,

$$x_1 = \frac{1400}{69} \cong 20.290, \quad x_2 = \frac{1480}{69} \cong 21.449, \quad \text{and} \quad x_3 = \frac{1430}{69} \cong 20.725$$

are the values of the unknowns.

Double checking our results

An important aspect of the solution process is to verify that the results we obtained are correct. We will verify that the values for x_1, x_2, and x_3 satisfy system (1.13), that is, we must verify that the equations are correct.

$$20\frac{1400}{69} + 21\frac{1480}{69} + 19\frac{1430}{69} = 1250,$$

$$11\frac{1400}{69} + 12\frac{1480}{69} + 13\frac{1430}{69} = 750, \tag{1.23}$$

$$9\frac{1400}{69} + 8\frac{1480}{69} + 8\frac{1430}{69} = 520.$$

Performing calculations in each case, one verifies that the equations are correct.

Exercise 1.1.2. Consider the same processing conditions as in Example 1.1.4, except that the demand is twice as that in the example. Is there a solution for this case? Can we find a solution if the quantities of gasoline, diesel, and lubricant oil are 5000, 3000, and 5000 gallons, respectively?

After solving a system of linear equations in three variables, a question arises. Can we solve any system of linear equations in three or more variables? Before trying to give an answer, it might be interesting to discuss a system of two linear equations in two variables to illustrate geometric aspects.

1.2 Systems of linear equations and their geometric representation

Recall from analytic geometry that a linear equation in two variables represents a line in the coordinate plane, thus a system of linear equations represents a collection of lines.

For instance, equations $x + y = 1$ and $2x - 3y + 4 = 0$ represent two lines as shown in Figure 1.3.

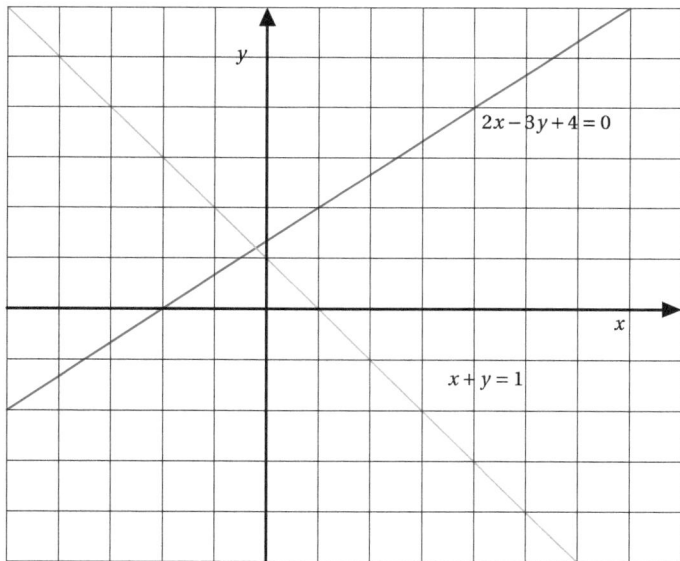

Figure 1.3: Graphical representation of equations $x + y = 1$ and $2x - 3y + 4 = 0$.

For the case of linear equations in two variables, to find values for the variables that satisfy the system, one can equivalently find the coordinates of a point which belongs to both lines.

Recall that a straight line l, represented by a linear equation of the form $ax + by = c$, intersects the two coordinate axes if and only if $ab \neq 0$. From the algebraic point of view, the condition $ab = 0$, with one of a or b not zero, implies that the equation $ax + by = c$ reduces to one of the form $ax = c$ or $by = c$. From either of these equations, it is straightforward to find the value of x or y, and the next step would be to substitute the found value of x or y into the remaining equation and find the value of the other variable.

From what we just discussed, if we consider an equation of the form $ax + by = c$, we may assume that $ab \neq 0$. From this, we will discuss a system of linear equations

$$\begin{aligned} a_1 x + b_1 y &= c_1, \\ a_2 x + b_2 y &= c_2, \end{aligned}$$

(1.24)

with the condition $a_1 a_2 b_1 b_2 \neq 0$.

We notice that system (1.24) is completely determined by the coefficients of the variables and by the independent terms, c_1 and c_2. This means that (1.24) can be represented by a rectangular array using the coefficients and the independent terms. More precisely, each row in the array represents an equation of the system,

$$\left[\begin{array}{cc|c} a_1 & b_1 & c_1 \\ a_2 & b_2 & c_2 \end{array} \right].$$

The vertical segment is for "separating" the coefficients of the variables from the independent terms.

Example 1.2.1. Consider the system

$$x + y = 1,$$

(1.25)

$$2x - 3y = -4;$$

(1.26)

for this case, the array representation of the system is

$$\left[\begin{array}{cc|c} 1 & 1 & 1 \\ 2 & -3 & -4 \end{array} \right].$$

Using this representation of the system, we will proceed to apply successively Properties 1 and 2 stated above; see page 9.

Multiplying row 1 by -2 and adding it to row 2 gives

$$\left[\begin{array}{cc|c} 1 & 1 & 1 \\ 2 & -3 & -4 \end{array} \right] \sim \left[\begin{array}{cc|c} 1 & 1 & 1 \\ 0 & -5 & -6 \end{array} \right].$$

(1.27)

Here the symbol ~ is used to denote what will be called row equivalence of matrices. We notice that the first row has been used as a *pivot* and is not changed. The rectangular array on the right of (1.27) represents the system

$$x + y = 1, \tag{1.28}$$
$$0x - 5y = -6. \tag{1.29}$$

If we multiply equation (1.29) by $-\frac{1}{5}$, then we obtain the value of y. This process is represented in the last array of (1.30):

$$\begin{bmatrix} 1 & 1 & | & 1 \\ 2 & -3 & | & -4 \end{bmatrix} \sim \begin{bmatrix} 1 & 1 & | & 1 \\ 0 & -5 & | & -6 \end{bmatrix} \sim \begin{bmatrix} 1 & 1 & | & 1 \\ 0 & 1 & | & \frac{6}{5} \end{bmatrix}. \tag{1.30}$$

The last array in (1.30) represents the system

$$x + y = 1,$$
$$0x + y = \frac{6}{5}. \tag{1.31}$$

If in the system (1.31) we subtract the second equation from the first and represent all steps using the corresponding arrays then we have

$$\begin{bmatrix} 1 & 1 & | & 1 \\ 2 & -3 & | & -4 \end{bmatrix} \sim \begin{bmatrix} 1 & 1 & | & 1 \\ 0 & -5 & | & -6 \end{bmatrix} \sim \begin{bmatrix} 1 & 1 & | & 1 \\ 0 & 1 & | & \frac{6}{5} \end{bmatrix} \sim \begin{bmatrix} 1 & 0 & | & -\frac{1}{5} \\ 0 & 1 & | & \frac{6}{5} \end{bmatrix}.$$

The last array represents the system

$$x + 0y = -\frac{1}{5},$$
$$0x + y = \frac{6}{5},$$

where the values of x and y are known.

Example 1.2.2. Recall that a linear function is of the form $f(x) = ax + b$, where a and b are constants. Assume that a linear function $f(x)$ satisfies $f(1) = 2$ and $f(-2) = -1$. Which function is $f(x)$?

To answer the question, we need to find the values of a and b, in the representation of $f(x) = ax + b$. From the given conditions, we have $f(1) = a(1) + b = 2$ and $f(-2) = a(-2) + b = -1$. That is, we have the equations $a + b = 2$ and $-2a + b = -1$.

Writing these equations in the form

$$a + b = 2,$$
$$-2a + b = -1,$$

the array representation of the system is

$$\left[\begin{array}{cc|c} 1 & 1 & 2 \\ -2 & 1 & 1 \end{array}\right]. \tag{1.32}$$

We invite the reader to apply "operations" in (1.32), in order to obtain the values $a = b = 1$. From this, we have that the function that satisfies the required conditions is $f(x) = x+1$.

Exercise 1.2.1. This exercise has several items, we encourage the reader to approach them all.
1. For each system of linear equations, use an array to represent it. Then use the method used in Example 1.2.1 to solve the system. Finally, represent the system geometrically:

$$\begin{array}{cccc} x + 3y = 3; & x + y = -2; & x - 5y = 7; & x - 3y = 6; \\ 3x + y = 4; & x + 3y = 2; & x - 2y = -1; & -2x + 6y = 7. \end{array}$$

2. Find all the values of a for which the following system has a solution:

$$\begin{aligned} ax + y &= 3, \\ 3x - ay &= 4. \end{aligned}$$

3. Consider the system whose equations are given by (1.24). Use the method described above to solve it.
4. How would you represent a system of 4 linear equations in 5 variables? Can your idea be extended to any number of equations, in any number of variables?

1.3 Fundamental concepts and the Gauss–Jordan reduction method

Before we present a general method to approach the solutions of a system of linear equations, it is important to introduce some terminology.

Definition 1.3.1. By a linear equation in the variables $x_1, x_2, \ldots, x_n$ we mean an equation of the form

$$a_1 x_1 + a_2 x_2 + \cdots + a_n x_n = b, \tag{1.33}$$

where $a_1, a_2, \ldots, a_n$, and b are independent of $x_1, x_2, \ldots, x_n$.

Example 1.3.1. The examples that follow are linear equations in x, y and z:
- $3x + 6y - z = 0$,
- $x + y = 4z$,
- $x + t^2 y + z = t^3$, t independent of x, y, z.

Exercise 1.3.1. Decide in which variables the given system is linear:
- $x^2 y + t^3 z = w,$
- $ax + by = w, b = x^2,$
- $t^3 + x + y + z = 0.$

Definition 1.3.2. A system of linear equations is a collection of equations of type (1.33). More precisely, a system of m linear equations in the variables $x_1, x_2, \ldots, x_n$ is a collection of equations of the form

$$a_{11}x_1 + a_{12}x_2 + \cdots + a_{1n}x_n = b_1,$$
$$a_{21}x_1 + a_{22}x_2 + \cdots + a_{2n}x_n = b_2,$$
$$\vdots$$
$$a_{m1}x_1 + a_{m1}x_1 + \cdots + a_{mn}x_n = b_m,$$

(1.34)

where a_{ij} and b_i are real numbers independent of $x_1, x_2, \ldots, x_n$, for $i \in \{1, 2, \ldots, m\}$ and $j \in \{1, 2, \ldots, n\}$.

Given the system (1.34), it can be represented by a rectangular array

$$\left[\begin{array}{cccc|c} a_{11} & a_{12} & \cdots & a_{1n} & b_1 \\ a_{21} & a_{21} & \cdots & a_{2n} & b_2 \\ \vdots & \vdots & \ddots & \vdots & \vdots \\ a_{m1} & a_{m2} & \cdots & a_{mn} & b_m \end{array} \right],$$

where the ith row represents the ith equation. For instance, the first row represents equation 1. Notice that in this representation the vertical line separates the coefficients of the variables and the independent terms.

The array representation of a system of linear equations contains all the information of the system, then it is reasonable to give it a name. It is called the *augmented matrix* of the system. The last column of the augmented matrix will be called the *vector of constant terms* of the system. When the vector of constants is omitted, the array is called the *coefficient matrix* of the system and it is given by

$$\left[\begin{array}{cccc} a_{11} & a_{12} & \cdots & a_{1n} \\ a_{21} & a_{21} & \cdots & a_{2n} \\ \vdots & \vdots & \ddots & \vdots \\ a_{m1} & a_{m2} & \cdots & a_{mn} \end{array} \right].$$

Before continuing the discussion, it is important to state some terminology that will be used. The element a_{ij} is referred as the entry in row i and column j.

Definition 1.3.3. An n-tuple of real numbers $(c_1, c_2, \ldots, c_n)$ is said to be a solution of the system (1.34) if $a_{i1}c_1 + a_{i2}c_2 + \cdots + a_{in}c_n = b_i$, for every $i \in \{1, 2, \ldots, n\}$. The collection of all solutions of (1.34) is called the solution set of the system. When the system has a solution, it is called *consistent*, otherwise it will be called *inconsistent*.

If $b_i = 0$ for every $i \in \{1, 2, \ldots, m\}$ in system (1.34), then $(0, 0, \ldots, 0)$ is a solution, that is, the system is consistent.

Example 1.3.2. Consider the system

$$
\begin{aligned}
x + 2y &= 3, \\
2x + 4y &= 1.
\end{aligned}
\tag{1.35}
$$

Geometrically, the equations in (1.35) represent parallel lines and the algebraic meaning is that the system has no solutions as the following argument shows. If (c_1, c_2) were a solution, then $c_1 + 2c_2 = 3$ and $2c_1 + 4c_2 = 1$ would hold, however, if we multiply $c_1 + 2c_2 = 3$ by -2 and add the result to $2c_1 + 4c_2 = 1$, we obtain $0 = -5$, which is not true.

When discussing possible solutions in Example 1.2.1, p. 12, the process was essentially described by the following "operations":
1. Interchange two equations.
2. Multiply an equation by a nonzero constant.
3. Multiply an equation by a constant, and add the result to another equation.

The above operations, when applied to solve a system of linear equations, represented by the augmented matrix, are called *elementary row operations* and are stated below.

Definition 1.3.4. The elementary row operations applied to a matrix are:
1. Swap any two rows.
2. Multiply a row by a nonzero constant.
3. Multiply a row by a constant and add the result to another row.

Notice that elementary row operations have inverses. For instance, if row i has been interchanged with row j, then the inverse of this operation is to interchange row j with row i.

Definition 1.3.5. Two systems of equations are called equivalent if they have the same set of solutions.

Theorem 1.3.1 below is one of the results guaranteeing that to solve a system of linear equations it is enough to apply elementary row operations on the augmented matrix of the system.

Theorem 1.3.1. *If a system of linear equations is obtained from another, by a sequence of elementary row operations, then the systems are equivalent.*

Proof. It is enough to prove that, when applying one elementary operation to a system of linear equations, the new system is equivalent to the original one.

Let

$$a_{11}x_1 + a_{12}x_2 + \cdots + a_{1n}x_n = b_1,$$
$$a_{21}x_1 + a_{22}x_2 + \cdots + a_{2n}x_n = b_2,$$
$$\vdots$$
$$a_{m1}x_1 + a_{m1}x_1 + \cdots + a_{mn}x_n = b_m$$

be a system of linear equations. It is clear that if the operation consists of interchanging two rows, the two systems are equivalent, since interchanging equations does not change the set of solutions. It is also clear that if the operation consists of multiplying one row by a nonzero constant, the two systems have the same solution set. Hence, it remains to prove that if a system of equations is obtained from another by multiplying one row by a constant and adding it to another row, then the systems have the same set of solutions. Assuming that equation i is multiplied by a and the result is added to equation j, then equation j in the new system is

$$(a_{j1} + aa_{i1})x_1 + (a_{j2} + aa_{i2})x_2 + \cdots + (a_{jn} + aa_{in})x_n = b_j + ab_i, \tag{1.36}$$

and all other equations remain the same.

If $(c_1, c_2, \ldots, c_n)$ is a solution of the original system, then

$$a_{k1}c_1 + a_{k2}c_2 + \cdots + a_{kn}c_n = b_k, \quad \text{for every } k \in \{1, 2, \ldots, m\}, \tag{1.37}$$

in particular,

$$a_{i1}c_1 + a_{i2}c_2 + \cdots + a_{in}c_n = b_i \tag{1.38}$$

and

$$a_{j1}c_1 + a_{j2}c_2 + \cdots + a_{jn}c_n = b_j \tag{1.39}$$

are true.

Multiplying equation (1.38) by a, adding it to (1.39), and grouping, one obtains equation (1.36) evaluated at $(c_1, c_2, \ldots, c_n)$. We conclude from this that $(c_1, c_2, \ldots, c_n)$ is a solution of the new system.

Conversely, if $(c_1, c_2, \ldots, c_n)$ is a solution of the new system, then

$$a_{k1}c_1 + a_{k2}c_2 + \cdots + a_{kn}c_n = b_k, \quad \text{for every } k \neq j, \tag{1.40}$$

and

$$(a_{j1} + aa_{i1})c_1 + (a_{j2} + aa_{i2})c_2 + \cdots + (a_{jn} + aa_{in})c_n = b_j + ab_i. \tag{1.41}$$

Taking $k = i$ in (1.40), multiplying by a, and subtracting from (1.41), we have

$$a_{j1}c_1 + a_{j2}c_2 + \cdots + a_{jn}c_n = b_j, \tag{1.42}$$

concluding that $(c_1, c_2, \ldots, c_n)$ is a solution of the original system of equations. $\quad\square$

In the following lines, we will represent a system of linear equations by a matrix and then apply row operations on the matrix. To see how this idea works, we will use a matrix to represent the system of linear equations given by (1.13), p. 8.

Recall that the system is

$$20x_1 + 21x_2 + 19x_3 = 1250,$$
$$11x_1 + 12x_2 + 13x_3 = 750,$$
$$9x_1 + 8x_2 + 8x_3 = 520,$$

from where the augmented matrix is

$$\left[\begin{array}{ccc|c} 20 & 21 & 19 & 1250 \\ 11 & 12 & 13 & 750 \\ 9 & 8 & 8 & 520 \end{array}\right].$$

Next, we will apply row operations on the augmented matrix to solve the system. We ask the reader to identify which elementary row operations were applied:

$$\left[\begin{array}{ccc|c} 20 & 21 & 19 & 1250 \\ 11 & 12 & 13 & 750 \\ 9 & 8 & 8 & 520 \end{array}\right] \sim \left[\begin{array}{ccc|c} 20 & 21 & 19 & 1250 \\ 2 & 4 & 5 & 230 \\ 9 & 8 & 8 & 520 \end{array}\right] \sim \left[\begin{array}{ccc|c} 0 & -19 & -31 & -1050 \\ 2 & 4 & 5 & 230 \\ 1 & -8 & -12 & -400 \end{array}\right]$$

$$\sim \left[\begin{array}{ccc|c} 0 & 19 & 31 & 1050 \\ 0 & 20 & 29 & 1030 \\ 1 & -8 & -12 & -400 \end{array}\right] \sim \left[\begin{array}{ccc|c} 0 & 19 & 31 & 1050 \\ 0 & 1 & -2 & -20 \\ 1 & -8 & -12 & -400 \end{array}\right] \sim \left[\begin{array}{ccc|c} 0 & 0 & 69 & 1430 \\ 0 & 1 & -2 & -20 \\ 1 & 0 & -28 & -560 \end{array}\right]$$

$$\sim \left[\begin{array}{ccc|c} 0 & 0 & 1 & 1430/69 \\ 0 & 1 & -2 & -20 \\ 1 & 0 & -28 & -560 \end{array}\right] \sim \left[\begin{array}{ccc|c} 0 & 0 & 1 & 1430/69 \\ 0 & 1 & 0 & 1480/69 \\ 1 & 0 & 0 & 1400/69 \end{array}\right] \sim \left[\begin{array}{ccc|c} 1 & 0 & 0 & 1400/69 \\ 0 & 1 & 0 & 1480/69 \\ 0 & 0 & 1 & 1430/69 \end{array}\right].$$

The solution of the system is read off from the last matrix, which is the same as that used in the checking in (1.23), p. 10.

1.3.1 Gauss–Jordan reduction method

In this section we present a general method to solve a system of linear equations, which will allow us to answer two fundamental questions that arise when solving linear equations:

1. When does a system of linear equations have a solution?
2. If the system has a solution, how do we find it?

We start the discussion of the questions above by considering some examples.

Example 1.3.3. The matrix $\begin{bmatrix} 1 & 0 & 0 & 4 \\ 0 & 1 & 0 & -1 \\ 0 & 0 & 1 & 3 \end{bmatrix}$ represents a solved system of linear equations. The interpretation is as follows. If the variables are x, y, and z, the first, second, and third row represent equations $x + 0y + 0z = 4$, $0x + y + 0z = -1$, and $0x + 0y + z = 3$, respectively, in other words, the solution of the system is $(x, y, z) = (4, -1, 3)$.

Example 1.3.4. If the matrix in Example 1.3.3 is changed to $\begin{bmatrix} 1 & 0 & 0 & 4 \\ 0 & 1 & 0 & -1 \\ 0 & 0 & 0 & 1 \end{bmatrix}$, then the new system has no solution, since the last row in the matrix would represent the inconsistency $0a + 0b + 0c = 1$, for any triple (a, b, c) of real numbers.

Example 1.3.5. The following matrix represents a system which has infinitely many solutions (why?):

$$\begin{bmatrix} 1 & 0 & 0 & 4 \\ 0 & 1 & 1 & -1 \\ 0 & 0 & 0 & 0 \end{bmatrix}.$$

The examples above show that the augmented matrix of a system contains all the information of the system. Then to decide if a system of linear equations has a solution or not, it is enough to know properties of the augmented matrix, after applying elementary row operations. Notice that matrices from the examples above have something in common, they are "reduced", this property needs to be made precise.

Definition 1.3.6. A matrix R is said to be in *row reduced form*, or *row reduced echelon form*, abbreviated RREF, if it satisfies:
1. The rows of R consisting only of zeroes, if there are any, are located at the bottom.
2. The first nonzero entry in a row of R, from left to right, must be 1. This first entry is called the main entry of that row.
3. The column of R that contains a main entry of a row must have only zeros, except for the main entry.
4. If the main entries of nonzero rows of R are located in columns numbered $k_1, k_2, \ldots, k_r$, then $k_1 < k_2 < \cdots < k_r$.

Exercise 1.3.2. The matrix $A = \begin{pmatrix} 1 & 0 & 0 & 0 & 4 \\ 0 & 0 & 1 & 4 & 7 \\ 0 & 3 & 3 & 0 & 0 \\ 0 & 0 & 0 & 1 & 2 \end{pmatrix}$ is not in row reduced form. Which conditions fail to hold? Apply row operations on the matrix to transform it to row reduced form.

Exercise 1.3.3. Assume that we have a system of n linear equations in n variables. Furthermore, assume that the augmented matrix of the system has been brought to RREF by elementary row operations and the reduced matrix does not have a row consisting only of zeros. Does the system have a solution?

In the process of solving a system of linear equations, it is important to answer the following questions:

1. What conditions should the rows of the row reduced matrix satisfy in order for the system to have a solution?
2. How do we know if a system of linear equations has more than one solution or a unique one?
3. Is it possible to transform, by elementary row operations, any nonzero matrix into one which is in row reduced form?

Exercise 1.3.4. By providing examples of matrices, show that there are several possible answers to the posed questions. Compare your answers with the conclusion of Theorem 1.3.2.

Definition 1.3.7. A matrix A is said to be the zero matrix, if all its entries are zero.

Theorem 1.3.2 (Gauss–Jordan reduction method). *Every nonzero matrix A can be transformed, by elementary row operations, to a matrix in row reduced form.*

Proof. Since A is not zero, $a_{ij} \neq 0$ for some i and j. We may assume that j is the least integer which satisfies the condition, for some i. This means that the columns, if any exist, with index less than j are zero. From this, the matrix A looks like

$$A = \begin{bmatrix} 0 & \cdots & 0 & a_{1j} & \cdots & a_{1n} \\ 0 & \cdots & 0 & a_{2j} & \cdots & a_{2n} \\ \vdots & \ddots & \vdots & \vdots & \ddots & \vdots \\ 0 & \cdots & 0 & a_{mj} & \cdots & a_{mn} \end{bmatrix}.$$

By swapping rows, we may assume that $i = 1$, then multiplying the first row by a_{1j}^{-1}, the element that appears in position $(1, j)$ is 1. Now, for $i = 2, \ldots, m$, multiply the first row by $-a_{ij}$ and add to row i, obtaining

$$A_1 = \begin{bmatrix} 0 & \cdots & 1 & b_{1\,j+1} & \cdots & b_{1n} \\ 0 & \cdots & 0 & b_{2\,j+1} & \cdots & b_{2n} \\ \vdots & \ddots & \vdots & \vdots & \ddots & \vdots \\ 0 & \cdots & 0 & b_{m\,j+1} & \cdots & b_{mn} \end{bmatrix}.$$

The next step is to examine the submatrix of A_1 whose entries are b_{ik} where $i \geq 2$ and $k > j$.

If the submatrix is zero, we have finished, otherwise there are $k > j$ and $i \geq 2$ so that $b_{ik} \neq 0$. We may assume that k is the minimum index with that property.

Again, by swapping rows, we may assume that $i = 2$, multiply row 2 of A_1 by b_{2k}^{-1}. Now, for $i > 2$, multiply row 2 by $-b_{ik}$ and add to row i, obtaining the matrix

$$A_2 = \begin{bmatrix} 0 & \cdots & 1 & \cdots & 0 & c_{1\,k+1} & \cdots & c_{1n} \\ 0 & \cdots & 0 & \cdots & 1 & c_{2\,k+1} & \cdots & c_{2n} \\ \vdots & \ddots & \vdots & \ddots & \vdots & \vdots & \ddots & \vdots \\ 0 & \cdots & 0 & \cdots & 0 & c_{m\,k+1} & \cdots & c_{mn} \end{bmatrix}.$$

We notice that the columns of A_2 are in "reduced" form. Continuing the process with A_2, one concludes that it finishes in a finite number of steps, producing a matrix in row reduced form, completing the proof of the theorem. □

Example 1.3.6. Solve the system

$$x + 2y + 3z = -1,$$
$$2x + 3z = 4,$$
$$-x + y + 3z = 0.$$

Discussion. First, we will transform the augmented matrix to row reduced form, after this we can determine which type of solutions the system has.

The augmented matrix is

$$B = \begin{bmatrix} 1 & 2 & 3 & -1 \\ 2 & 0 & 3 & 4 \\ -1 & 1 & 3 & 0 \end{bmatrix}.$$

Applying row operations, we have

$$\begin{bmatrix} 1 & 2 & 3 & -1 \\ 2 & 0 & 3 & 4 \\ -1 & 1 & 3 & 0 \end{bmatrix} \sim \begin{bmatrix} 1 & 2 & 3 & -1 \\ 0 & -4 & -3 & 6 \\ 0 & 3 & 6 & -1 \end{bmatrix} \sim \begin{bmatrix} 1 & 2 & 3 & -1 \\ 0 & -4 & -3 & 6 \\ 0 & -1 & 3 & 5 \end{bmatrix}$$

$$\sim \begin{bmatrix} 1 & 0 & 9 & 9 \\ 0 & 0 & -15 & -14 \\ 0 & -1 & 3 & 5 \end{bmatrix} \sim \begin{bmatrix} 1 & 0 & 9 & 9 \\ 0 & 0 & 1 & 14/15 \\ 0 & -1 & 3 & 5 \end{bmatrix} \sim \begin{bmatrix} 1 & 0 & 0 & 3/5 \\ 0 & 0 & 1 & 14/15 \\ 0 & -1 & 0 & 11/5 \end{bmatrix}$$

$$\sim \begin{bmatrix} 1 & 0 & 0 & 3/5 \\ 0 & 1 & 0 & -11/5 \\ 0 & 0 & 1 & 14/15 \end{bmatrix}.$$

According to the row reduced form, we have that the system has only one solution which is $C = \left(\frac{3}{5}, \frac{-11}{5}, \frac{14}{15}\right)$, as can be verified:

$$3/5 + 2(-11/5) + 3(14/15) = -1,$$
$$2(3/5) + 3(14/15) = 4,$$
$$-3/5 - 11/5 + 3(14/15) = 0.$$

1.3.2 Analysis of the solutions of a system of linear equations

If we know the row reduced form of the augmented matrix of a system of linear equations, it is not difficult to determine if the system:
1. has no solution,
2. has infinitely many solutions, or
3. has a unique solution.

Let

$$
A = \begin{bmatrix}
a_{11} & a_{12} & \cdots & a_{1n} & b_1 \\
a_{21} & a_{21} & \cdots & a_{2n} & b_2 \\
\vdots & \vdots & \ddots & \vdots & \vdots \\
a_{m1} & a_{m2} & \cdots & a_{mn} & b_m
\end{bmatrix}
\tag{1.43}
$$

be the augmented matrix of a system of linear equations, and let R be the row reduced form of A. In what follows, we will determine the type of solutions that the system has, according to the structure of R.

The system is inconsistent
If the main entry of a nonzero row appears in the $(n+1)$th column, then the system has no solution, since in this case such a row will represent an inconsistent equation.

The system is consistent
Let us assume that the main entry of the last nonzero row of R is not in the $(n+1)$th column, then the system has at least one solution. More precisely, if R has l nonzero rows, with main entries in columns $k_1 < k_2 < \cdots < k_l$, then the equations represented by R are of the form:

$$
x_{k_1} + \sum_{j=1}^{n-l} c_{1j} u_j = d_1,
$$

$$
x_{k_2} + \sum_{j=1}^{n-l} c_{2j} u_j = d_2,
\tag{1.44}
$$

$$
\vdots
$$

$$
x_{k_l} + \sum_{j=1}^{n-l} c_{lj} u_j = d_l,
$$

where the variables $u_1, u_2, \ldots, u_{n-l}$, called *free variables*, are a rearrangement of the original ones, which are different from $x_{k_1}, x_{k_2}, \ldots, x_{k_l}$. All the solutions of the system represented by matrix in (1.43) are obtained by assigning values to $u_1, u_2, \ldots, u_{n-l}$ and solving for $x_{k_1}, \ldots x_{k_l}$ in (1.44).

Example 1.3.7. Let $A = \begin{bmatrix} 1 & 0 & 0 & 0 & 3 & | & 4 \\ 0 & 1 & 2 & 0 & 2 & | & 5 \\ 0 & 0 & 0 & 1 & -1 & | & 6 \end{bmatrix}$ be the augmented matrix of a system of linear equations in the variables x_1, x_2, x_3, x_4, and x_5. Write the system represented by A and decide if the system has solutions. In the affirmative case, find all solutions.

Discussion. The system represented by A is

$$x_1 + 0x_2 + 0x_3 + 0x_4 + 3x_5 = 4,$$
$$0x_1 + x_2 + 2x_3 + 0x_4 + 2x_5 = 5, \qquad (1.45)$$
$$0x_1 + 0x_2 + 0x_3 + x_4 - x_5 = 6.$$

Notice that matrix A is already in row reduced form; the main entry of the last nonzero row is located before the last column of A, so the system is consistent. The free variables are x_3 and x_5 (why?). Therefore, all solutions are obtained by solving for x_1, x_2, and x_4 from system (1.45) and making x_3 and x_5 to take any value in the real numbers.

Unique solution

From the discussion above, we have that if the system has a solution, then it is unique if and only if there are no free variables. This holds if and only if the number of nonzero rows is equal to the number of variables. Particularly, if $m = n$, the system has a unique solution if and only if the main entries of the row reduced form of A appear in the (i, i) position for every $i \in \{1, 2, 3, 4, \ldots, n\}$.

Homogeneous system

A very important case of a system of linear equations occurs when the constant terms are zero, that is, the system looks as follows:

$$a_{11}x_1 + a_{12}x_2 + \cdots + a_{1n}x_n = 0,$$
$$a_{21}x_1 + a_{22}x_2 + \cdots + a_{2n}x_n = 0,$$
$$\vdots \qquad (1.46)$$
$$a_{m1}x_1 + a_{m1}x_1 + \cdots + a_{mn}x_n = 0.$$

We notice that the system of linear equations (1.46) is consistent since it has the zero solution, hence it is important to know when the system has a unique solution. The conditions under which (1.46) has a unique solution are established in Theorem 1.3.3.

Theorem 1.3.3. *Let A be the augmented matrix of a system of m homogeneous linear equations in n variables. Then the system has a unique solution, if and only if the number of nonzero rows of the row reduced form of A is m. Under this condition, $n \leq m$.*

Proof. When we say that a system is in n variables, this means that each variable has a nonzero coefficient in at least one equation. This is equivalent to saying that none of the first n columns of A are zero.

We notice that if a matrix has a nonzero column, then, by means of elementary row operations, this column never transforms to zero. Thus, the row reduced form of A has n nonzero columns.

From (1.44), p. 22, we have that the system represented by A has a unique solution if and only if there are no free variables and $d_1 = d_2 = \cdots = d_l = 0$, that is, the only solution is $x_1 = x_2 = \cdots = x_{k_l} = 0$, then necessarily $l = n$.

If $m < n$, then by the pigeonhole principle, Principle A.1.1, p. 263, at least one row of the row reduced form of A contains two nonzero entries, that is, the system has at least one free variable. From this we conclude that the system has infinitely many solutions, a contradiction, thus $n \leq m$. $\qquad\square$

General solution of a nonhomogeneous system of linear equations

Consider the ith equation represented by row i of matrix A, as in (1.43), more precisely we are considering equation

$$a_{i1}x_1 + a_{i2}x_2 + \cdots + a_{in}x_n = b_i. \tag{1.47}$$

Assume that the system represented by A has a solution and denote it by $C = (c_1, c_2, \ldots, c_n)$. From the definition of a solution, one has

$$a_{i1}c_1 + a_{i2}c_2 + \cdots + a_{in}c_n = b_i, \tag{1.48}$$

for every $i \in \{1, 2, \ldots, m\}$.

Let A_1 be the matrix that has its first n columns equal to those of A, and the last column equal to zero, then A_1 represents a homogeneous system of linear equations whose ith equation is

$$a_{i1}x_1 + a_{i2}x_2 + \cdots + a_{in}x_n = 0. \tag{1.49}$$

Let $D = (d_1, d_2, \ldots, d_n)$ be any solution of the system represented by A_1, then

$$a_{i1}d_1 + a_{i2}d_2 + \cdots + a_{in}d_n = 0, \tag{1.50}$$

for every $i \in \{1, 2, \ldots, m\}$.

Adding equations (1.48) and (1.50), one has that $C + D = (c_1 + d_1, c_2 + d_2, \ldots, c_n + d_n)$ is a solution of the system represented by A.

If $K = (k_1, k_2, \ldots, k_n)$ is another solution of the system represented by A, then

$$a_{i1}k_1 + a_{i2}k_2 + \cdots + a_{in}k_n = b_i, \tag{1.51}$$

for every $i \in \{1, 2, \ldots, m\}$. Subtracting equation (1.48) from equation (1.51) and grouping, we have

$$a_{i1}(k_1 - c_1) + a_{i2}(k_2 - c_2) + \cdots + a_{in}(k_n - c_n) = 0, \tag{1.52}$$

that is, $K - C = (k_1 - c_1, k_2 - c_2, \ldots, k_n - c_n)$ is a solution of the system of linear equations represented by A_1.

The discussion above is the proof of the following theorem.

Theorem 1.3.4. *Let A be the augmented matrix of a system of m linear equations in n variables, and let A_1 be the matrix whose n first columns are the same as those of A and the last is zero. If the system represented by A has a solution, say $C = (c_1, c_2, \ldots, c_n)$, and W is the solution set of the homogeneous system represented by A_1, then the solution set of the system represented by A is $\{w + C \mid w \in W\}$ and denoted by $W + C$.*

Example 1.3.8. Decide if the system

$$x_1 + 2x_2 + 3x_3 + 4x_4 + 5x_5 = 10,$$
$$2x_1 + 4x_2 + 6x_3 + 8x_4 + 10x_5 = 20,$$
$$-x_1 - 2x_2 - 2x_3 - x_4 + 3x_5 = 4$$

is consistent. If it is so, find the solution set of the system.

Discussion. The augmented matrix of the system is

$$\left(\begin{array}{ccccc|c} 1 & 2 & 3 & 4 & 5 & 10 \\ 2 & 4 & 6 & 8 & 10 & 20 \\ -1 & -2 & -2 & -1 & 3 & 4 \end{array}\right).$$

Applying row operations, one has

$$\left(\begin{array}{ccccc|c} 1 & 2 & 3 & 4 & 5 & 10 \\ 2 & 4 & 6 & 8 & 10 & 20 \\ -1 & -2 & -2 & -1 & 3 & 4 \end{array}\right) \sim \left(\begin{array}{ccccc|c} 1 & 2 & 3 & 4 & 5 & 10 \\ 0 & 0 & 0 & 0 & 0 & 0 \\ 0 & 0 & 1 & -1 & 8 & 14 \end{array}\right)$$

$$\sim \left(\begin{array}{ccccc|c} 1 & 2 & 3 & 4 & 5 & 10 \\ 0 & 0 & 1 & -1 & 8 & 14 \\ 0 & 0 & 0 & 0 & 0 & 0 \end{array}\right) \sim \left(\begin{array}{ccccc|c} 1 & 2 & 0 & -5 & 19 & -32 \\ 0 & 0 & 1 & -1 & 8 & 14 \\ 0 & 0 & 0 & 0 & 0 & 0 \end{array}\right).$$

The last matrix represents the system

$$x_1 + 2x_2 + 0x_3 - 5x_4 + 19x_5 = -32,$$
$$0x_1 + 0x_2 + x_3 - x_4 + 8x_5 = 14,$$
$$0x_1 + 0x_2 + 0x_3 + 0x_4 + 0x_5 = 0.$$

It can be verified that $C = (-32, 0, 14, 0, 0)$ is a solution, thus, according to Theorem 1.3.4, all the solutions of the system are obtained by adding $C = (-32, 0, 14, 0, 0)$ to the solutions of the system

$$x_1 + 2x_2 + 0x_3 - 5x_4 + 19x_5 = 0,$$
$$0x_1 + 0x_2 + x_3 - x_4 + 8x_5 = 0.$$

This system has three free variables, x_2, x_4, and x_5, then the coordinates of its solutions satisfy

$$x_2 = x_2,$$
$$x_4 = x_4,$$
$$x_5 = x_5,$$
$$x_1 = -2x_2 + 5x_4 - 19x_5,$$
$$x_3 = x_4 - 8x_5.$$

Written as a set,

$$W = \{(-2x_2 + 5x_4 - 19x_5, x_2, x_4 - 8x_5, x_4, x_5) \mid x_2, x_3, x_5 \in \mathbb{R}\}.$$

Summarizing, the solutions of the system

$$x_1 + 2x_2 + 3x_3 + 4x_4 + 5x_5 = 10,$$
$$2x_1 + 4x_2 + 6x_3 + 8x_4 + 10x_5 = 20,$$
$$-x_1 - 2x_2 - 2x_3 - x_4 + 3x_5 = 4$$

are given by

$$W = \{(-2x_2 + 5x_4 - 19x_5, x_2, x_4 - 8x_5, x_4, x_5) \mid x_2, x_3, x_5 \in \mathbb{R}\} + (-32, 0, 14, 0, 0)$$
$$= \{(-2x_2 + 5x_4 - 19x_5 - 32, x_2, x_4 - 8x_5 + 14, x_4, x_5) \mid x_2, x_3, x_5 \in \mathbb{R}\}.$$

Exercise 1.3.5. In each case, the given matrix is the augmented matrix of a system of linear equations. Determine if the system has: a unique solution, infinitely many, or no solutions:

1. $\left(\begin{array}{cc|c} 2 & -1 & 2 \\ -4 & 1 & 5 \end{array} \right)$;

2. $\left(\begin{array}{ccc|c} 1 & 2 & -1 & 0 \\ -2 & 3 & 2 & -1 \\ 5 & 2 & 0 & 4 \end{array} \right)$;

3. $\left(\begin{array}{cccc|c} -1 & 1 & 1 & -1 & 2 \\ 1 & 0 & 3 & -3 & 1 \\ 2 & 1 & 0 & 1 & 0 \\ -1 & 2 & 0 & 1 & 0 \end{array} \right)$;

4. Determine the values of a in such a way that the system represented by the matrix
   $\left(\begin{array}{ccc|c} a & -1 & 1 & 0 \\ 1 & 1-a & -1 & 1 \\ -1 & 1 & 1 & 2 \end{array} \right)$, has: (1) a unique solution; (2) infinitely many solutions, or (3) has no solution at all.

1.3.3 Row operations using SageMath

SageMath is a free open-source mathematics software system licensed under the GPL. We think that when performing calculation, the use of a digital tool can help to concentrate in the understanding of methods and concepts and the calculations can be done by the use of a tool. For instance, using SageMath, one can perform row operations on a matrix. This can be done by applying one row operation at the time; it could be a little bit tedious, however, it is better than doing the computations by hand. At this point, we think that it is a good exercise to perform row operations on a matrix to get a better understanding of the reduction process for a matrix. For each row operation, using SageMath, there are two methods which produce different results. One changes the matrix "in-place" while the other produces a new matrix which is a modified version of the original. This is an important difference that we encourage the reader to distinguish. Consider the row operation that swaps two rows. There are two matrix methods to do this with SageMath, a "with" version that will produce a new matrix different from that on which you have applied the operation, which you will likely want to save, and a plain version of the operation that will change the matrix it operates on "in-place". The copy() function, which is a general-purpose command, is a way to make a copy of a matrix before you make changes to it. We should remember that row and column indices in SageMath start at 0.

It is important to understand the following commands:

```
A1=matrix(QQ,2,3,[1,2,3,4,5,6])
B=A1
C1=copy(A1)
D1=A1.with_swapped_rows(0,1)
```

The results of these commands are:

$$A1 = \begin{pmatrix} 1 & 2 & 3 \\ 4 & 5 & 6 \end{pmatrix},$$

$$B = \begin{pmatrix} 1 & 2 & 3 \\ 4 & 5 & 6 \end{pmatrix},$$

$$C1 = \begin{pmatrix} 1 & 2 & 3 \\ 4 & 5 & 6 \end{pmatrix},$$

$$D1 = \begin{pmatrix} 4 & 5 & 6 \\ 1 & 2 & 3 \end{pmatrix}.$$

Explain why the command

```
A1.swap_rows(0,1)
```

changes $A1$ to $A2 = \begin{pmatrix} 4 & 5 & 6 \\ 1 & 2 & 3 \end{pmatrix}$.

The following example shows how to use SageMath in the process of applying row operations on the rows of a matrix to obtain the row reduced form.

Define a matrix in SageMath with the use of the following command:

```
A=matrix([[3,2,0,4],[1,4,sqrt(2),3],[2,3,4,5]])
```

This defines A to be the matrix

$$A = \begin{pmatrix} 3 & 2 & 0 & 4 \\ 1 & 4 & \sqrt{2} & 3 \\ 2 & 3 & 4 & 5 \end{pmatrix}.$$

The next commands produce elementary row operations on the defined matrix. For example, the command

```
A.swap_rows(0,1)
```

swaps rows 1 and 2 in matrix A, obtaining

$$A_1 = \begin{pmatrix} 1 & 4 & \sqrt{2} & 3 \\ 3 & 2 & 0 & 4 \\ 2 & 3 & 4 & 5 \end{pmatrix}.$$

The following command produces a matrix which is obtained from A_1 by adding -3 times row 1 to row 2

```
A.add_multiple_of_row(1,0,-3)
```

Let us call this new matrix A_2, that is,

$$A_2 = \begin{pmatrix} 1 & 4 & \sqrt{2} & 3 \\ 0 & -10 & -3\sqrt{2} & -5 \\ 2 & 3 & 4 & 5 \end{pmatrix}.$$

Now, the command

```
A.add_multiple_of_row(2,0,-2)
```

represents the process of multiplying row 1 of A_2 by -2 and adding the result to row 3. The new matrix is

$$A_3 = \begin{pmatrix} 1 & 4 & \sqrt{2} & 3 \\ 0 & -10 & -3\sqrt{2} & -5 \\ 0 & -5 & -2\sqrt{2}+4 & -1 \end{pmatrix}.$$

The command

```
A.add_multiple_of_row(1,2,-2)
```

multiplies row 3 of A_3 by –2 and adds the result to row 2, obtaining

$$A_4 = \begin{pmatrix} 1 & 4 & \sqrt{2} & 3 \\ 0 & 0 & \sqrt{2}-8 & -3 \\ 0 & -5 & -2\sqrt{2}+4 & -1 \end{pmatrix}.$$

The process of multiplying row 3 of A_4 by –1/5 is obtained with the command

```
A.rescale_row(2,-1/5)
```

which produces

$$A_5 = \begin{pmatrix} 1 & 4 & \sqrt{2} & 3 \\ 0 & 0 & \sqrt{2}-8 & -3 \\ 0 & 1 & \frac{2}{5}\sqrt{2}-\frac{4}{5} & \frac{1}{5} \end{pmatrix}.$$

The command

```
A.add_multiple_of_row(0,2,-4)
```

multiplies row 3 of A_5 by –4 and adds the result to row 1, obtaining

$$A_6 = \begin{pmatrix} 1 & 0 & -\frac{3}{5}\sqrt{2}+\frac{16}{5} & \frac{11}{5} \\ 0 & 0 & \sqrt{2}-8 & -3 \\ 0 & 1 & \frac{2}{5}\sqrt{2}-\frac{4}{5} & \frac{1}{5} \end{pmatrix}.$$

Since A must be transformed to row reduced form, we need to swap rows 2 and 3 which is obtained from A_6 by using the command

```
A.swap_rows(1,2)
```

whose result is

$$A_7 = \begin{pmatrix} 1 & 0 & -\frac{3}{5}\sqrt{2}+\frac{16}{5} & \frac{11}{5} \\ 0 & 1 & \frac{2}{5}\sqrt{2}-\frac{4}{5} & \frac{1}{5} \\ 0 & 0 & \sqrt{2}-8 & -3 \end{pmatrix}.$$

It should be noticed that all operations were applied to matrix A, but it was changing with each row operation.

We can obtain matrix A_7 with the use of only one command and call the result R, namely

`A.rref()`

whose result is

$$R = \begin{pmatrix} 1 & 0 & -\frac{3}{5}\sqrt{2} + \frac{16}{5} & \frac{11}{5} \\ 0 & 1 & \frac{2}{5}\sqrt{2} - \frac{4}{5} & \frac{1}{5} \\ 0 & 0 & \sqrt{2} - 8 & -3 \end{pmatrix}.$$

We encourage the reader to continue applying row operations on A_7 by using SageMath up to the point where the last matrix is in row reduced form. When you have finished this process, you can directly answer the questions:

1. What is the system of linear equations that A represents?
2. Which system is represented by the row reduced form of A?
3. Is the system represented by A consistent?

1.4 Exercises

In the discussion of some of the following exercises you are invited to use SageMath.

1. Represent each system of linear equations by an augmented matrix and apply row operations on the matrix to solve the systems:

$$\begin{aligned} x + y &= 3; & x - y &= 3; \\ 3x + y &= 0; & 3x + 2y &= 2; \end{aligned} \qquad \begin{aligned} x - y + z &= 3; & 2x - 3y + 8z &= 4; \\ 3x + 2y - z &= 2; & 3x + 2y - 2z &= 5; \\ x - y + 2z &= 0; & 4x - y + 2z &= -1. \end{aligned}$$

2. Assume that $A = \begin{bmatrix} 1 & 0 & \lambda & 1 \\ 0 & \lambda & \lambda & 0 \\ \lambda & 2 & 1 & -1 \end{bmatrix}$ is the augmented matrix of a system of linear equations. Determine the values of λ in such a way that the system:
   (a) has a unique solution,
   (b) has infinitely many solutions,
   (c) has no solution.

3. Assume that each of the following matrices is the augmented matrix of a system of linear equations. Determine if the system is consistent or not. Explain

$$\left(\begin{array}{cccc|c} 1 & 3 & 0 & 5 & -5 \\ 0 & 1 & -6 & 9 & 0 \\ 0 & 0 & 2 & 7 & 1 \\ 0 & 0 & 1 & 4 & -2 \end{array}\right); \qquad \left(\begin{array}{cccc|c} 1 & 3 & 0 & 5 & -5 \\ 0 & 3 & -6 & 9 & 0 \\ 0 & 0 & 2 & 0 & 8 \\ 0 & 0 & 1 & 4 & -16 \end{array}\right);$$

$$\begin{pmatrix} 3 & 1 & -1 & 2 & | & 4 \\ 2 & -1 & 2 & 5 & | & 0 \\ 5 & 4 & 4 & -2 & | & 6 \\ 7 & 0 & 1 & -1 & | & 0 \end{pmatrix}.$$

4. Let $a, b, c,$ and d be fixed real numbers. Find necessary and sufficient conditions in order that the system of linear equations

$$ax + bz = 1,$$
$$ay + bw = 0,$$
$$cx + dz = 0,$$
$$cx + dw = 1$$

has a unique solution

5. (Geometry of elementary operations) Consider the system of linear equations

$$x - y = -2,$$
$$2x + y = 3, \tag{1.53}$$

and let a be a parameter. Multiplying the first equation of (1.53) by a and adding the result to the second equation, we have

$$(2 + a)x + (1 - a)y = 3 - 2a. \tag{1.54}$$

For which values of a is the line represented by (1.54) parallel to each axis? What is the geometric meaning of changing the values of a continuously? Are there values of a for which the line represented by (1.54) coincides with each of the lines represented by (1.53). Explain.

6. Extend the ideas and questions from Exercise 5 to a system of three linear equations in three variables and use GeoGebra, or another geometry system, to represent your examples.

7. Decide if the following statements are true or false. Provide examples to convince yourself about the correctness of your answers:
   (a) Given a linear system of equations in the variables $x, y,$ and z, the solutions, if there are any, are triples of numbers.
   (b) A system of linear equations is inconsistent if it has more than one solution.
   (c) There are systems of linear equations with exactly two solutions.
   (d) In order to solve a system of linear equations, it is necessary that the number of variables and equations coincide.
   (e) If the augmented matrix of a system of linear equations has in its last column only zeros, the system is inconsistent.
   (f) If the row reduced form of the augmented matrix of a system of linear equations has a nonzero row, whose main entry is located in the last column, then the system has a unique solution.

(g) If the number of variables in a system of linear equations is greater than the number of equations, then the system has infinitely many solutions.

(h) If the coefficients of a system of linear equations are rational numbers and it has solutions, then the entries of the solutions are rational numbers.

(i) If a system of linear equations has a unique solution, then the number of equations is equal to the number of variables.

8. In the following matrices determine the values of h and k so that the augmented matrices that represent systems of linear equations satisfy: (a) the system has a unique solution; (b) the system has infinitely many solutions; (c) the system has no solutions:

(a) $\left(\begin{smallmatrix} 1 & h & | & 1 \\ 2 & 3 & | & k \end{smallmatrix} \right)$,

(b) $\left(\begin{smallmatrix} 1 & -3 & | & 1 \\ 2 & -h & | & k \end{smallmatrix} \right)$,

(c) $\left(\begin{smallmatrix} k & 4 & | & 2 \\ 5 & -h & | & 1 \end{smallmatrix} \right)$.

9. Write a program in your favorite language to solve a system of linear equations in two variables.

10. To solve this exercise it is recommended to review Example 1.2.2, p. 13. Determine the coefficients of a polynomial function of degree 3 whose graph passes through the points $(1, 2)$, $(2, 0)$, $(-1, 0)$, and $(-2, 4)$. Explain why it is necessary to know four points for polynomials of degree three; 5 points for a polynomial of degree 4. How many points do we need to uniquely determine a polynomial function of degree n that passes through them?

11. The reader is invited to review Example 1.1.2, p. 5, and solve the system of linear equations given by (1.8). From this, answer the question posed there.

12. There is a zone in the world which is divided in three regions denoted by R_1, R_2, and R_3. In the zone the population growth is zero, and each year people migrate from one region to the others according to the following scheme: from region R_1, 5 % migrate to R_2 and 10 % migrate to R_3; from region R_2, 7 % migrate to region R_1 and 6 % migrate to region R_3; from region R_3, 7 % migrate to region R_2 and 8 % migrate to region R_1. At the beginning of the observation, in regions R_1, R_2, and R_3 there are 15, 20, and 25 million people, respectively. How many people will there be in each region after 1, 2, 3, 4, and 5 years? How the population will be distributed in each region after $k > 5$ years?

13. Assume that a mining company manages three mines and each produces three minerals: gold, silver, and copper. From each ton of ore, the mines produce gold, silver, and copper according to Table 1.5. What is the total production of gold, silver, and copper if mine 1, mine 2, and mine 3 process 1000, 2500, and 3000 tons of ore, respectively? How many tons of ore must mines 1, 2, and 3 process to have a total production of 0.2095, 5.4275, and 27.967 tons of gold, silver, and copper, respectively?

14. A clothing factory produces jackets, pants, t-shirts, skirts, and blouses using five resources: cotton, plastic, metal, paper, and electricity. Table 1.6 shows how much of each resource is used in making each product. How much of each resource was

Table 1.5: Mining production (in kg) from each ton of ore.

	Mine 1	Mine 2	Mine 3
Gold	0.025	0.03	0.031
Silver	0.7	0.75	0.80
Copper	3.8	3.9	4.0

used to produce 60 jackets, 100 pants, 40 t-shirts, 100 skirts, and 150 blouses? The company is planning to produce 70 jackets, 85 pants, 75 t-shirts, 60 skirts, and 100 blouses. How much of each resource is needed to produce what is required?

Table 1.6: Requirements for the production of jackets, pants, t-shirts, skirts, and blouses.

	cotton	plastic	metal	paper	electricity
jacket	0.6	0.4	0.2	0.15	1.2
pants	0.4	0.3	0.1	0.2	1.1
T-shirt	0.4	0.5	0.2	0.3	1.0
skirt	0.3	0.7	0.2	0.1	1.2
blouse	0.5	0.6	0.1	0.1	1.0

15. Let n be a positive integer and assume that the augmented matrix of a linear system of equations is of type $n \times (n+1)$, and its row reduced form has all nonzero rows. Is the system consistent? Provide examples for the cases $n = 2, 3, 4, 5$.

16. Let $A = \left[\begin{smallmatrix} a & b & | & c \\ a_1 & b_1 & | & c_1 \end{smallmatrix} \right]$ be the augmented matrix of a system of linear equations, where all the entries of A are whole numbers. Under which conditions on the entries of A, are the solutions of the system whole numbers?

17. Let A be the coefficient matrix of a homogeneous system of n linear equations in n variables. Assume that the row reduced form of A has only nonzero rows. Can A be the coefficient matrix of an inconsistent nonhomogeneous system of linear equations? Provide examples.

18. Assume that there is an inconsistent linear system of m equations in n variables. Transforming the system into a homogeneous one by equating the constant terms to zero, are there infinitely many solutions for the new system?

19. Let

$$
\begin{aligned}
a_{11}x_1 + a_{12}x_2 + \cdots + a_{1n}x_n &= b_1, \\
a_{21}x_1 + a_{22}x_2 + \cdots + a_{2n}x_n &= b_2, \\
&\ \vdots \\
a_{m1}x_1 + a_{m1}x_1 + \cdots + a_{mn}x_n &= b_m
\end{aligned}
\tag{1.55}
$$

be a consistent system of linear equations and assume that $m < n$. Is it possible that (1.55) has a unique solution? Explain by providing examples.

20. Assume that the augmented matrix of a system of linear equations A has n rows and $n+1$ columns. What conditions should the rows of A satisfy in order that the system has a unique solution?

21. Assume that in a system of two linear equations in two variables, the coefficients of the variables are irrational numbers, while the independent terms are integers. Could $(\sqrt{2}, \sqrt{2})$ be a solution of the system?

22. Consider the system of homogeneous linear equations given in (1.46), p. 23. Prove the following:

    (a) if C and C_1 are solutions of (1.46), then $C + C_1$ is a solution,

    (b) if C is a solution and $r \in \mathbb{R}$, then rC is also a solution.

2 Matrices

In mathematics, number systems have been proposed and developed when solving different types of concrete problems. For instance, complex numbers were proposed by Girolamo Cardano (1545), when he was solving equations of degree 3. This happens within mathematics itself or from other disciplines. For the case of matrices, many great mathematicians, among them Cramer, Bezout, Vandermonde, Lagrange, Gauss, Cauchy, Cayley, Jordan, Sylvester, Frobenius, and others, initiated the study and use of matrices to solve important problems in mathematics and other scientific disciplines.

In this chapter we present basic properties of matrices, starting from their use to represent and "operate" with data. This will lead us to consider matrices not only associated to systems of linear equations, as was done in Chapter 1, but also as an important algebraic system, which will allow us to attain a better understanding of their properties and applications.

2.1 Sum and product of matrices

In this section we present the properties of a sum and product of matrices. Concerning the sum, it is defined "naturally", entry by entry; however, the definition of the product seems a little bit artificial. This idea has motivated that the definition of matrix product be presented from a motivation in which matrices are used to represent production processes. We hope that this approach of motivating the product of matrices would help students make sense out of the definition of the product.

2.1.1 Sum of matrices

Assume that there are two companies which manufacture five different products each. The value of the production is calculated monthly, and it is registered during a period of six months. The companies will be labeled as A and B, while the products will be labeled as 1, 2, 3, 4, and 5. To represent the value of each product, we use double indices. For instance, the values of the products that come from company A, corresponding to the first month will be denoted by $a_{11}, a_{21}, a_{31}, a_{41}$, and a_{51}. Likewise, for the second month the values of the same products are represented by $a_{12}, a_{22}, a_{32}, a_{42}$, and a_{52}. Notice that the first index corresponds to the product number, while the second corresponds to the month number.

One efficient way to represent all this information is by means of a table. That is, the production of company A during six months can be represented by Table 2.1.

A similar representation is used for company B, Table 2.2.

https://doi.org/10.1515/9783111135915-002

Table 2.1: Production of company A in a period of six months.

Product No.	Month 1	Month 2	Month 3	Month 4	Month 5	Month 6
1	a_{11}	a_{12}	a_{13}	a_{14}	a_{15}	a_{16}
2	a_{21}	a_{22}	a_{23}	a_{24}	a_{25}	a_{26}
3	a_{31}	a_{32}	a_{33}	a_{34}	a_{35}	a_{36}
4	a_{41}	a_{42}	a_{43}	a_{44}	a_{45}	a_{46}
5	a_{51}	a_{52}	a_{43}	a_{54}	a_{55}	a_{56}

Table 2.2: Production of company B in the same period of six months.

Product No.	Month 1	Month 2	Month 3	Month 4	Month 5	Month 6
1	b_{11}	b_{12}	b_{13}	b_{14}	b_{15}	b_{16}
2	b_{21}	b_{22}	b_{23}	b_{24}	b_{25}	b_{26}
3	b_{31}	b_{32}	b_{33}	b_{34}	b_{35}	b_{36}
4	b_{41}	b_{42}	b_{43}	b_{44}	b_{45}	b_{46}
5	b_{51}	b_{52}	b_{43}	b_{54}	b_{55}	b_{56}

Using this representation of the production, it is clear how to compute the total monthly production: just add the corresponding entries in the tables. For example, the total amount of product 2 in month 4 is $a_{24} + b_{24}$.

The tables can be transformed into rectangular arrays as

$$
\begin{pmatrix}
a_{11} & a_{12} & a_{13} & a_{14} & a_{15} & a_{16} \\
a_{21} & a_{22} & a_{23} & a_{24} & a_{25} & a_{26} \\
a_{31} & a_{32} & a_{33} & a_{34} & a_{35} & a_{36} \\
a_{41} & a_{42} & a_{43} & a_{44} & a_{45} & a_{46} \\
a_{51} & a_{52} & a_{53} & a_{54} & a_{55} & a_{56}
\end{pmatrix}
\text{ and }
\begin{pmatrix}
b_{11} & b_{12} & b_{13} & b_{14} & b_{15} & b_{16} \\
b_{21} & b_{22} & b_{23} & b_{24} & b_{25} & b_{26} \\
b_{31} & b_{32} & b_{33} & b_{34} & b_{35} & b_{36} \\
b_{41} & b_{42} & b_{43} & b_{44} & b_{45} & b_{46} \\
b_{51} & b_{52} & b_{53} & b_{54} & b_{55} & b_{56}
\end{pmatrix}.
$$

Hence these arrays represent the production of companies A and B. Then, to obtain the total monthly production for each product, just add the corresponding entries in the arrays. The discussion above can be generalized to m products and n months. In this case, the production of the companies can be represented by rectangular arrays of m rows and n columns. An array of this type is called an $m \times n$ matrix. We also say that A and B are $m \times n$ matrices.

The above example can be used to define the sum of two $m \times n$ matrices.

Definition 2.1.1. Given two $m \times n$ matrices A and B, the sum $A + B$ is denoted by $A + B = C$, where the entries of C are given by $c_{ij} = a_{ij} + b_{ij}$, for $i \in \{1, 2, \ldots, m\}$ and $j \in \{1, 2, \ldots, n\}$.

Example 2.1.1. Given matrices

$$
A = \begin{pmatrix} 1 & -2 & 4 & 5 \\ 4 & 5 & 0 & \pi \\ 4 & 0 & 3 & 4 \end{pmatrix} \text{ and } B = \begin{pmatrix} 0 & -5 & 4 & 5 \\ -4 & 15 & 10 & -\pi \\ -3 & 1 & -4 & 3 \end{pmatrix},
$$

their sum is

$$A + B = \begin{pmatrix} 1 & -2 & 4 & 5 \\ 4 & 5 & 0 & \pi \\ 4 & 0 & 3 & 4 \end{pmatrix} + \begin{pmatrix} 0 & -5 & 4 & 5 \\ -4 & 15 & 10 & -\pi \\ -3 & 1 & -4 & 3 \end{pmatrix} = \begin{pmatrix} 1 & -7 & 8 & 10 \\ 0 & 20 & 10 & 0 \\ 1 & 1 & -1 & 7 \end{pmatrix}.$$

Example 2.1.2. If A and B are given by

$$A = \begin{pmatrix} a_{11} & a_{12} & a_{13} & a_{14} \\ a_{21} & a_{22} & a_{23} & a_{24} \\ a_{31} & a_{32} & a_{33} & a_{34} \end{pmatrix} \quad \text{and} \quad B = \begin{pmatrix} b_{11} & b_{12} & b_{13} & b_{14} \\ b_{21} & b_{22} & b_{23} & b_{24} \\ b_{31} & b_{32} & b_{33} & b_{34} \end{pmatrix},$$

then $A + B = C$ is

$$C = \begin{pmatrix} a_{11} + b_{11} & a_{12} + b_{12} & a_{13} + b_{13} & a_{14} + b_{14} \\ a_{21} + b_{21} & a_{22} + b_{22} & a_{23} + b_{23} & a_{24} + b_{24} \\ a_{31} + b_{31} & a_{32} + b_{32} & a_{33} + b_{33} & a_{34} + b_{34} \end{pmatrix}.$$

2.1.2 Product of matrices

The sum of matrices was defined entry by entry, following a "natural" way. From that definition, it is not difficult to verify that the sum shares properties analogous to the sum of real numbers.

For the product of matrices, it might be interesting to mentioned why it is defined as it appears in textbooks. First of all, there are several ways to define a product of matrices. The "natural" procedure to define the product, leads to what is called *Hadamard product of matrices*. To have an idea of the properties that Hadamard product shares, see Exercise 18, p. 65. The product of matrices that appears in textbooks does not have a very well-known history. According to Agarwal and Sen [1, p. 235], Binet "discovered the multiplication rule for matrices in 1812, which was used by Eisenstein to simplify the process of making substitutions in linear systems and was formalized by Cayley". Concerning the previous reference, the first time that the product of matrices appeared the way it is known today occurred in Cayley's 1858 paper [9]. Leaving this little historical note aside, we will explore, via an example, a way to motivate the product of matrices.

Example 2.1.3. Three companies produce smartphones, tablets, and smartwatches. For each million dollars invested, the production is as follows: company C_1 produces 0.60 million dollars worth of smartphones, 0.35 million dollars worth of tablets, and 0.20 million dollars worth of smartwatches. Analogously, company C_2 produces 0.58 million dollars worth of smartphones, 0.32 million dollars worth of tablets, and 0.20 million dollars worth of smartwatches. Similarly, company C_3 produces 0.62 million dollars worth of smartphones, 0.33 million dollars worth of tablets, and 0.21 million dollars worth of smartwatches. Furthermore, suppose that the investments in companies C_1, C_2, and

C_3 are 7, 6, and 4 million dollars, respectively. What is the production of smartphones, tablets, and smartwatches?

First of all, we will represent the information by means of a table.

Table 2.3: Production worth per one million of investment for smartphones, tablets, and smartwatches.

	Company C_1	Company C_2	Company C_3
Smartphones	0.6	0.58	0.62
Tablets	0.35	0.32	0.33
Smartwatches	0.20	0.20	0.21

From Table 2.3, we can compute the total price of smartphones produced by the three companies. The price is $7(0.6) + 6(0.58) + 4(0.62) = 10.16$ million dollars. In the same way, we can compute the total price of tablets and smartwatches, which are $7(0.35) + 6(0.32) + 4(0.33) = 5.69$ and $7(0.20) + 6(0.20) + 4(0.21) = 3.44$ million dollars, respectively.

Pushing the example a little further, assume that the production scheme is kept as before, but now the investments in companies C_1, C_2, and C_3 are x, y, and z dollars, respectively. What is the total production price of smartphones, tablets, and smartwatches produced by the three companies?

Let us use p_1, p_2, p_3 to represent the total production prices of smartphones, tablets, and smartwatches, respectively. Then as before, we have that the total production price of smartphones is $p_1 = 0.6x + 0.58y + 0.62z$. Analogously, we have that $p_2 = 0.35x + 0.32y + 0.33z$ and $p_3 = 0.20x + 0.20y + 0.21z$ are the total production prices of tablets and smartwatches, respectively. These equations can be written in a more compact form as follows:

$$0.6x + 0.58y + 0.62z = p_1,$$
$$0.35x + 0.32y + 0.33z = p_2,$$
$$0.20x + 0.20y + 0.21z = p_3,$$

or in "matrix form"

$$\underbrace{\begin{bmatrix} 0.6 & 0.58 & 0.62 \\ 0.35 & 0.32 & 0.33 \\ 0.20 & 0.20 & 0.21 \end{bmatrix}}_{\text{Production data}} \underbrace{\begin{bmatrix} x \\ y \\ z \end{bmatrix}}_{\text{Investment}} = \underbrace{\begin{bmatrix} p_1 \\ p_2 \\ p_3 \end{bmatrix}}_{\text{Production in dollars}}.$$

Exercise 2.1.1. With a production scheme as given in Table 2.3, what would be the investments in companies C_1, C_2, and C_3 in order to have a production of 27.1, 14.9, and 9.2 million dollars worth of smartphones, tablets, and smartwatches, respectively?

The discussion in Example 2.1.3 can be pushed further, this is done in Example 2.1.4.

Example 2.1.4. Consider a consortium with n factories, and assume that each factory produces m different products. Furthermore, assume that all factories produce the same products. Under these assumptions, the production is described below.

For each million dollars invested in factory 1, the value of the production is as follows:

$$\text{value of product 1 is } a_{11} \text{ dollars,}$$
$$\text{value of product 2 is } a_{21} \text{ dollars,}$$
$$\vdots$$
$$\text{value of product } m \text{ is } a_{m1} \text{ dollars.}$$

In general, for each million dollars that is invested in the jth factory, production is described as follows:

$$\text{value of product 1 is } a_{1j} \text{ dollars,}$$
$$\text{value of product 2 is } a_{2j} \text{ dollars,}$$
$$\vdots$$
$$\text{value of product } m \text{ is } a_{mj} \text{ dollars.}$$

Data above can be represented by Table 2.4, where E_j and P_i represent factory j and product i, respectively.

Table 2.4: Production representation for each factory.

	E_1	E_2	E_3	E_4	$\cdots$	E_n
P_1	a_{11}	a_{12}	a_{13}	a_{14}	$\cdots$	a_{1n}
P_2	a_{21}	a_{22}	a_{23}	a_{24}	$\cdots$	a_{2n}
P_3	a_{31}	a_{32}	a_{33}	a_{34}	$\cdots$	a_{3n}
$\vdots$	$\vdots$	$\vdots$	$\vdots$	$\vdots$	$\ddots$	$\vdots$
P_m	a_{m1}	a_{m2}	a_{m3}	a_{m4}	$\cdots$	a_{mn}

If in each factory the investment is one million dollars, then the total value of product 1 is

$$a_{11} + a_{12} + \cdots + a_{1n}.$$

In general, the total value of product i is

$$a_{i1} + a_{i2} + \cdots + a_{in}.$$

If we denote by x_j the amount of money invested in factory j, then the total value of product i is

$$a_{i1}x_1 + a_{i2}x_2 + \cdots + a_{in}x_n = c_i.$$

This information can be represented in the form

$$
\underbrace{\begin{bmatrix}
a_{11} & a_{12} & \cdots & a_{1n} \\
a_{21} & a_{22} & \cdots & a_{2n} \\
a_{31} & a_{32} & \cdots & a_{3n} \\
\vdots & \vdots & \ddots & \vdots \\
a_{m1} & a_{m2} & \cdots & a_{mn}
\end{bmatrix}}_{\text{Production data}}
\underbrace{\begin{bmatrix}
x_1 \\ x_2 \\ x_3 \\ \vdots \\ x_n
\end{bmatrix}}_{\text{Investment}}
\implies
\underbrace{\begin{bmatrix}
c_1 \\ c_2 \\ c_3 \\ \vdots \\ c_m
\end{bmatrix}}_{\text{Production in dollars}}
$$

or in the form

$$
\underbrace{\begin{bmatrix}
a_{11} & a_{12} & \cdots & a_{1n} \\
a_{21} & a_{22} & \cdots & a_{2n} \\
a_{31} & a_{32} & \cdots & a_{3n} \\
\vdots & \vdots & \ddots & \vdots \\
a_{m1} & a_{m2} & \cdots & a_{mn}
\end{bmatrix}}_{\text{Production data}}
\underbrace{\begin{bmatrix}
x_1 \\ x_2 \\ x_3 \\ \vdots \\ x_n
\end{bmatrix}}_{\text{Investment}}
=
\underbrace{\begin{bmatrix}
c_1 \\ c_2 \\ c_3 \\ \vdots \\ c_m
\end{bmatrix}}_{\text{Production in dollars}} ,
$$

where equality is interpreted as

$$c_i = a_{i1}x_1 + a_{i2}x_2 + \cdots + a_{in}x_n, \quad \text{for every } i \in \{1, 2, \ldots, m\}. \tag{2.1}$$

With this notation, we have

$$
\begin{bmatrix}
a_{11} & a_{12} & \cdots & a_{1n} \\
a_{21} & a_{22} & \cdots & a_{2n} \\
a_{31} & a_{32} & \cdots & a_{3n} \\
\vdots & \vdots & \ddots & \vdots \\
a_{m1} & a_{m2} & \cdots & a_{mn}
\end{bmatrix}
\begin{bmatrix}
x_1 \\ x_2 \\ x_3 \\ \vdots \\ x_n
\end{bmatrix}
=
\begin{bmatrix}
a_{11}x_1 + a_{12}x_2 + \cdots + a_{1n}x_n \\
a_{21}x_1 + a_{22}x_2 + \cdots + a_{2n}x_n \\
a_{31}x_1 + a_{32}x_2 + \cdots + a_{3n}x_n \\
\vdots \\
a_{m1}x_1 + a_{m2}x_2 + \cdots + a_{mn}x_n
\end{bmatrix}
=
\begin{bmatrix}
c_1 \\ c_2 \\ c_3 \\ \vdots \\ c_m
\end{bmatrix} .
$$

The equality above is the definition of the product of matrices A and X where

$$
A = \begin{bmatrix}
a_{11} & a_{12} & \cdots & a_{1n} \\
a_{21} & a_{22} & \cdots & a_{2n} \\
a_{31} & a_{32} & \cdots & a_{3n} \\
\vdots & \vdots & \ddots & \vdots \\
a_{m1} & a_{m2} & \cdots & a_{mn}
\end{bmatrix}
\quad \text{and} \quad
X = \begin{bmatrix}
x_1 \\ x_2 \\ x_3 \\ \vdots \\ x_n
\end{bmatrix} .
$$

In a more compact notation, we write $AX = C$, where the entries of C are $c_1, c_2, \ldots, c_m$, as described in (2.1).

Assume that the consortium manages n factories and it is required to keep records of the production during a period of p years, keeping the production conditions fixed. Can we use matrix notation to analyze the information during that period of time?

We will start by introducing notation to represent what is invested yearly in each company. For instance, let us denote by b_{11} the amount of money invested in factory 1 in the first year; let b_{12} denote the investment in the same company during the second year. In general, let b_{1j} denote the amount of money invested in company 1 in the jth year. Following these ideas and notation, matrix (2.2) represents the investments of the companies during p years. The interpretation is as follows: row 1 represents the investments in company 1 in the period of p years. Likewise, row 2 represents the investments in company 2 in the same period, etc.,

$$B = \begin{bmatrix} b_{11} & b_{12} & \cdots & b_{1p} \\ b_{21} & b_{22} & \cdots & b_{2p} \\ b_{31} & b_{32} & \cdots & b_{3p} \\ \vdots & \vdots & \ddots & \vdots \\ b_{n1} & b_{n2} & \cdots & b_{np} \end{bmatrix}. \tag{2.2}$$

Given that each year m consumer items will be produced, let c_{11} denote the total value of product 1 in year 1, c_{12} the total value of product 1 in year 2. In general, let c_{1j} denote the total value of product 1 in the jth year. From this, we have that the total value of the ith product in the jth year is denoted by c_{ij}.

With this notation, the value matrix during that period is

$$C = \begin{bmatrix} c_{11} & c_{12} & \cdots & c_{1p} \\ c_{21} & c_{12} & \cdots & c_{2p} \\ c_{31} & c_{32} & \cdots & c_{3p} \\ \vdots & \vdots & \ddots & \vdots \\ c_{m1} & a_{m2} & \cdots & c_{mp} \end{bmatrix}.$$

If

$$A = \begin{bmatrix} a_{11} & a_{12} & \cdots & a_{1n} \\ a_{21} & a_{122} & \cdots & a_{2n} \\ a_{31} & a_{32} & \cdots & a_{3n} \\ \vdots & \vdots & \ddots & \vdots \\ a_{m1} & a_{m2} & \cdots & a_{mn} \end{bmatrix}$$

represents the production conditions, then the production process can be represented by the product $AB = C$ which is specified in

$$\begin{bmatrix} a_{11} & a_{12} & \cdots & a_{1n} \\ a_{21} & a_{122} & \cdots & a_{2n} \\ a_{31} & a_{32} & \cdots & a_{3n} \\ \vdots & \vdots & \ddots & \vdots \\ a_{m1} & a_{m2} & \cdots & a_{mn} \end{bmatrix} \begin{bmatrix} b_{11} & b_{12} & \cdots & b_{1p} \\ b_{21} & b_{122} & \cdots & b_{2p} \\ b_{31} & b_{32} & \cdots & b_{3p} \\ \vdots & \vdots & \ddots & \vdots \\ b_{n1} & b_{p2} & \cdots & b_{np} \end{bmatrix} = \begin{bmatrix} c_{11} & c_{12} & \cdots & c_{1p} \\ c_{21} & c_{12} & \cdots & c_{2p} \\ c_{31} & c_{32} & \cdots & c_{3p} \\ \vdots & \vdots & \ddots & \vdots \\ c_{m1} & a_{m2} & \cdots & c_{mp} \end{bmatrix}.$$

The previous symbols mean, according to the production conditions:

$$c_{11} = a_{11}b_{11} + a_{12}b_{21} + \cdots + a_{1n}b_{n1},$$
$$c_{12} = a_{11}b_{12} + a_{12}b_{22} + \cdots + a_{1n}b_{n2}.$$

In general,

$$c_{ij} = a_{i1}b_{1j} + a_{i2}b_{2j} + \cdots + a_{in}b_{nj}, \tag{2.3}$$

for every $i \in \{1, \ldots, m\}$ and $j \in \{1, \ldots, p\}$.

The previous discussion can be used to define the product of two matrices of "appropriate" sizes.

Definition 2.1.2. If A and B are matrices of size $m \times n$ and $n \times p$, respectively, their product is defined by $AB = C$, where C is a matrix of size $m \times p$ whose entries are defined by equation (2.3).

With this terminology, we can formulate the problem of solving a system of linear equations in a very convenient form. Given the system

$$a_{11}x_1 + a_{12}x_2 + \cdots + a_{1n}x_n = b_1,$$
$$a_{21}x_1 + a_{22}x_2 + \cdots + a_{2n}x_n = b_2,$$
$$\vdots \tag{2.4}$$
$$a_{m1}x_1 + a_{m1}x_1 + \cdots + a_{mn}x_n = b_m,$$

and defining A, X, and B as

$$A = \begin{bmatrix} a_{11} & a_{12} & \cdots & a_{1n} \\ a_{21} & a_{22} & \cdots & a_{2n} \\ a_{31} & a_{32} & \cdots & a_{3n} \\ \vdots & \vdots & \ddots & \vdots \\ a_{m1} & a_{m2} & \cdots & a_{mn} \end{bmatrix}, \quad X = \begin{bmatrix} x_1 \\ x_2 \\ x_3 \\ \vdots \\ x_n \end{bmatrix}, \quad \text{and} \quad B = \begin{bmatrix} b_1 \\ b_2 \\ b_3 \\ \vdots \\ b_m \end{bmatrix},$$

the system (2.4) can be represented in the form $AX = B$. This representation has several advantages. For instance, we can "solve" for X by "dividing" by A. This actually means that we can obtain X by multiplying by the *inverse* of A. This formulation has one advantage, when A is a square matrix, solving the system is reduced to deciding when A has an inverse and if so, $X = A^{-1}B$.

Exercise 2.1.2. With a production scheme as that given in Table 2.3, what would be the investment in companies C_1, C_2, and C_3 during two consecutive years to have a total production according to Table 2.5?

Table 2.5: Production of smartphones, tablets, and smartwatches during a two-year period.

Total Production	Smartphones	Tablets	Smartwatches
First year	37.76	21.09	12.78
Second year	45.78	25.37	15.5

2.1.3 Properties of the sum and product of matrices

The number systems (integers, rationals, and real numbers), usually studied in high school mathematics courses, have two operations, addition and multiplication. These operations have several properties such as associativity, commutativity, and distributivity of the product with respect to the sum. Taking this into account, it is natural to ask if the product and sum of matrices satisfy some of those properties.

In the definition of the sum of matrices, it has been emphasized that to add two matrices it is needed that the matrices have the same size. The set of matrices of size $m \times n$ with entries in the real numbers will be denoted by $\mathcal{M}_{m\times n}(\mathbb{R})$.

To multiply two matrices A and B, it is necessary that the number of columns of the left-hand side factor be equal to the number of rows of the right-hand side factor. With the above notation, AB is defined only when $A \in \mathcal{M}_{m\times n}(\mathbb{R})$ and $B \in \mathcal{M}_{n\times p}(\mathbb{R})$. Under these conditions, we might ask which properties of the product and sum are satisfied.

Properties of the sum of matrices

To analyze properties of the sum of matrices, we will assume that A, B, and C are elements of $\mathcal{M}_{m\times n}(\mathbb{R})$. Since the justification of the properties of the sum is straightforward, we do not state them as a theorem.

1. If A and B are matrices, with entries a_{ij} and b_{ij}, respectively, for $i, j \in \{1, 2, \ldots, n\}$, then a typical entry of $A + B$ has the form $a_{ij} + b_{ij}$. Now, since a_{ij} and b_{ij} are real numbers and the sum of real numbers is commutative, we get $a_{ij} + b_{ij} = b_{ij} + a_{ij}$, so $A + B = B + A$, i. e., the *commutativity of the sum*.

2. There is a matrix that will be denoted by 0, whose entries are all zero. This matrix satisfies $A + 0 = 0 + A = A$, for every A, showing the *existence of addition identity*.

3. Given the matrix A, there is $-A$ such that $A + (-A) = 0$. If the entries of A are a_{ij} with $i, j \in \{1, 2, \ldots, n\}$, then the entries of $-A$ are $-a_{ij}$. It is clear that $a_{ij} + (-a_{ij}) = 0$, establishing the *existence of an additive inverse*.

4. If A, B, and C are matrices, then $(A + B) + C = A + (B + C)$. Equality is obtained by writing the elements of $(A + B) + C$ and using the associativity property of the real numbers, yielding the *associativity of the sum*.

Properties of the product of matrices

In Theorem 2.1.1, we state the properties of the product of matrices, proving only the associativity property, inviting the reader to provide the proof of the others.

Theorem 2.1.1. *Assume that $A \in \mathcal{M}_{m \times n}(\mathbb{R})$, then the following holds:*

Associativity: *If $B \in \mathcal{M}_{n \times p}(\mathbb{R})$ and $C \in \mathcal{M}_{p \times q}(\mathbb{R})$, then $(AB)C = A(BC)$.*

Distributivity on the right: *If $B \in \mathcal{M}_{m \times n}(\mathbb{R})$ and $C \in \mathcal{M}_{n \times p}(\mathbb{R})$, then $(A+B)C = AC+BC$.*

Distributivity on the left: *If $B \in \mathcal{M}_{m \times n}(\mathbb{R})$ and $C \in \mathcal{M}_{p \times m}(\mathbb{R})$, then $C(A + B) = CA + CB$.*

Identity on the left: *There is a matrix of size $m \times m$ denoted I_m, whose entries c_{ij} satisfy*

$$c_{ij} = \begin{cases} 0 & \text{if } i \neq j, \\ 1 & \text{if } i = j, \end{cases} \text{ such that } I_m A = A.$$

Identity on the right: *There is a matrix of size $n \times n$, denoted I_n, whose entries c_{ij}, with*

$$i, j \in \{1, 2, \ldots, n\}, \text{ satisfy } c_{ij} = \begin{cases} 0 & \text{if } i \neq j, \\ 1 & \text{if } i = j, \end{cases} \text{ such that } AI_n = A.$$

Proof. To prove the first property, we notice that the products $(AB)C$ and $A(BC)$ are defined and are matrices of size $m \times q$. Set $(AB)C = D$ and $A(BC) = E$, with entries d_{ij} and e_{ij}, respectively. We will prove that $d_{ij} = e_{ij}$ for every $i \in \{1, 2, \ldots, m\}$ and for every $j \in \{1, 2, \ldots, q\}$.

We recall how the product of matrices is defined. For instance, the element in row i and column j in AB is obtained by multiplying the ith row of A and the jth column of B. Such an element is given by $c_{ij} = a_{i1}b_{1j} + a_{i2}b_{2j} + \cdots + a_{in}b_{nj}$. Hence the elements of row i in AB are obtained by taking the row i of A and letting j vary in the set $\{1, 2, \ldots, p\}$.

From this, the ith row in AB is

$$[a_{i1}b_{11} + a_{i2}b_{21} + \cdots + a_{in}b_{n1} \quad a_{i1}b_{12} + a_{i2}b_{22} + \cdots + a_{in}b_{n2} \quad \cdots \quad a_{i1}b_{1p} + a_{i2}b_{2p} + \cdots + a_{in}b_{np}].$$

Using this and the definition of the product of matrices, one has

$$d_{ij} = (a_{i1}b_{11} + a_{i2}b_{21} + \cdots + a_{in}b_{n1})c_{1j} + (a_{i1}b_{12} + a_{i2}b_{22} + \cdots + a_{in}b_{n2})c_{2j}$$
$$+ \cdots + (a_{i1}b_{1p} + a_{i2}b_{2p} + \cdots + a_{in}b_{np})c_{pj}.$$

Likewise, to obtain e_{ij}, we take the ith row of A and the jth column of BC, the latter being

$$
\begin{bmatrix}
b_{11}c_{1j} + b_{12}c_{2j} + \cdots + b_{1p}c_{pj} \\
b_{21}c_{1j} + b_{22}c_{2j} + \cdots + b_{2p}c_{pj} \\
\vdots \\
b_{n1}c_{1j} + b_{n2}c_{2j} + \cdots + b_{np}c_{pj}
\end{bmatrix}.
$$

Hence, applying the definition of $A(BC)$, we have

$$
e_{ij} = a_{i1}(b_{11}c_{1j} + b_{12}c_{2j} + \cdots + b_{1p}c_{pj}) + a_{i2}(b_{21}c_{1j} + b_{22}c_{2j} + \cdots + b_{2p}c_{pj})
$$
$$
+ \cdots + a_{in}(b_{n1}c_{1j} + b_{n2}c_{2j} + \cdots + b_{np}c_{pj}).
$$

Now, the equality $d_{ij} = e_{ij}$ is obtained by comparing terms in their representations. □

Remark 2.1.1. We notice that if A and B are matrices such that AB is defined, then the product of B and A is not necessarily defined. It might happen that AB and BA are defined and $AB \neq BA$. Another possibility is that $AB = 0$, and neither A nor B is zero. Actually, this property occurs when solving systems of homogeneous linear equations.

2.2 Elementary matrices and the inverse of a matrix

In this section we will present a discussion of the identification of elementary operations on the rows of a matrix, and its relation with some special matrices called elementary matrices. Also, we will develop a method to decide when a square matrix has an inverse and, in the affirmative case, to compute it.

Recall that a system of linear equations can be represented in the form $AX = B$. If A has an inverse, A^{-1}, then $A^{-1}AX = A^{-1}B$. From this and $I_nX = X$, we have $X = A^{-1}B$, that is, the system has been solved.

When using the Gauss–Jordan reduction method, we identify three types of row operation on a matrix:
1. Swap any two rows.
2. Multiply a row by a nonzero real number.
3. Multiply a row by a real number and add the result to another row.

In what follows, we will identify each of these operations with matrices obtained from the identity matrix.

Assume that we have an $m \times n$ matrix

$$
A = \begin{pmatrix}
a_{11} & a_{12} & a_{13} & \cdots & a_{1n} \\
a_{21} & a_{22} & a_{23} & \cdots & a_{2n} \\
\vdots & \vdots & \vdots & \ddots & \vdots \\
a_{m1} & a_{m2} & a_{m3} & \cdots & a_{mn}
\end{pmatrix},
$$

and let I_m denote the $m \times m$ identity matrix.

Swapping rows. Let E_{ik} be the matrix that is obtained from I_m by swapping rows i and k. What is the result when computing $E_{ik}A$?

Multiplying a row. Let $r \neq 0$ and let $E_i(r)$ be the matrix obtained from I_m when multiplying row i by r. What is the result when computing $E_i(r)A$?

Multiplying a row and adding it to another row. Let $E_{ik}(r)$ be the matrix that is obtained from I_m when multiplying row k by r and adding the result to row i. What is the result when computing $E_{ik}(r)A$?

Examples to help identify possible answers to the posed questions

1. Take $m = 2$ and $n = 3$ in the previous questions. If $I_2 = \left(\begin{smallmatrix} 1 & 0 \\ 0 & 1 \end{smallmatrix}\right)$ and $A = \left(\begin{smallmatrix} a_{11} & a_{12} & a_{13} \\ a_{21} & a_{22} & a_{23} \end{smallmatrix}\right)$, then $E_{12} = \left(\begin{smallmatrix} 0 & 1 \\ 1 & 0 \end{smallmatrix}\right)$ and $E_{12}A = \left(\begin{smallmatrix} 0 & 1 \\ 1 & 0 \end{smallmatrix}\right)\left(\begin{smallmatrix} a_{11} & a_{12} & a_{13} \\ a_{21} & a_{22} & a_{23} \end{smallmatrix}\right) = \left(\begin{smallmatrix} a_{21} & a_{22} & a_{23} \\ a_{11} & a_{12} & a_{13} \end{smallmatrix}\right)$. Perform the calculations for the cases 2 and 3 of the posed questions.

2. Take $m = 3$ and $n = 4$ and proceed as before. Notice that in these cases you have to be careful when defining the matrices E_{ij}, more precisely, we have the following matrices:

$$E_{12} = \begin{pmatrix} 0 & 1 & 0 \\ 1 & 0 & 0 \\ 0 & 0 & 1 \end{pmatrix}, \quad E_{13} = \begin{pmatrix} 0 & 0 & 1 \\ 0 & 1 & 0 \\ 1 & 0 & 0 \end{pmatrix}, \quad \text{and} \quad E_{23} = \begin{pmatrix} 1 & 0 & 0 \\ 0 & 0 & 1 \\ 0 & 1 & 0 \end{pmatrix}.$$

If

$$A = \begin{pmatrix} a_{11} & a_{12} & a_{13} & a_{14} \\ a_{21} & a_{22} & a_{23} & a_{24} \\ a_{31} & a_{32} & a_{33} & a_{34} \end{pmatrix}$$

then we have

$$E_{12}A = \begin{pmatrix} 0 & 1 & 0 \\ 1 & 0 & 0 \\ 0 & 0 & 1 \end{pmatrix} \begin{pmatrix} a_{11} & a_{12} & a_{13} & a_{14} \\ a_{21} & a_{22} & a_{23} & a_{24} \\ a_{31} & a_{32} & a_{33} & a_{34} \end{pmatrix} = \begin{pmatrix} a_{21} & a_{22} & a_{23} & a_{24} \\ a_{11} & a_{12} & a_{13} & a_{14} \\ a_{31} & a_{32} & a_{33} & a_{34} \end{pmatrix},$$

$$E_{13}A = \begin{pmatrix} 0 & 0 & 1 \\ 0 & 1 & 0 \\ 1 & 0 & 0 \end{pmatrix} \begin{pmatrix} a_{11} & a_{12} & a_{13} & a_{14} \\ a_{21} & a_{22} & a_{23} & a_{24} \\ a_{31} & a_{32} & a_{33} & a_{34} \end{pmatrix} = \begin{pmatrix} a_{31} & a_{32} & a_{33} & a_{34} \\ a_{21} & a_{22} & a_{23} & a_{24} \\ a_{11} & a_{12} & a_{13} & a_{14} \end{pmatrix},$$

$$E_{23}A = \begin{pmatrix} 1 & 0 & 0 \\ 0 & 0 & 1 \\ 0 & 1 & 0 \end{pmatrix} \begin{pmatrix} a_{11} & a_{12} & a_{13} & a_{14} \\ a_{21} & a_{22} & a_{23} & a_{24} \\ a_{31} & a_{32} & a_{33} & a_{34} \end{pmatrix} = \begin{pmatrix} a_{11} & a_{12} & a_{13} & a_{14} \\ a_{31} & a_{32} & a_{33} & a_{34} \\ a_{21} & a_{22} & a_{23} & a_{24} \end{pmatrix}.$$

3. Do you have any general conclusion to answer the posed questions? Explain.

Definition 2.2.1. An $n \times n$ matrix A is said to have an inverse if there is an $n \times n$ matrix B such that $AB = BA = I_n$.

Remark 2.2.1. The inverse of a matrix is unique, that is, if there are matrices B and C such that $AB = BA = I_n = AC = CA$, then $C = CI_n = CAB = (CA)B = I_nB = B$, proving that $B = C$. The inverse of A is denoted by A^{-1}.

Exercise 2.2.1. Prove that the elementary matrices have inverses.

In the language of matrices, Theorem 1.3.2 can be stated as follows:

Theorem 2.2.1. *If A is a nonzero matrix, then there exist elementary matrices $E_1, E_2, \ldots, E_t$ so that $E_t \cdots E_2 E_1 A = R$, where R is the row reduced form of A.*

Proof. By Theorem 1.3.2, there are finitely many elementary row operations $O_1, O_2, \ldots,$ O_k so that, when applied to A, it is transformed to a row reduced form, R. Since the effect on A of each elementary row operation is obtained by multiplying A by the corresponding elementary matrix, there are elementary matrices $E_1, E_2, \ldots, E_k$ so that $E_k \cdots E_1 A = R$, finishing the proof. $\qquad \square$

A very important case is when A is a square matrix, that is, $m = n$. Under this assumption, we will find conditions when A has an inverse. Also, we will develop a method to compute A^{-1}. With this and Remark 2.2.1 in mind, we will have a method to solve a square linear system of equations.

Assume that A is an $n \times n$ matrix, and let $E_1, \ldots, E_t$ be elementary matrices so that $E_t \cdots E_1 A = R$ is the row reduced form of A. Recall that R is a matrix whose nonzero rows have the main entry equal to 1. The following remark will be important to decide if A has an inverse.

Remark 2.2.2. Let R be the row reduced form of an $n \times n$ matrix A, then R has no zero rows if and only if $R = I_n$.

Proof. In fact, if $R = I_n$, it is clear that all the rows of R are not zero. Conversely, if R has only nonzero rows, then R has n nonzero rows and the condition on R, being row reduced, implies $R = I_n$. $\qquad \square$

There is one additional important remark.

Remark 2.2.3.
1. If two matrices A and B have inverses, then AB has an inverse and $(AB)^{-1} = B^{-1}A^{-1}$.
2. If A has a zero row and B is another matrix so that AB is defined, then AB has a zero row, particularly, if A is a square matrix, then A has no inverse. What is the result if B has a zero column?

From Remarks 2.2.2 and 2.2.3, we derive the following important result:

Theorem 2.2.2. *Let A be an $n \times n$ matrix, and let $E_1, \ldots, E_t$ be elementary matrices so that $E_t \cdots E_1 A = R$ is the row reduced form of A. Then A has an inverse if and only if $R = I_n$, and in this case $A^{-1} = E_t \cdots E_1$.*

Proof. If $R = I_n$, it is clear that the inverse of A is $E_t \cdots E_1$. Assume that A has an inverse. From Remark 2.2.3(1) and the fact that elementary matrices have inverses, Exercise 2.2.1, one concludes that $R = E_t \cdots E_1 A$ has an inverse. Now applying Remark 2.2.3(2), we have that the rows of R are nonzero. The final conclusion follows from Remark 2.2.2. □

Theorem 2.2.2 is the foundation of one of the most interesting methods to compute the inverse of a matrix. It is known as the method to compute the inverse by means of elementary operations. In the following subsection, we will present a practical application of Theorem 2.2.2.

2.2.1 Computing the inverse of a matrix

Assume that A is an $n \times n$ matrix and consider the $n \times 2n$ matrix whose first n columns are equal to those of A and the others are equal to those of I_n. Abusing the notation, the new matrix can be represented by $[A : I_n]$.

If $E_1, \ldots, E_t$ are elementary matrices so that $E_t \cdots E_1 A = R$, then applying row operations, represented by elementary matrices $E_1, \ldots, E_t$, on the rows of $[A : I_n]$, we will end up with the matrix $[R : E_1 \cdots E_t]$. From Theorem 2.2.2, A has an inverse if and only if $R = I_n$ and A^{-1} is the matrix that appears on the right in $[R : E_1 \cdots E_t]$, that is, $A^{-1} = E_1 \cdots E_t$.

Example 2.2.1. Consider the matrix

$$A := \begin{bmatrix} 1 & 4 & 7 \\ 2 & 5 & 8 \\ 3 & 6 & 9 \end{bmatrix}.$$

Decide if A has inverse and, if so, find it.

Discussion. We apply the method described above.

Starting from A and the identity I_3, form the matrix

$$A_1 := \left[\begin{array}{ccc|ccc} 1 & 4 & 7 & 1 & 0 & 0 \\ 2 & 5 & 8 & 0 & 1 & 0 \\ 3 & 6 & 9 & 0 & 0 & 1 \end{array} \right].$$

Applying row operations to A_1, we have

$$A_1 := \left[\begin{array}{ccc|ccc} 1 & 4 & 7 & 1 & 0 & 0 \\ 2 & 5 & 8 & 0 & 1 & 0 \\ 3 & 6 & 9 & 0 & 0 & 1 \end{array} \right] \rightarrow A_2 := \left[\begin{array}{ccc|ccc} 1 & 4 & 7 & 1 & 0 & 0 \\ 0 & -3 & -6 & -2 & 1 & 0 \\ 0 & -6 & -12 & -3 & 0 & 1 \end{array} \right]$$

$$\rightarrow A_3 := \left[\begin{array}{ccc|ccc} 1 & 4 & 7 & 1 & 0 & 0 \\ 0 & -3 & -6 & 2 & 1 & 0 \\ 0 & 0 & 0 & 1 & -2 & 1 \end{array} \right].$$

At this point we can stop the calculations and conclude that A has no inverse, since the last row of R will be zero, and, according to Remark 2.2.2 and Theorem 2.2.2, A has no inverse.

We can verify the calculations using SageMath. In order to use SageMath to verify the calculations, we need to apply a sequence of commands. The first and second define the matrix A and the identity, respectively, the third defines the matrix where row operations will be performed, and the last is the one that gives the row reduced echelon form, from where we can decide if A has and inverse and, if so, identify it.

```
A=matrix(QQ, [[1,4,7],[2,5,8],[3,6,9]])
B=identity_matrix(3)
C=A.augment(B)
D=C.rref()
```

$$D = \begin{pmatrix} 1 & 0 & -1 & 0 & -2 & \frac{5}{3} \\ 0 & 1 & 2 & 0 & 1 & -\frac{2}{3} \\ 0 & 0 & 0 & 1 & -2 & 1 \end{pmatrix}.$$

Exercise 2.2.2. Choose several 4×4 matrices with rational entries and decide if they have an inverse or not. Use SageMath to verify your answers.

Example 2.2.2. Determine if the matrix

$$A := \begin{bmatrix} 1 & 2 & 0 \\ 2 & 5 & 8 \\ 3 & 6 & 9 \end{bmatrix}$$

has an inverse and, if so, find it.

Discussion. Applying the method, we consider the matrix

$$A_1 := \begin{bmatrix} 1 & 2 & 0 & 1 & 0 & 0 \\ 2 & 5 & 8 & 0 & 1 & 0 \\ 3 & 6 & 9 & 0 & 0 & 1 \end{bmatrix}.$$

Applying row operations successively on the rows we obtain

$$A_1 := \begin{bmatrix} 1 & 2 & 0 & 1 & 0 & 0 \\ 2 & 5 & 8 & 0 & 1 & 0 \\ 3 & 6 & 9 & 0 & 0 & 1 \end{bmatrix} \rightarrow A_2 := \begin{bmatrix} 1 & 2 & 0 & 1 & 0 & 0 \\ 0 & 1 & 8 & -2 & 1 & 0 \\ 3 & 6 & 9 & 0 & 0 & 1 \end{bmatrix} \rightarrow$$

$$A_2 := \begin{bmatrix} 1 & 2 & 0 & 1 & 0 & 0 \\ 0 & 1 & 8 & -2 & 1 & 0 \\ 3 & 6 & 9 & 0 & 0 & 1 \end{bmatrix} \rightarrow A_3 := \begin{bmatrix} 1 & 2 & 0 & 1 & 0 & 0 \\ 0 & 1 & 8 & -2 & 1 & 0 \\ 0 & 0 & 1 & -\frac{1}{3} & 0 & \frac{1}{9} \end{bmatrix}.$$

At this point we can say that A has an inverse (why?), and we can find it. Continue applying row operations up to the point where the left-hand side block is the identity. On the right-hand side, the block that we have is the inverse of A,

$$A_6 := \begin{bmatrix} 1 & 0 & 0 & | & -\frac{1}{3} & -2 & \frac{16}{9} \\ 0 & 1 & 0 & | & \frac{2}{3} & 1 & -\frac{8}{9} \\ 0 & 0 & 1 & | & -\frac{1}{3} & 0 & \frac{1}{9} \end{bmatrix},$$

which is

$$A^{-1} = \begin{bmatrix} -\frac{1}{3} & -2 & \frac{16}{9} \\ \frac{2}{3} & 1 & -\frac{8}{9} \\ -\frac{1}{3} & 0 & \frac{1}{9} \end{bmatrix}.$$

Using SageMath to verify the answer

```
X=matrix(QQ, [[1,2,0],[2,5,8],[3,6,9]])
Y=identity_matrix(3)
W=X.augment(Y)
E=W.rref()
```

$$E = \begin{pmatrix} 1 & 0 & 0 & -\frac{1}{3} & -2 & \frac{16}{9} \\ 0 & 1 & 0 & \frac{2}{3} & 1 & -\frac{8}{9} \\ 0 & 0 & 1 & -\frac{1}{3} & 0 & \frac{1}{9} \end{pmatrix}.$$

From E, we identify the inverse of A.

Using SageMath, there is a direct way to compute A^{-1}:

```
X=matrix(QQ, [[1,2,0],[2,5,8],[3,6,9]])
Y=X.inverse()
```

$$X^{-1} = \begin{pmatrix} -\frac{1}{3} & -2 & \frac{16}{9} \\ \frac{2}{3} & 1 & -\frac{8}{9} \\ -\frac{1}{3} & 0 & \frac{1}{9} \end{pmatrix}.$$

Example 2.2.3. Solve the following system of linear equations by finding the inverse of the coefficient matrix:

$$x + 2y = 3,$$
$$2x + 5y + 8z = 1,$$
$$3x + 6y + 9z = 2.$$

Discussion. The system can be represented in the form $AX = B$, where

$$A = \begin{bmatrix} 1 & 2 & 0 \\ 2 & 5 & 8 \\ 3 & 6 & 9 \end{bmatrix}, \quad X = \begin{bmatrix} x \\ y \\ z \end{bmatrix}, \quad \text{and} \quad B = \begin{bmatrix} 3 \\ 1 \\ 2 \end{bmatrix}.$$

Since A is the matrix in Example 2.2.2, and we know it has an inverse, the equation $AX = B$ is equivalent to $X = A^{-1}B$. Using the result in the mentioned example, we have

$$X = \begin{bmatrix} x \\ y \\ z \end{bmatrix} = \begin{bmatrix} -\frac{1}{3} & -2 & \frac{16}{9} \\ \frac{2}{3} & 1 & -\frac{8}{9} \\ -\frac{1}{3} & 0 & \frac{1}{9} \end{bmatrix} \begin{bmatrix} 3 \\ 1 \\ 2 \end{bmatrix} = \begin{bmatrix} \frac{5}{9} \\ \frac{11}{9} \\ -\frac{7}{9} \end{bmatrix}.$$

This matrix equation is equivalent to $x = \frac{5}{9}, y = \frac{11}{9}$, and $z = -\frac{7}{9}$, that is, the given system has a unique solution $\left(\frac{5}{9}, \frac{11}{9}, -\frac{7}{9} \right)$.

Exercise 2.2.3. In each of the following cases, determine if A has an inverse and, if so, find it:

- Let a and b be different real numbers and $A = \left(\begin{smallmatrix} 1 & a \\ 1 & b \end{smallmatrix} \right)$.
- Let a, b, and c be three pairwise different real numbers and $A = \left(\begin{smallmatrix} 1 & a & a^2 \\ 1 & b & b^2 \\ 1 & c & c^2 \end{smallmatrix} \right)$.

Exercise 2.2.4. Let $A = \left(\begin{smallmatrix} a_{11} & a_{12} \\ a_{21} & a_{22} \end{smallmatrix} \right)$ be such that $\Delta = a_{11}a_{22} - a_{12}a_{21} \neq 0$. Determine if A^{-1} exists and, if so, obtain it.

When studying solutions of systems of linear equations, or criteria to decide if a matrix has an inverse, several terms are needed. In this line of ideas, the following definition and theorem are in order.

Definition 2.2.2. Let A and B be two $m \times n$ matrices. We say that A is row equivalent to B if there are elementary matrices $E_1, E_2, \ldots, E_t$ so that $A = E_1 E_2 \cdots E_t B$.

Theorem 2.2.3 is one of the most important results in linear algebra, this will be evolving through the text to become a theorem with about 13 equivalent statements.

Theorem 2.2.3. *If A is an $n \times n$ matrix, then the following conditions are equivalent:*
1. *The matrix A has an inverse.*
2. *The matrix A is row equivalent to I_n.*
3. *The matrix A is a product of elementary matrices.*
4. *The system of linear equations $AX = 0$ has a unique solution.*
5. *For every $B = \begin{bmatrix} b_1 \\ b_2 \\ \vdots \\ b_n \end{bmatrix}$, the equation $AX = B$ has a unique solution, where $X = \begin{bmatrix} x_1 \\ x_2 \\ \vdots \\ x_n \end{bmatrix}$.*

Note 1. As an excellent exercise, we ask the reader to provide the details of the proof of Theorem 2.2.3; see Exercise 3, p. 64.

The following definitions are important to classify different types of matrix.

Definition 2.2.3. If A is a square matrix that satisfies any of the conditions of Theorem 2.2.3, A is called nonsingular. Otherwise, A is called singular.

Definition 2.2.4. Let $A = (a_{ij})$ be an $m \times n$ matrix. The transpose of A is defined by $A^t = (b_{ij})$, where $b_{ij} = a_{ji}$ for $i \in \{1, 2, \dots, n\}$ and $j \in \{1, 2, \dots, m\}$. If A is a matrix such that $A^t = A$ ($A^t = -A$), we say that A is symmetric (antisymmetric). Notice that in this case necessarily A is a square matrix. When $m = n$, we define the *trace* of A, denoted by $\text{tr}(A) = \sum_{i=1}^{n} a_{ii}$.

Definition 2.2.5. An $n \times n$ matrix A is called lower (upper) triangular if $a_{ij} = 0$ for every $i < j \le n$ ($a_{ij} = 0$ for every $j < i \le n$). If $a_{ij} = 0$ for every $i \ne j$, A is called a diagonal matrix and denoted by $A = \text{diag}\{a_{11}, a_{22}, \dots, a_{nn}\}$.

Remark 2.2.4. If A is antisymmetric, then the elements a_{ii} are zero. If A is any square matrix, then the matrices $B = A^t + A$ and $C = A^t - A$ are symmetric and antisymmetric, respectively. It is straightforward to verify that $A = \frac{1}{2}(B - C)$.

Example 2.2.4. If $A = \begin{pmatrix} a_{11} & a_{12} & a_{13} & a_{14} \\ a_{21} & a_{22} & a_{23} & a_{24} \\ a_{31} & a_{32} & a_{33} & a_{34} \end{pmatrix}$, then $A^t = \begin{pmatrix} a_{11} & a_{21} & a_{31} \\ a_{12} & a_{22} & a_{32} \\ a_{13} & a_{23} & a_{33} \\ a_{14} & a_{24} & a_{34} \end{pmatrix}$.

Some properties of the trace and transpose of a matrix

The following statements are related to the transpose and trace of a matrix. It is important to discuss them; we ask the reader to provide an answer.

1. Let A and B be matrices so that their product is defined. Prove that $(AB)^t = B^t A^t$.
2. Assume that A has an inverse, is it true that A^t has an inverse?
3. For a square matrix A, when is $\text{tr}(AA^t) = 0$?
4. If A and B are square matrices, compute $\text{tr}(AB)$ and $\text{tr}(BA)$. Is there any relationship between those numbers?
5. Is it true that an invertible matrix has trace zero?
6. Is $\text{tr}(A + B) = \text{tr}(A) + \text{tr}(B)$ true?
7. Are there $n \times n$ matrices A and B, so that $AB - BA = I_n$?

2.3 Hypothetical applications of linear equations

In this section we present some hypothetical applications where a system of linear equations plays an important role.

Deformation of beams

In what follows, we need to recall Hooke's elasticity law, whose statement is:

Hook's law. The magnitude of the deformation of a rigid body under the action of a force F is directly proportional to the magnitude of F.

Assume that there is an elastic beam subject to the action of several weights of magnitudes $W_1, W_2, \ldots, W_n$ applied at points $P_1, P_2, \ldots, P_n$ on the beam, respectively. See Figure 2.1. Let us denote by y_i the magnitude of the deformation at point P_i due to the action of W_i. Assume also that the beam obeys Hooke's elasticity law for elastic materials. According to such a law, W_j produces a displacement equal to $d_{ij} W_j$ at point P_i, where d_{ij} is a constant that depends only on the material and the distance from P_i to P_j.

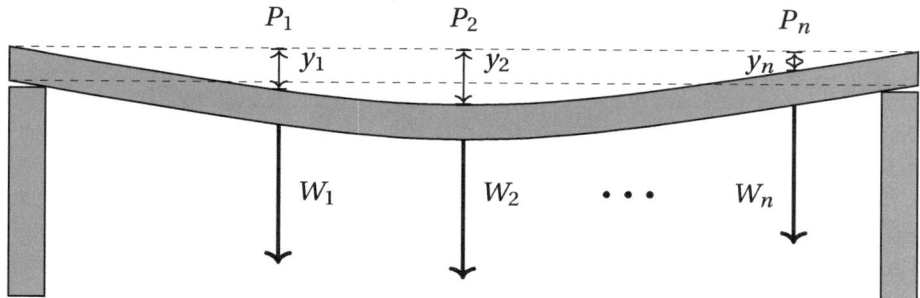

Figure 2.1: Deformation of a beam under the action of n weights.

From these considerations, we have that y_i is the sum of the displacements caused by each weight, that is, we have

$$y_i = d_{i1} W_1 + d_{i2} W_2 + \cdots + d_{in} W_n, \quad i \in \{1, 2, \ldots, n\}. \tag{2.5}$$

In matrix form, equation (2.5) can be represented by

$$DF = Y, \quad \text{where } Y = \begin{bmatrix} y_1 \\ y_2 \\ \vdots \\ y_n \end{bmatrix} \text{ and } F = \begin{bmatrix} W_1 \\ W_2 \\ \vdots \\ W_n \end{bmatrix}. \tag{2.6}$$

Matrix D is called the *flexibility matrix* and its columns are interpreted as follows.

Assume that at point P_i a weight of magnitude 1 is applied, and at the other points there is no weight, then

$$F = \begin{bmatrix} 0 \\ \vdots \\ 1 \\ \vdots \\ 0 \end{bmatrix},$$

from where

$$Y = \begin{bmatrix} d_{11} & \cdots & d_{1n} \\ d_{21} & \cdots & d_{2n} \\ \vdots & \ddots & \vdots \\ d_{n1} & \cdots & d_{nn} \end{bmatrix} \begin{bmatrix} 0 \\ \vdots \\ 1 \\ \vdots \\ 0 \end{bmatrix} = \begin{bmatrix} d_{1i} \\ d_{2i} \\ \vdots \\ d_{ni} \end{bmatrix}.$$

The meaning of the above equation is that column i of matrix D represents the deformation when a unitary weight is applied at point P_i, that is, a weight of magnitude 1, applied at point i, produces displacements $d_{1i}, d_{2i}, \ldots, d_{ni}$ at the corresponding points $P_1, P_2, \ldots, P_n$.

When D has an inverse, from the length of the deformations, $y_1, y_2, \ldots, y_n$, we can obtain the values of weights $W_1, W_2, \ldots, W_n$, since from equation $DF = Y$ we obtain $F = D^{-1}Y$. Columns of D^{-1} are interpreted as follows. When at point P_i we measure a deformation of length 1 and zero at the other points, we have

$$Y = \begin{bmatrix} 0 \\ \vdots \\ 1 \\ \vdots \\ 0 \end{bmatrix}.$$

Thus, when calculating $D^{-1}Y$, the result is the ith column of D^{-1} whose entries are magnitudes of the weights $W_1, W_2, \ldots W_n$, which produce a deformation of length 1 at the point P_i and no deformation at point P_j, for all $j \neq i$.

Electrical circuits

Applying Kirchhoff's circuit laws and Ohm's law, a wide variety of electrical circuits can be solved using systems of linear equations. For convenience, we state these laws right below.

Kirchhoff's circuits law for electric current. The algebraic sum of all currents entering and exiting a node must be zero.

Kirchhoff's law for voltage. The algebraic sum of all voltages in a loop must be zero.

Ohm's law. The current through a conductor between two points is directly proportional to the potential difference across the two points.

If V, I, and R denote the potential difference, current, and resistance, respectively, then Ohm's law can be expressed by the equation

$$V = RI. \tag{2.7}$$

Let us assume that there is a circuit as illustrated in Figure 2.2.

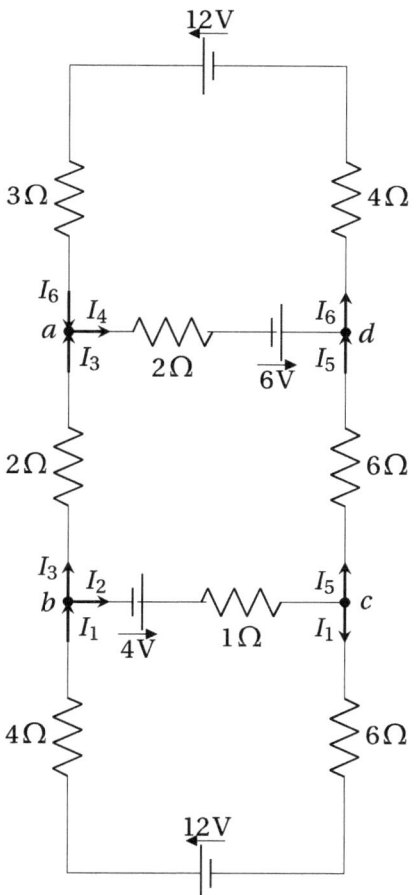

Figure 2.2: Electric circuit with three branches.

Determine the electric current at each node of the circuit.

We notice that in the circuit to be analyzed there are three loops and four nodes. We start by applying Kirchhoff's law to each of the four nodes: a, b, c, and d. In node a, currents labeled I_3 and I_6 go in, while current I_4 goes out, thus from Kirchhoff's law for electric current we have $I_4 = I_3 + I_6$; in node b, I_1 goes in, while I_2 and I_3 go out, thus by a reasoning as before, $I_1 = I_2 + I_3$; for node c, I_1 and I_5 go out, and I_2 goes in, thus $I_2 = I_1 + I_5$; finally, in node d, I_4 and I_5 go in and I_6 goes out, therefore $I_6 = I_4 + I_5$.

Now, analyzing each loop in the circuit, we have: in the upper loop, there are currents I_4 and I_6 and, when passing through resistors of magnitude 3Ω and 2Ω, applying Ohm's law, there will be a potential drop equal to $2I_4 + 4I_6 + 3I_6$. Now, by Kirchhoff's law for voltage, this drop of potential is equal to the sum of voltages due to the two batteries, thus we have $18 = 2I_4 + 4I_6 + 3I_6$. Continuing the analysis with the same sort of arguments in the middle and lower loops, we have equations $10 = I_2 + 2I_3 + 2I_4 + 6I_5$ and $16 = 10I_1 + I_2$.

Collecting all equations and rearranging terms, we have a linear system of equations:

$$
\begin{aligned}
I_1 - I_2 - I_3 &= 0, \\
I_1 - I_2 + I_5 &= 0, \\
I_4 + I_5 - I_6 &= 0, \\
I_3 - I_4 + I_6 &= 0, \\
2I_4 + 7I_6 &= 18, \\
10I_1 + I_2 &= 16, \\
I_2 + 2I_3 + 2I_4 + 6I_5 &= 10.
\end{aligned}
\tag{2.8}
$$

It should be noticed that in the system (2.8) there are seven equations in six variables, and that its coefficient matrix is

$$
E = \begin{pmatrix}
1 & -1 & -1 & 0 & 0 & 0 \\
1 & -1 & 0 & 0 & 1 & 0 \\
0 & 0 & 0 & 1 & 1 & -1 \\
0 & 0 & 1 & -1 & 0 & 1 \\
0 & 0 & 0 & 2 & 0 & 7 \\
10 & 1 & 0 & 0 & 0 & 0 \\
0 & 1 & 2 & 2 & 6 & 0
\end{pmatrix}.
\tag{2.9}
$$

We will solve the system (2.8) using SageMath. To start the solving process, we define, in SageMath, the matrix E and the vector of constant terms w, namely

```
E=matrix(QQ,[[1,-1,-1,0,0,0],
            [1,-1,0,0,1,0],
            [0,0,0,1,1,-1],
            [0,0,1,-1,0,1],
```

```
            [0,0,0,2,0,7],
            [10,1,0,0,0,0],
            [ 0,1,2,2,6,0]])
w=vector(QQ,[0,0,0,0,18,16,10])
X1=E.solve_right(w)
E
```

$$E = \begin{pmatrix} 1 & -1 & -1 & 0 & 0 & 0 \\ 1 & -1 & 0 & 0 & 1 & 0 \\ 0 & 0 & 0 & 1 & 1 & -1 \\ 0 & 0 & 1 & -1 & 0 & 1 \\ 0 & 0 & 0 & 2 & 0 & 7 \\ 10 & 1 & 0 & 0 & 0 & 0 \\ 0 & 1 & 2 & 2 & 6 & 0 \end{pmatrix},$$

$$X1 = \left(\frac{221}{166}, \frac{223}{83}, -\frac{225}{166}, \frac{157}{166}, \frac{225}{166}, \frac{191}{83} \right).$$

With the notation for currents, we have $(I_1, I_2, I_3, I_4, I_5, I_6) = \left(\frac{221}{166}, \frac{223}{83}, -\frac{225}{166}, \frac{157}{166}, \frac{225}{166}, \frac{191}{83} \right).$

Temperatures distribution on thin plates

To analyze the heat conduction on an isotropic and homogeneous medium, it is necessary to use the heat equation

$$U_t = k(U_{xx} + U_{yy} + U_{zz}), \tag{2.10}$$

where k is a constant that depends on the thermal conductivity, density, and thermal capacity of the material; U_{xx}, U_{yy}, and U_{zz} are the second partial derivatives of U with respect to x, y, and z, respectively, and U_t is the partial derivative of U with respect to t, which measures the change of temperature with respect to time t.

The example that we are discussing posses different conditions.

Assume that we have a thin rectangular plate as shown in Figure 2.3. Additionally, assume that we know the temperature on the border of the plate, and the temperature at each interior node of the plate is the average of the temperature of the closest points on the net. For instance, if we denote by t_i the temperature at node i, for the case $i = 1$ we have $t_1 = \frac{1}{4}(40 + t_2 + t_4 + 15)$.

Continuing this way, for each node we have an equation and altogether a system of linear equations in nine variables, as described below:

$$t_1 = \frac{1}{4}(40 + t_2 + t_4 + 15),$$

$$t_2 = \frac{1}{4}(40 + t_1 + t_3 + t_5),$$

$$t_3 = \frac{1}{4}(40 + t_2 + t_6 + 10), \tag{2.11}$$

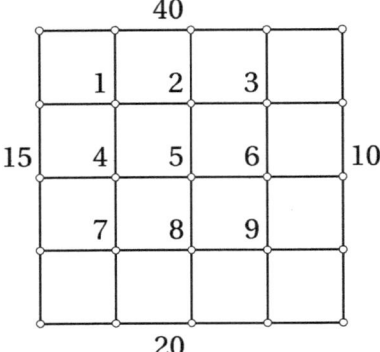

Figure 2.3: Temperature distribution on a thin plate.

$$t_4 = \frac{1}{4}(15 + t_1 + t_5 + t_7),$$

$$t_5 = \frac{1}{4}(t_2 + t_4 + t_6 + t_8),$$

$$t_6 = \frac{1}{4}(10 + t_3 + t_5 + t_9),$$

$$t_7 = \frac{1}{4}(15 + 20 + t_4 + t_8),$$

$$t_8 = \frac{1}{4}(20 + t_5 + t_7 + t_9),$$

$$t_9 = \frac{1}{4}(10 + 20 + t_6 + t_8).$$

We will solve the system (2.11) using SageMath. Make sure that the code given below is clear to you. We will denote by $X = (t_1, t_2, \ldots, t_9)$ the vector of temperatures.

```
H=matrix(QQ,[[4,-1,0,-1,0,0,0,0,0],
            [-1,4,-1,0,-1,0,0,0,0],
            [0,-1,4,0,0,-1,0,0,0],
            [-1,0,0,4,-1,0,-1,0,0],
            [0,-1,0,-1,4,-1,0,-1,0],
            [0,0,-1,0,-1,4,0,0,-1],
            [0,0,0,-1,0,0,4,-1,0],
            [0,0,0,0,-1,0,-1,4,-1],
            [0,0,0,0,0,-1,0,-1,4]])
B=vector(QQ,[55,40,50,15,0,10,35,20,30])
X=H.solve_right(B)
H
```

$$H = \begin{pmatrix} 4 & -1 & 0 & -1 & 0 & 0 & 0 & 0 & 0 \\ -1 & 4 & -1 & 0 & -1 & 0 & 0 & 0 & 0 \\ 0 & -1 & 4 & 0 & 0 & -1 & 0 & 0 & 0 \\ -1 & 0 & 0 & 4 & -1 & 0 & -1 & 0 & 0 \\ 0 & -1 & 0 & -1 & 4 & -1 & 0 & -1 & 0 \\ 0 & 0 & -1 & 0 & -1 & 4 & 0 & 0 & -1 \\ 0 & 0 & 0 & -1 & 0 & 0 & 4 & -1 & 0 \\ 0 & 0 & 0 & 0 & -1 & 0 & -1 & 4 & -1 \\ 0 & 0 & 0 & 0 & 0 & -1 & 0 & -1 & 4 \end{pmatrix},$$

$$X = \left(\frac{180}{7}, \frac{3105}{112}, \frac{335}{14}, \frac{2255}{112}, \frac{85}{4}, \frac{2015}{112}, \frac{130}{7}, \frac{2145}{112}, \frac{235}{14} \right).$$

Forest preservation

Forest preservation and exploitation must be done with strict observation of the law, since otherwise, the ecological system would be at great risk. The hypothetical example that we are discussing has one purpose – to illustrate the possibility of formulating a model to explore a way of forest wood production.

One company manages three regions to produce pine, cedar, and oak wood. The company has signed a contract with a sawmill for a monthly supply of 2,000 cubic meters of pine wood; 1,500 cubic meters of oak wood; and 800 cubic meters of cubic meters of cedar wood. The exploitation volume, per region, is described in Table 2.6.

Table 2.6: Wood exploitation per region.

Region	Exploitation/ha	%Pine	%Oak	%Cedar
Region 1	$310\,m^3/ha$	80	10	10
Region 2	$350\,m^3/ha$	10	80	10
Region 3	$280\,m^3/ha$	10	10	80

How many hectares of land should be exploited, per region, in order to satisfy the sawmill demand?

First of all, the information in Table 2.6 is interpreted as follows. In Region 1 it is allowed to obtain 310 cubic meters per hectare, from which 80 % is of pine wood, 10 % of oak, and the other 10 % of cedar. The interpretation for other regions is similar. Let us denote by x, y, and z the number of hectares that can be exploited in regions 1, 2, and 3, respectively. With this notation, the amounts of pine wood allowed in regions 1, 2, and 3 are $0.8(310)x$, $0.1(350)y$, and $0.1(280)z$, respectively. Since the demand is 2,000 cubic meters of pine wood, one has

$$0.8(310)x + 0.1(350)y + 0.1(280)z = 2000. \tag{2.12}$$

Likewise, for the oak and cedar wood, we have

$$0.1(310)x + 0.8(350)y + 0.1(280)z = 1500 \tag{2.13}$$

and

$$0.01(310)x + 0.1(350)y + 0.8(280)z = 800, \tag{2.14}$$

respectively.

With those equations, we have the system

$$\begin{aligned}
0.8(310)x + 0.1(350)y + 0.1(280)z &= 2000, \\
0.1(310)x + 0.8(350)y + 0.1(280)z &= 1500, \\
0.1(310)x + 0.1(350)y + 0.8(280)z &= 800.
\end{aligned} \tag{2.15}$$

Multiplying each equation in (2.15) by 10 and introducing the changes $s = 350y$, $t = 310x$, and $u = 280z$, (2.15) becomes

$$\begin{aligned}
8s + t + u &= 15000, \\
s + 8t + u &= 20000, \\
s + t + 8u &= 8000,
\end{aligned} \tag{2.16}$$

which helps to perform calculations.

Using Gauss–Jordan reduction method to solve (2.16), we obtain

$$s = \frac{10700}{7},$$
$$t = \frac{15700}{7},$$
$$u = \frac{3700}{7}.$$

Finally, using these values in $s = 350y$, $t = 310x$, $u = 280z$, and simplifying, we obtain

$$x = \frac{1570}{217},$$
$$y = \frac{1070}{245}, \tag{2.17}$$
$$z = \frac{370}{196}.$$

2.4 Integral matrices

The objective of this section is to analyze under which conditions on the coefficient matrix of a system of linear equations whose entries are integers we get solutions whose entries are also integers. We start by presenting an example.

A factory produces shirts, t-shirts, and jackets using three types of sewing machine. The production by the week is as shown in Table 2.7. At the end of a week, the production supervisor found in stock the following data: 3910 shirts, 2790 t-shirts, and 1880 jackets. How many sewing machines have worked during this week?

Table 2.7: Production by each machine.

	Machine type 1	Machine type 2	Machine type 3
shirts	100	120	90
t-shirts	70	80	70
jackets	50	60	40

Let x represent the number of type 1 sewing machines that worked during the week, then its production was $100x$ shirts, $70x$ t-shirts, and $50x$ jackets. In the same way, if y denotes the number of type 2 sewing machines that worked in this week, the production by these machines was $120y$ shirts, $80y$ t-shirts, and $60y$ jackets. Likewise, if z represents the number of type 3 sewing machines that worked during the week, then the production was: $90z$ shirts, $70z$ t-shirts, and $40z$ jackets.

From this information, one has that the total number of shirts produced weekly by the factory is

$$\text{total number of shirts/week:} \quad 100x + 120y + 90z.$$

Similar expressions are obtained for t-shirts and jackets; that is,

$$\text{total number of t-shirts/week:} \quad 70x + 80y + 70z,$$
$$\text{total umber of jackets/week:} \quad 50x + 60y + 40z.$$

In order to answer the posed question, it is necessary to solve the following system:

$$100x + 120y + 90z = 3910,$$
$$70x + 80y + 70z = 2790,$$
$$50x + 60y + 40z = 1880,$$

which is equivalent to

$$10x + 12y + 9z = 391,$$
$$7x + 8y + 7z = 279, \tag{2.18}$$
$$5x + 6y + 4z = 188.$$

To solve (2.18), we will start by considering the augmented matrix

$$A = \begin{bmatrix} 10 & 12 & 9 & 391 \\ 7 & 8 & 7 & 279 \\ 5 & 6 & 4 & 188 \end{bmatrix}.$$

One property of matrix A is that all of its entries are integers. From the production conditions, it is necessary that the coordinates of the solutions, if any, must be nonnegative integers.

Recall that applying row operations on the augmented matrix, by Theorem 1.3.1, p. 16, the resulting matrix represents an equivalent linear system of equations. Taking this as a starting point and given that we are searching for integer solutions, it is important to apply only row operations that guarantee that the resulting system has integral solutions.

Given that one of the row operations consists in multiplying a row by a nonzero constant and the row operations have inverses, the only allowed constants are ± 1, since those are the only integers that have a multiplicative inverse.

In other words, the type of row operations that are allowed are:
1. interchange any two rows,
2. multiply a row by ± 1, and
3. multiply a row by an integer and add to another row.

We will apply these row operations to A in order to find out the type of solutions that the system (2.18) has. The reader should verify the process.

$$\begin{bmatrix} 10 & 12 & 9 & 391 \\ 7 & 8 & 7 & 279 \\ 5 & 6 & 4 & 188 \end{bmatrix} \sim \begin{bmatrix} 0 & 0 & 1 & 15 \\ 2 & 2 & 3 & 91 \\ 5 & 6 & 4 & 188 \end{bmatrix} \sim \begin{bmatrix} 2 & 2 & 3 & 91 \\ 5 & 6 & 4 & 188 \\ 0 & 0 & 1 & 15 \end{bmatrix} \sim \begin{bmatrix} 2 & 2 & 3 & 91 \\ 1 & 2 & -2 & 6 \\ 0 & 0 & 1 & 15 \end{bmatrix}$$

$$\sim \begin{bmatrix} 0 & -2 & 7 & 79 \\ 1 & 2 & -2 & 6 \\ 0 & 0 & 1 & 15 \end{bmatrix} \sim \begin{bmatrix} 0 & -2 & 7 & 79 \\ 1 & 0 & 5 & 85 \\ 0 & 0 & 1 & 15 \end{bmatrix} \sim \begin{bmatrix} 0 & -2 & 0 & -26 \\ 1 & 0 & 0 & 10 \\ 0 & 0 & 1 & 15 \end{bmatrix} \sim \begin{bmatrix} 1 & 0 & 0 & 10 \\ 0 & 1 & 0 & 13 \\ 0 & 0 & 1 & 15 \end{bmatrix}.$$

From the last matrix, we have that the solution is $x = 10$, $y = 13$, and $z = 15$.

It should be guessed that there are systems of linear equations whose coefficient matrices have integer entries and the solutions have no integer coordinates. For example, if we change system (2.18) to

$$10x + 12y + 9z = 391,$$
$$7x + 8y + 7z = 278, \tag{2.19}$$
$$5x + 6y + 4z = 188,$$

then the solution of (2.19) is $\left(7, \frac{31}{2}, 15\right)$ or $x = 7, y = \frac{31}{2}$, and $z = 15$.

We notice that this solution has no meaning in the context where it comes from. However, something important to point out is that the operations that were applied can take the original matrix to one where the main entry in a nonzero row is a positive integer (if it were negative, multiplying by –1 will transform it into a positive).

The above examples lead to formulate some questions concerning such systems of linear equations. Equations with integral or rational coefficients are called *diophantine equations*, honoring the great Greek mathematician Diophantus who lived approximately in the period 200–284 AD and was one of the first to systematically study such equations.

To make the discussion more precise, some terms are required.

Definition 2.4.1. A matrix A is said to be integral, if all its entries are integers.

Considering integral matrices which represent systems of linear equations, we can ask several questions.

When does a system of linear equations have a solution? How do we find all solutions if there are any? Can the Gauss–Jordan method be adjusted to deal with integral matrices?

Discussing these questions leads to what is called the *theory of integral matrices*.

One of the most fundamental results concerning integral matrices is the following.

Theorem 2.4.1 (Smith normal form). *For every $m \times n$ integral matrix A, there exist invertible integral matrices L and R of size $m \times m$ and $n \times n$, respectively, so that $LAR = D = \operatorname{diag}\{d_1, d_2, \ldots, d_r, 0 \ldots, 0\}$, where d_i is a positive integer for $i \in \{1, 2, \ldots, r\}$ and d_i divides d_{i+1}, for $i \in \{1, 2, \ldots, r-1\}$.*

We strongly invite the reader to give the proof of the theorem for a 2×2 integral matrix A. One possible approach to prove the theorem consists in applying elementary operations on the rows and columns of A. The proof of the general case is out of the scope of this book. The interested reader can consult [21], where he/she will find the proof of the above result.

To close this section, a couple of definitions related to Theorem 2.4.1 are presented.

Definition 2.4.2. The matrix $D = \operatorname{diag}\{d_1, d_2, \ldots, d_r, 0 \ldots, 0\}$ in Theorem 2.4.1 is called the Smith normal form of A.

Definition 2.4.3. Two $m \times n$ matrices A and B are equivalent if there are nonsingular matrices P and Q so that $B = PAQ$.

2.5 Exercises

1. Assume that A and B are $n \times n$ matrices. Prove that AB is invertible if and only if A and B are invertible.

2. In this exercise you will apply row operations on the given matrices.
   (a) Let $A = \begin{bmatrix} 1 & 2 & 1 & 0 \\ -1 & 0 & 3 & 5 \\ 1 & -2 & 1 & 1 \end{bmatrix}$ be a matrix. Determine the row reduced form of A, call it R, and find an invertible matrix P so that $PA = R$.

    (b) For each of the following matrices, determine if their inverses exist and, if so, find them:

$$\begin{bmatrix} 2 & 5 & -1 \\ 4 & -1 & 2 \\ 6 & 4 & 1 \end{bmatrix}, \quad \begin{bmatrix} 1 & -1 & 2 \\ 3 & 2 & 4 \\ 0 & 1 & -2 \end{bmatrix}.$$

3.   Prove Theorem 2.2.3.

4.   Assume that A is an invertible matrix and $AB = 0$, for some matrix B. Prove that $B = 0$.

5.   Assume that the square matrix A has no inverse. Show that there is a matrix $B \neq 0$ so that $AB = 0$.

6.   Find two matrices A and B so that $AB = 0$ and $BA \neq 0$.

7.   Let A and B be nonzero square matrices so that $AB = 0$. Prove that A and B have no inverses.

8.   Let a, b, c, and d be real numbers and let $A = \begin{bmatrix} a & b \\ c & d \end{bmatrix}$. Prove that A has an inverse if and only if $ad - bc \neq 0$. Assume that the entries of A are integers, then A^{-1} has integer entries if and only if $ad - bc = \pm 1$.

9.   Let a be a real number and let $A = \begin{bmatrix} a & 0 \\ 1 & a \end{bmatrix}$. Prove that $A^k = \begin{bmatrix} a^k & 0 \\ ka^{k-1} & a^k \end{bmatrix}$ for every positive integer $k \geq 1$. If $a \neq 0$, find $(A^{-1})^k$ by using two different methods.

10. Let A and B be $m \times n$ and $n \times m$ matrices, respectively, so that $AB = I_m$ and $BA = I_n$. Prove that $n = m$.

11. Let A be a $n \times n$ matrix.
    (a) Assume that there exists a positive integer k so that $A^k = 0$. Does A have an inverse?
    (b) For each $n \in \{2, 3, 4, 5, 6\}$, construct examples of $n \times n$ matrices $A \neq 0$, so that $A^k = 0$ for some integer $k > 1$.
    (c) Let A be a square matrix and assume that there is an integer $k > 1$ so that $A^k = I_n$. Is it possible that A has no inverse?
    (d) Find examples of matrices that satisfy part (c) and are not the identity matrix.

12. Let A be a matrix that has no inverse. Prove that there is an $n \times 1$ matrix B so that the system $AX = B$ has no solution.

13. Prove that the elementary matrices have inverses.

14. Let λ represent a real number and let $A = \begin{bmatrix} 1 & 0 & \lambda \\ 0 & \lambda & \lambda \\ \lambda & 2 & 1 \end{bmatrix}$ be the coefficient matrix of a system of linear equations.
    Determine the values of λ so that the system $AX = \begin{bmatrix} 1 \\ 2 \\ 3 \end{bmatrix}$:
    (a) has a unique solution,
    (b) has infinitely many solutions,
    (c) has no solution.

15. Let A be an $n \times n$ matrix. Prove that A has no inverse if and only if there is a nonzero $n \times n$ matrix B so that $AB = 0$.

16. Let A be a 2×2 matrix with positive entries. Prove that there exists $c > 0$ so that the matrix $A - cI$ has no inverse. Does the same result hold for 3×3 matrices?

17. Let $p(x) = a_m x^m + a_{m-1} x^{m-1} + \cdots + a_1 x + a_0$ be a polynomial and let A be a square matrix. We define $p(A) = a_m A^m + a_{m-1} A^{m-1} + \cdots + a_1 A + a_0 I_m$. Given other polynomial $q(x)$, prove that $p(A)q(A) = q(A)p(A)$.

18. Let $\mathcal{M}_{m \times n}(\mathbb{R})$ represent the set of $m \times n$ real matrices. Assume that $A, B \in \mathcal{M}_{m \times n}(\mathbb{R})$ have entries a_{ij} and b_{ij}, respectively; define a product $A * B = C$ (Hadamard product) where the entries of C are $c_{ij} = a_{ij} b_{ij}$. Is this product commutative, associative? Is there an identity, an inverse for this product? Does this product distribute with respect to the usual sum of matrices?

19. Let A be a lower triangular matrix with nonzero entries on the diagonal. Prove that A has an inverse and it is also lower triangular. Is the same result true for upper triangular matrices? What are the necessary and sufficient conditions on a diagonal matrix to be invertible?

20. Let $X = \begin{bmatrix} x_1 \\ x_2 \\ \vdots \\ x_n \end{bmatrix}$ be such that $x_k \neq 0$ for some $k \in \{1, 2, \ldots, n\}$. We define $M_k = I_n - \Omega E_k$, where

$$\Omega = \begin{bmatrix} 0 \\ \vdots \\ 0 \\ \frac{x_{k+1}}{x_k} \\ \vdots \\ \frac{x_n}{x_k} \end{bmatrix},$$

and E_k is the matrix $1 \times n$ with zero entries except the kth entry whose value is 1. Compute $M_k X$. Does M_k have an inverse? The matrix M_k is called a *Gauss matrix*.

21. Let A be an $n \times n$ matrix so that $a_{11} \neq 0$. Find a Gauss matrix M so that MA has zeros in the first column, except at the entry $(1,1)$ whose value is a_{11}. Let A be a 2×2 matrix. Which conditions must A satisfy to guarantee that there exist Gauss matrices $M_1, \ldots, M_k$ so that $M_k \cdots M_1 A$ is upper triangular?

22. Let $A = \text{diag}\{A_1, A_2, \ldots, A_k\}$ be a matrix, with A_i square matrix for every $i \in \{1, 2, \ldots, k\}$. Prove that $A^n = \text{diag}\{A_1^n, A_2^n, \ldots, A_k^n\}$ for every $n \geq 1$. What are the conditions under which A has an inverse?

23. Let $A = \begin{bmatrix} A_{11} & A_{12} \\ 0 & A_{22} \end{bmatrix}$ be a matrix, with A_{11} and A_{22} square matrices. Prove that A is invertible if and only if A_{11} and A_{22} are invertible. In this case determine an expression for the inverse of A.

24. Let A and B be matrices of sizes $m \times n$ and $n \times p$, respectively. Let us denote by $A_1, A_2, \ldots, A_n$ the columns of A and by $B_1, B_2, \ldots, B_n$ the rows of B. Does the equation $AB = A_1 B_1 + A_2 B_2 + \cdots + A_n B_n$ have any meaning? Explain.

25. Assume that A is an $n \times n$ matrix. What conditions should A satisfy in order that there exist matrices X and D such that $AX = XD$, with D diagonal?

26. In this exercise we present the elements to justify Leontief's model, Example 1.1.2, p. 5. For a discussion from the economical point of view, see [19].

Suppose that an economical system consists of n sectors and denote by X the production vector, by B the consumer vector, and for A the input–output matrix in Leontief's model. Matrix A determines the demand of the sectors needed to satisfy the final demand. With this interpretation and assumption on the economical system, which is in equilibrium, the above data leads to

$$X = AX + B. \tag{2.20}$$

A natural question arises. Is there a production vector that satisfies the demand? Recall that one assumption is that the sum of the elements of each column of A is less than 1. In this exercise we will provide some hints that allow to answer the question.

Given a square matrix $A = (a_{ij})$, the norm of A is defined by $\|A\| := \max[\sum_{i=1}^{n} a_{ij}]_{j=1}^{n}$.

Prove:

(a) For each pair of matrices A and B, one has $\|AB\| \le \|A\|\|B\|$, in particular $\|A^n\| \le \|A\|^n$, for every $n \in \mathbb{N}$.

(b) Let A be a square matrix so that $\|A\| < 1$. Prove that $A^k \to 0$ when $k \to \infty$. Using the equality $I - A^k = (I - A)(I + A + A^2 + \cdots + A^{k-1})$ and the above result, conclude that the matrix $I - A$ has an inverse and $(I-A)^{-1} = I + A + A^2 + \cdots + A^{k-1} + \cdots$. From this, one obtains that equation (2.20) has a positive solution for every B with positive entries. An approximation of the solution is $X = (I + A + A^2 + \cdots + A^{k-1})B$.

27. Consider the equation $X = AX + B$, where

$$A = \begin{bmatrix} 0.1588 & 0.0064 & 0.0025 & 0.0304 & 0.0014 & 0.0083 & 0.1594 \\ 0.0057 & 0.2645 & 0.0436 & 0.0099 & 0.0083 & 0.0201 & 0.3413 \\ 0.0264 & 0.1506 & 0.3557 & 0.0139 & 0.0142 & 0.0070 & 0.0236 \\ 0.3299 & 0.0565 & 0.0495 & 0.3636 & 0.0204 & 0.0483 & 0.0649 \\ 0.0089 & 0.0081 & 0.0333 & 0.0295 & 0.03412 & 0.0237 & 0.0020 \\ 0.1190 & 0.0901 & 0.0996 & 0.1260 & 0.1722 & 0.2368 & 0.3369 \\ 0.0063 & 0.0126 & 0.0196 & 0.0098 & 0.0064 & 0.0132 & 0.0012 \end{bmatrix}, \quad B = \begin{bmatrix} 74000 \\ 56000 \\ 10500 \\ 25000 \\ 17500 \\ 196000 \\ 5000 \end{bmatrix}.$$

Use the previous results and SageMath to find an approximate solution up to three decimal places. (Data of matrix A is taken from [16, Exercise 13, p. 154].)

28. In Exercises 28–31, D will represent the flexibility matrix that appears in Example 2.3, p. 53. Assume that $D = \begin{bmatrix} 0.0090 & 0.0070 & 0.0040 & 0.0010 \\ 0.0070 & 0.0095 & 0.0070 & 0.0040 \\ 0.0040 & 0.0070 & 0.0095 & 0.0070 \\ 0.0010 & 0.0040 & 0.0070 & 0.0090 \end{bmatrix}$ is the flexibility matrix for an elastic beam under the action of four weights uniformly distributed. The units are in centimeters over newtons. Answer the following questions:

(a) What is the physical meaning of the condition $D = D^t$?

(b) Does the matrix D have an inverse? Use SageMath to answer this and the following question.

(c) What are the weights magnitudes that act on the beam to produce the deformations 0.3, 0.25, 0.4, 0.2 given in centimeters?

29. Under which physical conditions of a beam, the flexibility matrix is symmetric?
30. Can the matrix D be antisymmetric?
31. Assume that each square matrix is the sum of two matrices, one symmetric and the other antisymmetric. What is the physical meaning of this mathematical fact if the matrix D is not symmetric?
32. In Figure 2.4, two electric circuits are shown. Find the electric current at each node of the given circuits. After formulating a system of linear equations, it is advisable to use SageMath to solve it.

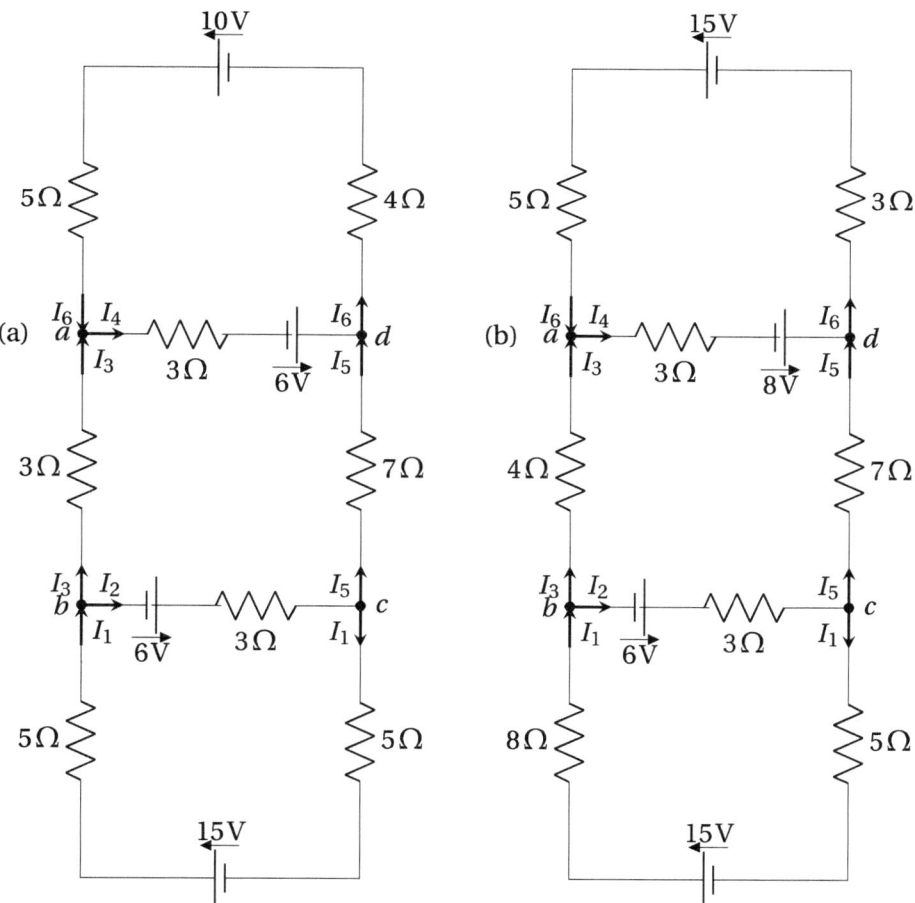

Figure 2.4: Each circuit has four nodes and three loops.

3 The vector spaces $\mathbb{R}^2$ and $\mathbb{R}^3$

The term "vector" comes from the Latin word "carrier" or "transporter". Thus a "Vector Control Program" could be understood as a program to control propagation of something, particularly a disease. No, we are not interested in those topics, we are interested in the term "vector" from the mathematical point of view. In this respect, historians claim that the term vector was coined by William Rowan Hamilton in one of his papers, in the year 1846. Hamilton used the term vector to describe what now are called triples of real numbers. However, the meaning of vector as a term has changed ever since 1846. Now, the term vector or, more precisely, *vector space* is used to name a wide range of mathematical objects sharing a few important properties – the axioms of a vector space. Before going into the discussion of these axioms, we start by mentioning that the term vector is used in different disciplines such as physics, engineering, and mathematics, to mention a few. In physics, fundamental concepts such as force, acceleration, and velocity can be geometrically represented by the use of what are called geometric vectors. These entities can be thought of as directed segments and are quite helpful to think in geometric terms of what a vector is. The representation of forces by directed segments plays a crucial role when analyzing the action of several forces on a physical object. Nowadays, a vector can be used to represent an ample range of information. Moreover, using vectors to represent information is of great utility, especially when transmitting and operating with it.

In this chapter we start by discussing geometric and algebraic representations of vectors in the plane and in the three-dimensional space. Also, some properties of geometric objects (lines and planes) are discussed using the structure of $\mathbb{R}^2$ and $\mathbb{R}^3$ as vector spaces. To start the discussion, we consider that each point P of $\mathbb{R}^2$ or $\mathbb{R}^3$ has been associated to a directed segment with initial point, the origin of coordinates, and P as the end point.

To illustrate how the geometric representation of a vector is related to the algebraic representation, consider the directed segment or vector in Figure 3.1. The represented vector has magnitude 2 on the horizontal axis and magnitude 4 on the vertical axis. This information is given by the pair $(2, 4)$.

The algebraic representation of a vector can help us to give a geometric meaning to the sum of vectors, as we will see below. The result of the sum of vectors α and β must be a vector whose components are obtained from those of α and β.

In order to formulate this idea in a precise way, we will introduce some terms and notation.

3.1 Vectors in $\mathbb{R}^2$ and $\mathbb{R}^3$

In this section we present the algebraic representation of the cartesian plane $\mathbb{R}^2$ and the cartesian space $\mathbb{R}^3$. From the analytic geometry point of view, the cartesian plane is

https://doi.org/10.1515/9783111135915-003

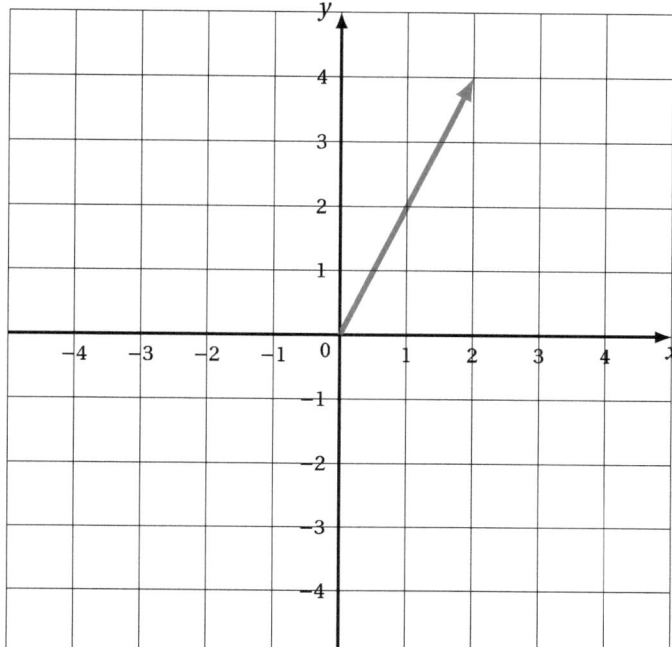

Figure 3.1: Geometric representation of a vector.

considered as the collection of pairs of real numbers (x,y) together with the euclidean distance, which is the core to define conic sections. To further explore important properties of the plane, we introduce a sum of elements in the plane, as well as the product of a real number and an element of the plane.

We start by considering that the cartesian plane is described algebraically as the set of ordered pairs of real numbers, and will be denoted by $\mathbb{R}^2$. More precisely,

$$\mathbb{R}^2 = \{(x,y) \mid x,y \in \mathbb{R}\}.$$

Definition 3.1.1. Given $\alpha = (a,b)$ and $\beta = (c,d)$ in $\mathbb{R}^2$, their sum is defined as $\alpha + \beta = (a+c, b+d)$. If $\lambda \in \mathbb{R}$, the product of λ and α is defined as $\lambda(a,b) = (\lambda a, \lambda b)$. The real number λ is called a scalar.

The sum and product with a scalar have a geometric meaning which are illustrated in Figures 3.2 and 3.4, respectively.

Sum. The geometric representation of the sum of two vectors is obtained by translating the origin point of one of them to the end point of the other, to form a parallelogram. The diagonal of this parallelogram that has one endpoint at the origin represents the sum. See Figure 3.2. The geometric representation of the sum is called the *parallelogram law*.

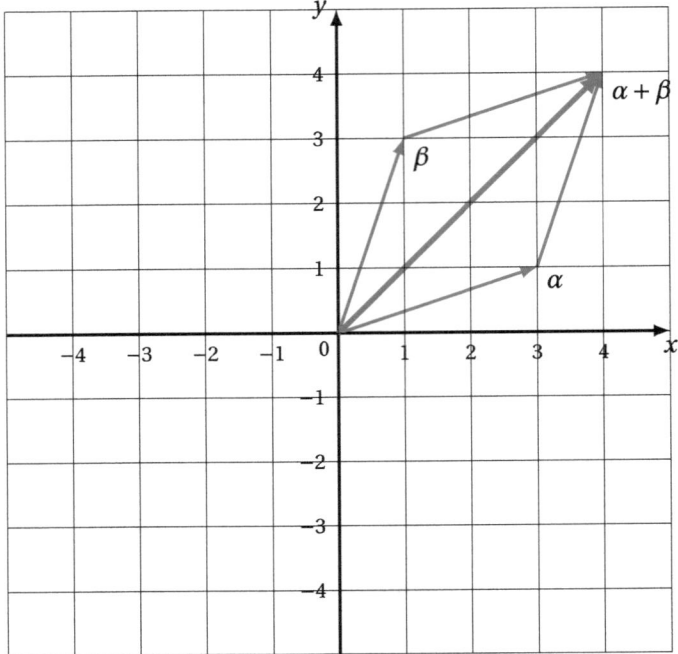

Figure 3.2: Geometric representation of the sum of two vectors – parallelogram law.

Example 3.1.1. Let (a, b) be a point on the circle C of radius 4, centered at the origin. Find all points (x, y) such that $(x, y) + (a, b) = (3, 2)$, when (a, b) moves along C.

Discussion. Since the point (a, b) lies on C, $a^2 + b^2 = 16$. Also, $(a, b) + (x, y) = (3, 2)$. From the definition of the sum in $\mathbb{R}^2$, we have $a + x = 3$ and $b + y = 2$. From these equations, we have

$$a^2 + b^2 = (3 - x)^2 + (2 - y)^2 = 16. \tag{3.1}$$

Hence, $(x - 3)^2 + (y - 2)^2 = 16$ represents the equation of a circle centered at $(3, 2)$ and radius 4, see Figure 3.3.

Exercise 3.1.1. In each of the exercises, sketch the graphical representations.
1. Represent geometrically the given vectors $(1, 2)$, $(-2, \sqrt{2})$, and $(\pi, -3)$.
2. Represent geometrically the sum of the vectors:
    – $u = (2, 3)$ and $v = (-1, -3)$;
    – $u = (-2, 4)$ and $v = (-\pi, -3)$;
    – $u = (0, 3)$ and $v = (-\sqrt{3}, -3)$.
3. Given a nonzero vector $u = (a, b)$, represent geometrically the vector $U = \frac{1}{\lambda} u$, where $\lambda = \sqrt{a^2 + b^2}$. Prove that the components of U satisfy the equation $x^2 + y^2 = 1$. The

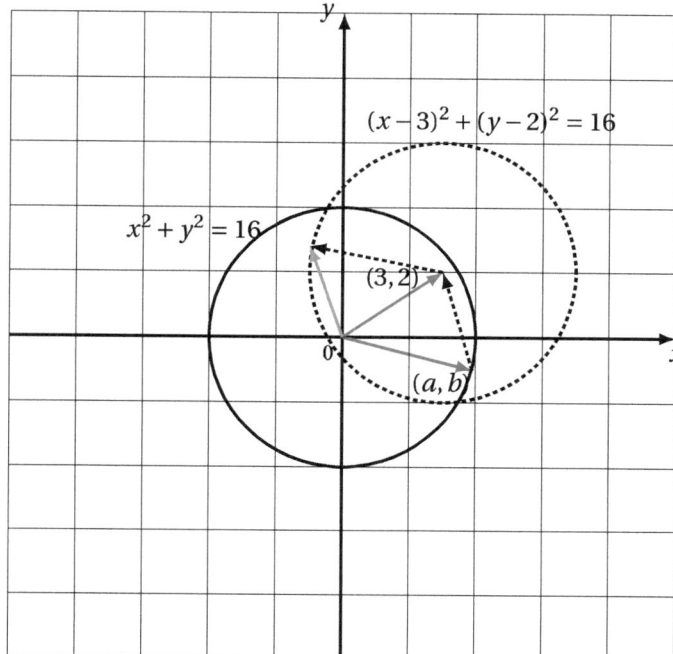

Figure 3.3: $(a, b) + (x, y) = (3, 2)$.

scalar λ is called the *norm* of u. Notice that λ is the distance from the origin to the end point that determines the vector.

4. Regarding Example 3.1.1, what is the locus of points (x, y) that satisfy $(x, y) + (a, b) = (1, 4)$ if (a, b) is on the ellipse given by equation $\frac{x^2}{9} + \frac{y^2}{4} = 1$?

Product with a scalar

The product of a vector and a scalar is interpreted as follows. If the scalar is positive and different from 1, the vector changes its norm; if the scalar is negative, then the vector changes its orientation and norm, except when the scalar is −1. In this case, the vector is reflected with respect to the origin. If the scalar is zero, the vector becomes zero. Figure 3.4 shows the case when the scalar is greater than 1.

The sum and product with a scalar that we have defined in $\mathbb{R}^2$ can be extended to $\mathbb{R}^3$; more generally, the sum and product with a scalar can be extended to $\mathbb{R}^n$. This will be defined in Chapter 4.

The geometric representation of $\mathbb{R}^3$ is by means of a system of three mutually perpendicular axes as shown in Figure 3.5. Such a representation can be interpreted as the construction of a perpendicular line to a horizontal coordinate plane. This can be used to describe the elements in the three-dimensional space as triples of real numbers. More precisely,

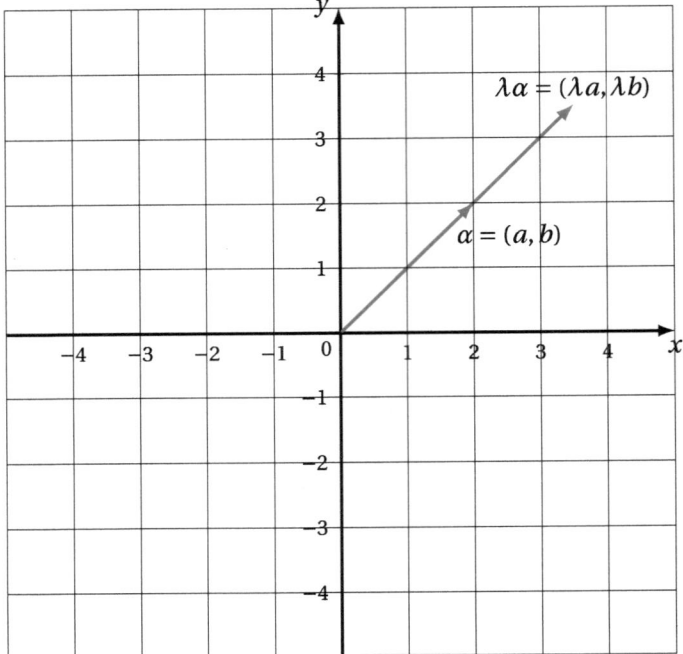

Figure 3.4: Graphical representation of a scalar multiplied by a vector.

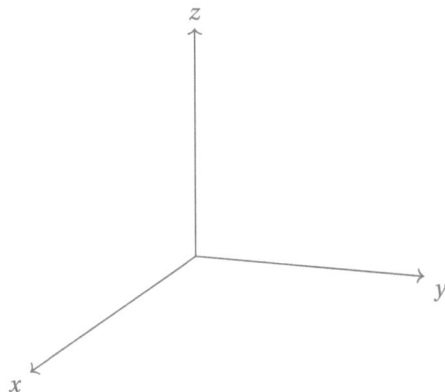

Figure 3.5: Graphical representation of a coordinate system in $\mathbb{R}^3$.

$$\mathbb{R}^3 := \{(x,y,z) \mid x,y,z \in \mathbb{R}\}.$$

To define the sum and product with scalars in $\mathbb{R}^3$, we mimic what is done in $\mathbb{R}^2$, that is, Definition 3.1.1 is extended to $\mathbb{R}^3$ in a natural way by adding a third coordinate.

3.2 Linear combinations and linear dependence

In the language of operations with vectors, we can give another geometric interpretation to a system of linear equations. Such an interpretation is useful from different points of view. For example, if we are interested in knowing the sign of the solutions without solving the system, the new interpretation might be useful.

Consider the system of linear equations in the variables x and y,

$$2x + y = 3,$$
$$x + 2y = 4. \tag{3.2}$$

Using vector notation, the system can be represented by

$$(2x + y, x + 2y) = (3, 4), \quad \text{or in equivalent form} \quad x(2, 1) + y(1, 2) = (3, 4).$$

Vectors of the form $x(2, 1) + y(1, 2)$ are called a *linear combination* of vectors $(2, 1)$ and $(1, 2)$. Notice that these vectors are obtained from the columns of the coefficient matrix of system (3.2). One way to interpret that system (3.2) has a solution is that the vector $(3, 4)$ can be expressed as a linear combination of $(2, 1)$ and $(1, 2)$.

For positive values of the scalars x and y, the expression on the left-hand side of the equation represents all vectors in the shaded region as shown in Figure 3.6.

When a system of linear equations is represented by means of a linear combination, the system has a solution if and only if the vector of constants is a linear combination of the columns of the coefficient matrix.

Since $(3, 4)$ is located in the shaded region, without solving the system we can say that the system has a solution with positive values.

Performing a more general analysis, we can conclude that the equation $(2x + y, x + 2y) = (a, b)$ has a solution for every (a, b). A geometric way to analyze the solutions of this system consists in noting that the vectors $(1, 2)$ and $(2, 1)$ determine the lines $L_1 : y = 2x$ and $L : y = \frac{x}{2}$, hence to arrive at point (a, b) starting from the origin, one needs to move along parallel lines to the given ones. See Figure 3.7.

According to Figure 3.7, the coordinate plane is divided into four regions that have been denoted by R_1, R_2, R_3, and R_4.

Without solving the equation $x(1, 2) + y(2, 1) = (a, b)$, we can determine the signs of the solutions. This is useful when solving systems of linear equations that represent production processes, since production and investment are related to nonnegative quantities:

1. For $(a, b) \in R_1$, the values of x and y are positive.
2. For $(a, b) \in R_2$, one has $x \leq 0$ and $0 \leq y$.
3. For $(a, b) \in R_3$, both x and y are negative.
4. For $(a, b) \in R_4$, one has $0 \leq x$ and $y \leq 0$.

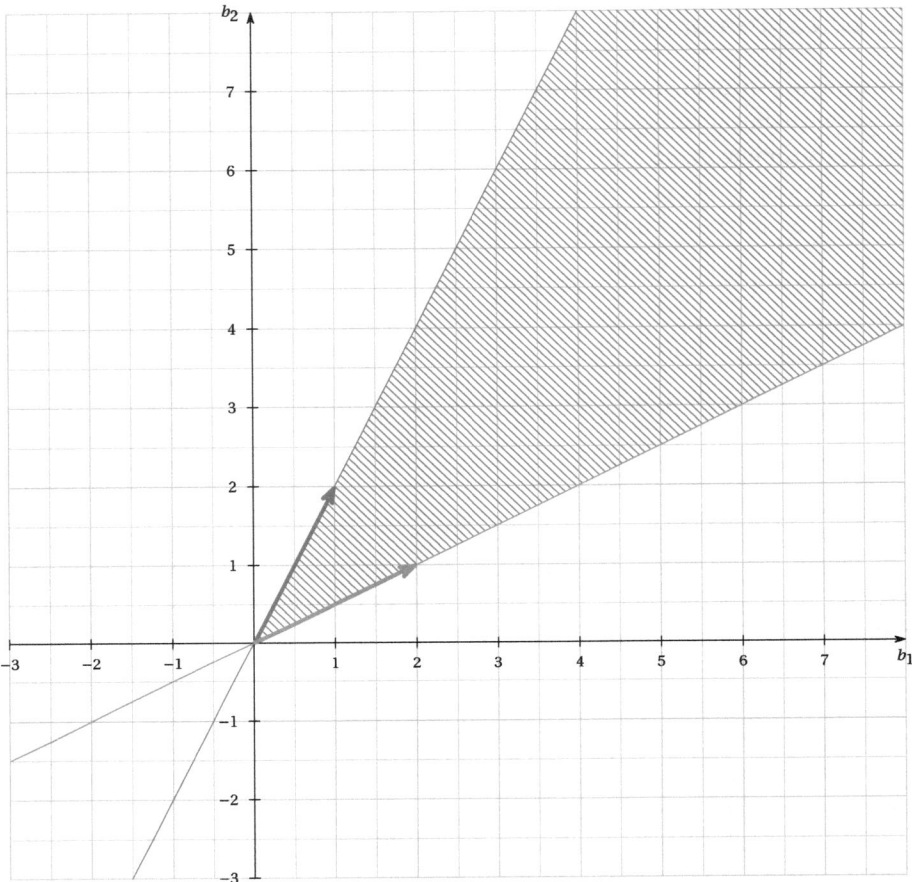

Figure 3.6: Representation of $x(1,2) + y(2,1)$, for $x, y \geq 0$.

Exercise 3.2.1. In the previous analysis, where is (a, b) located when one of x and y is zero?

From the above discussion, a system of linear equations has been represented as a single equation using vectors. This can be done for any system of linear equations. For instance, the system

$$a_1 x + b_1 y = c_1,$$
$$a_2 x + b_2 y = c_2, \tag{3.3}$$

can be represented as $x(a_1, a_2) + y(b_1, b_2) = (c_1, c_2)$, and this equation has a solution if and only if (c_1, c_2) is a linear combination of (a_1, a_2) and (b_1, b_2). Notice that if $c_1 = c_2 = 0$, the system has at least one solution. When the equation $x(a_1, a_2) + y(b_1, b_2) = (0, 0)$ has

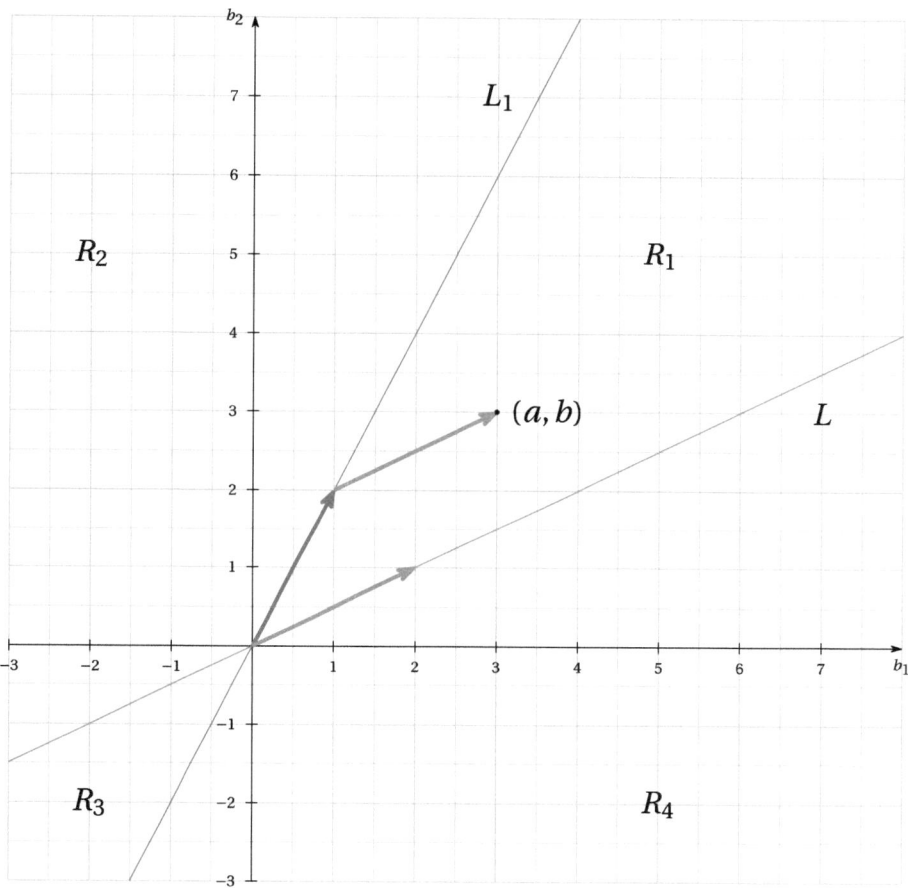

Figure 3.7: Representation of $x(1,2) + y(2,1) = (a,b)$.

more than one solution, the vectors (a_1, a_2) and (b_1, b_2) are called *linearly dependent*, otherwise we say that the vectors are *linearly independent*.

Given that this concept is one of the most important in linear algebra, we make it explicit.

Definition 3.2.1. The vectors $a_1, a_2, \ldots, a_k$, are linearly independent if the equation $x_1 a_1 + x_2 a_2 + \cdots + x_k a_k = 0$ has only the solution $x_1 = x_2 = \cdots = x_k = 0$. Otherwise, we say that the vectors are linearly dependent.

Remark 3.2.1. It should be noticed that if one of the vectors in Definition 3.2.1 is zero, then the vectors are linearly dependent.

If $k = 2$, then $a_1, a_2 \in \mathbb{R}^2$ are linearly dependent if and only if one is a multiple of the other, Figure 3.8. In fact, if there are scalars x and y, not both zero, so that $xa_1 + ya_2 = 0$,

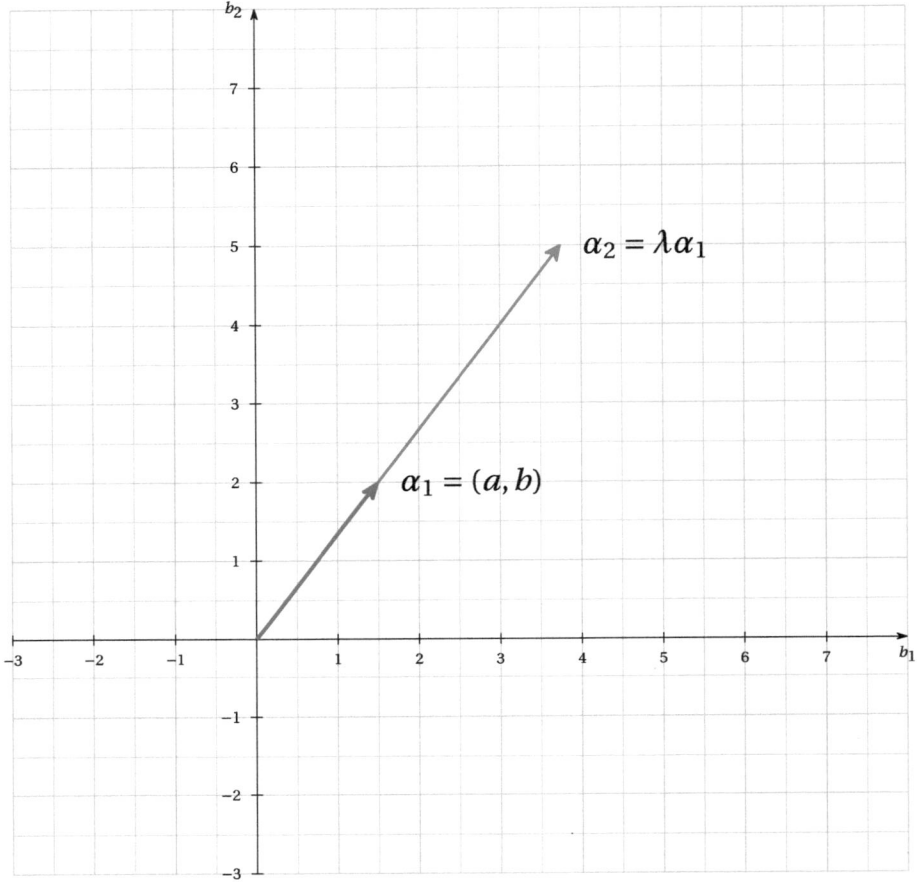

Figure 3.8: Linear dependence of two vectors.

we may assume that $x \neq 0$, and then $\alpha_1 = -\frac{y}{x}\alpha_2$. Conversely, if $\alpha_1 = c\alpha_2$, then $\alpha_1 - c\alpha_2 = 0$, proving that α_1, α_2 are linearly dependent.

Exercise 3.2.2. Prove the following:
1. Any three or more vectors in $\mathbb{R}^2$ are linearly dependent. This statement is equivalent to the following: given any three or more vectors in $\mathbb{R}^2$, at least one of them is a linear combination of the others.
2. Any four or more vectors in $\mathbb{R}^3$ are linearly dependent.

Exercise 3.2.3. Let A be a 3×3 matrix and let X and B be 3×1 matrices. Prove that the equation $AX = B$ has a solution if and only if there are scalars c_1, c_2, and c_3 so that $B = c_1 A_1 + c_2 A_2 + c_3 A_3$, where A_1, A_2, and A_3 are the columns of A.

3.2.1 Exercises

1. Given the vectors $\alpha = (1,2)$ and $\beta = (2,-3)$,
    (a) sketch the set $\{x\alpha + y\beta : x,y \geq 0\}$;
    (b) determine if any vector of $\mathbb{R}^2$ is a linear combination of α and β;
    (c) sketch the set $\{x\alpha + y\beta : x > 0, y \leq 0\}$;
    (d) explain, using linear combinations, the meaning that the system

    $$x + 2y = 1,$$
    $$2x - 3y = 6$$

    has a solution.
2. Represent the system

    $$2x + y + z = 1,$$
    $$x - y + 2z = \frac{1}{2},$$
    $$x + y - z = 3$$

    as a linear combination of vectors.
3. Represent any point of the square $C = \{(x,y) \mid 0 \leq x,y \leq 1\}$ as a linear combination of the vectors $(1,0)$ and $(0,1)$.
4. Determine which vectors are linear combinations of the columns of

    $$\begin{pmatrix} 2 & 0 \\ -1 & 2 \end{pmatrix}.$$

5. How do you represent a straight line that passes through the origin using vectors? What is the geometric representation of $t\alpha$, where $\alpha \in \mathbb{R}^2 \setminus \{0\}$ and $t < 0$?
6. If the vector $(1,2,3)$ can be expressed as a linear combination of the columns of the matrix

    $$\begin{pmatrix} 2 & 0 & 2 \\ 5 & 2 & -1 \\ 1 & 2 & 3 \end{pmatrix},$$

    how can this be interpreted in terms of solutions of equations?

3.3 Geometric properties of $\mathbb{R}^2$ and $\mathbb{R}^3$ as vector spaces

Before defining some geometric terms that will allow us to discuss geometric properties of $\mathbb{R}^2$ and $\mathbb{R}^3$, it is convenient to discuss the equations of a line and a plane.

From analytic geometry, we know that the general equation of a line is of the form $Ax + By + C = 0$, where A, B, and C are real constants and at least one of A and B is not zero. Considering a line as a subset of the coordinate plane, we can say that a line L consists of the points (x, y) which satisfy an equation of the form $Ax + By + C = 0$. This set of points can be described using vectors. *There is a nonzero vector a and a point $P \in L$ so that the points of L are precisely of the form $P + ta$, where t varies in* $\mathbb{R}$. This representation of L is called a *vector form or parametric form of L*. See Figure 3.9.

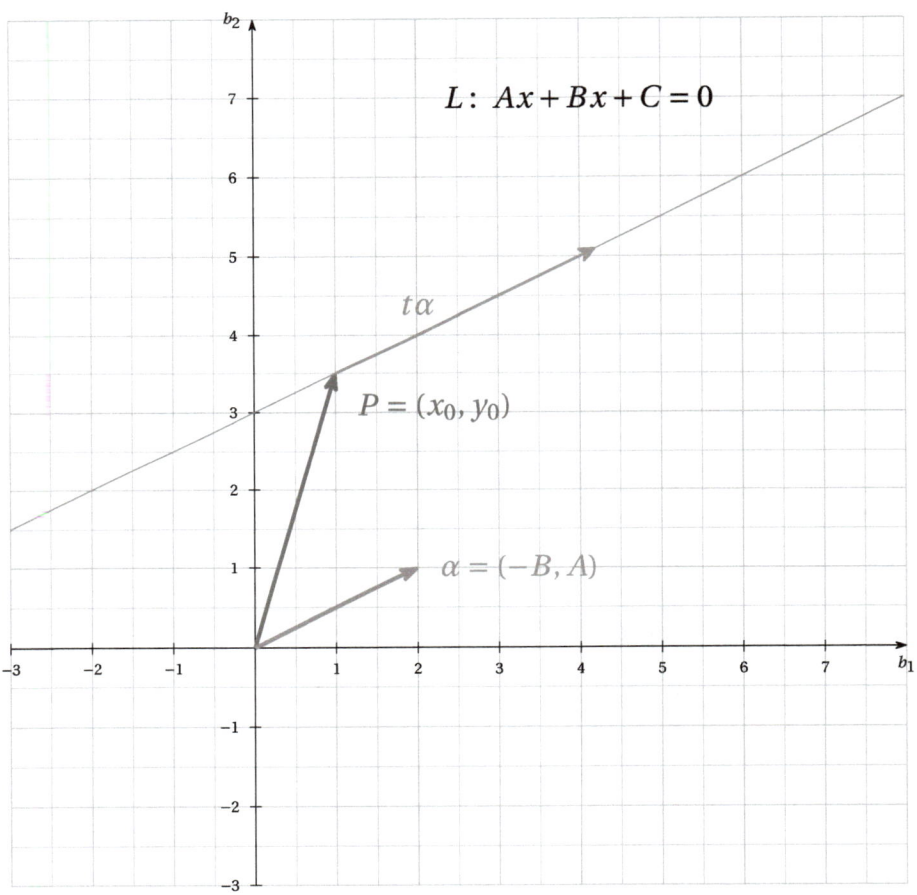

Figure 3.9: Vector representation of a line $L : Ax + By + C = 0$.

In fact, let $P = (x_0, y_0)$ be such that $Ax_0 + By_0 + C = 0$, then $-Ax_0 - By_0 = C$. Substituting C into $Ax + By + C = 0$ and collecting terms, we have $A(x - x_0) + B(y - y_0) = 0$, that is, the coordinates of point $Q = (x, y)$ satisfy the equation $Ax + By + C = 0$ if and only if $A(x - x_0) + B(y - y_0) = 0$. On the other hand, for each $t \in \mathbb{R}$ one has $A(-tB) + B(tA) = 0$, then, setting $x - x_0 = -tB$ and $y - y_0 = tA$, these equations can be represented as $(x, y) = (x_0, y_0) + t(-B, A)$.

Defining $\alpha = (-B, A)$ and $P = (x_0, y_0) \in L$, we have that (x, y) lies on L if and only if $(x, y) = (x_0, y_0) + t(-B, A)$.

Using the parametric representation of a line and taking into account that a plane is "generated" by lines, it is "reasonable" to define a plane in $\mathbb{R}^3$ as follows:

Definition 3.3.1. Given $P = (x_0, y_0, z_0) \in \mathbb{R}^3$ and two linearly independent vectors $\alpha, \beta \in \mathbb{R}^3$, the set $\{P + s\alpha + t\beta : s, t \in \mathbb{R}\}$ is called the plane generated by α and β that passes through P.

What would happen if in Definition 3.3.1 the linear independence condition were dropped? For instance, take the vectors $\alpha = (1, 1, 2)$, $\beta = (3, 3, 6)$, and point $P = (0, 0, 0)$. Sketch the points of the form $P + s\alpha + t\beta$, with $s, t \in \mathbb{R}$.

Exercise 3.3.1. Sketch the geometric representation of a plane according to Definition 3.3.1.

Exercise 3.3.2. Describe the plane that passes through $P = (1, 2, 0)$ and is generated by $\alpha = (1, 2, 0)$ and $\beta = (1, 1, 1)$.

Exercise 3.3.3. Describe the plane that contains the points $(1, 0, 0)$, $(0, 1, 0)$, and $(1, 1, 1)$.

Exercise 3.3.4. Determine if the following subsets of $\mathbb{R}^3$ are planes
-   $\{(1 + 2t + s, 3 + 4t + s, 0) \in \mathbb{R}^3 : s, t \in \mathbb{R}\}$,
-   $\{(t, s, 3) \in \mathbb{R}^3 : s, t \in \mathbb{R}\}$.

After finding the relationship between the general equation of a line and its parametric form, it is quite natural to ask if there is a general equation of a plane and its relationship with the parametric one. This will be discussed in Section 3.3.5, p. 90.

3.3.1 Norm and inner product in $\mathbb{R}^3$

Two fundamental concepts in euclidean geometry are the distance between two points and the angle defined by two segments or lines. These concepts can be defined in a vectors space under appropriate conditions such as in the case of $\mathbb{R}^2$ and $\mathbb{R}^3$. More generally, these concepts can be defined in $\mathbb{R}^n$. In order to introduce these ideas, we need to define the length or norm of a vector, as well as the angle between two vectors.

We start with a geometric idea of the length of a vector.

An application of the Pythagorean theorem in $\mathbb{R}^3$ allows us to define the distance from the origin to the point defining a vector $\alpha = (a, b, c)$. This is made precise in the following.

Definition 3.3.2. Given a vector $a = (a, b, c) \in \mathbb{R}^3$, the norm of a is defined by

$$\|a\| := \sqrt{a^2 + b^2 + c^2}.$$

The geometric meaning of the norm of a vector is that it represents the length of the segment associated to the vector a.

Given nonzero vectors $\alpha = (x, y)$ and $\beta = (a, b)$ in $\mathbb{R}^2$, they form an angle θ as illustrated in Figure 3.10.

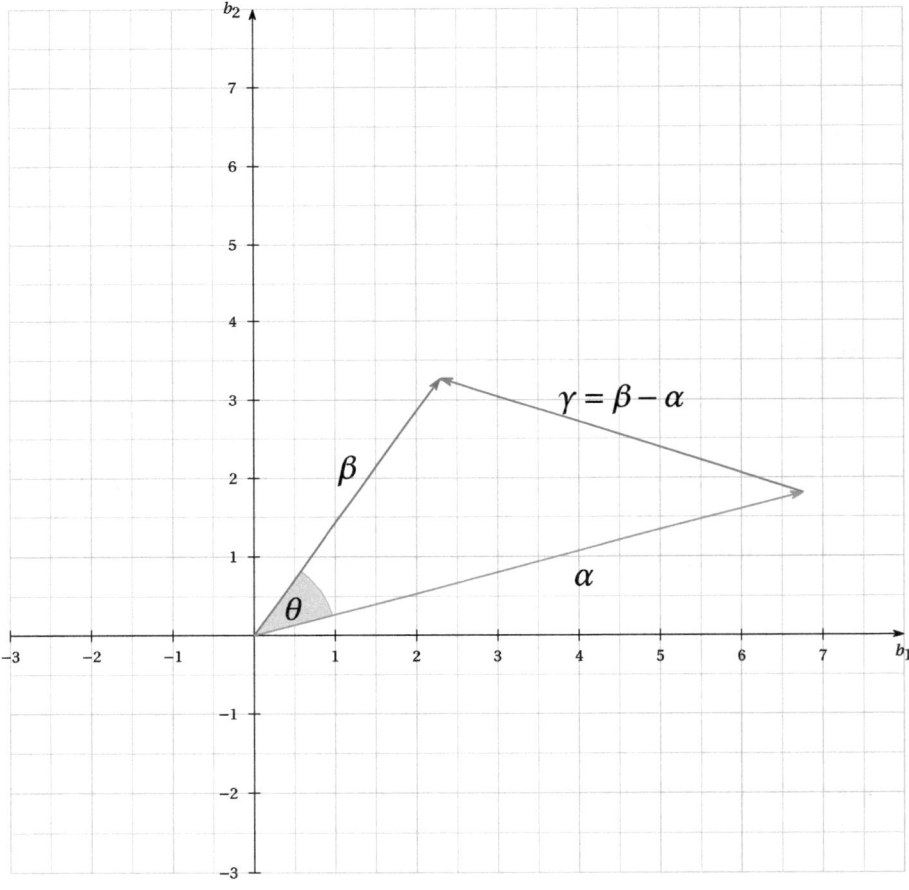

Figure 3.10: Angle between two vectors in $\mathbb{R}^2$.

Let $\gamma = \beta - \alpha = (a - x, b - y)$. According to the law of cosines, we have

$$\|\gamma\|^2 = \|\alpha\|^2 + \|\beta\|^2 - 2\|\alpha\|\|\beta\| \cos(\theta). \tag{3.4}$$

Using the coordinates of the vectors and the definition of the norm, this equation can be rewritten as

$$(a - x)^2 + (b - y)^2 = x^2 + y^2 + a^2 + b^2 - 2\sqrt{x^2 + y^2}\sqrt{a^2 + b^2}\cos(\theta). \qquad (3.5)$$

Expanding the squares and simplifying, equation (3.5) becomes

$$\cos(\theta) = \frac{ax + by}{\sqrt{x^2 + y^2}\sqrt{a^2 + b^2}}. \qquad (3.6)$$

We have that two vectors are perpendicular if the angle $\theta = \frac{\pi}{2}$. Also $\cos(\theta) = 0$ if and only if $\theta = \frac{\pi}{2}$. Thus, from equation (3.6) the vectors a and β are perpendicular if and only if $ax + by = 0$.

Definition 3.3.3. Given the vectors $a = (x, y)$ and $\beta = (a, b)$ in $\mathbb{R}^2$, we define the inner product of a and β as $\langle a, \beta \rangle := xa + yb$.

With this notation and using equation (3.6), the angle between vectors a and β is obtained from

$$\cos(\theta) = \frac{\langle a, \beta \rangle}{\|a\|\|\beta\|}. \qquad (3.7)$$

From the previous discussion, it is natural to ask if there is an equation similar to equation (3.7) for vectors in $\mathbb{R}^3$. That is, if $a = (x, y, z)$ and $\beta = (a, b, c)$ are elements in $\mathbb{R}^3$, how can we determine the angle between them?

Since $a = (x, y, z)$ and $\beta = (a, b, c)$ are vectors in $\mathbb{R}^3$, then those and $(0, 0, 0)$ belong to a plane in $\mathbb{R}^3$, see Figure 3.11.

We can use again the law of cosines, as in $\mathbb{R}^2$, and then

$$\|\beta - a\|^2 = \|a\|^2 + \|\beta\|^2 - 2\|a\|\|\beta\|\cos(\theta). \qquad (3.8)$$

Using coordinates of a and β, as well as the definition of the norm, equation (3.8) is equivalent to

$$(a - x)^2 + (b - y)^2 + (c - z)^2$$
$$= x^2 + y^2 + z^2 + a^2 + b^2 + c^2 - 2\sqrt{x^2 + y^2 + z^2}\sqrt{a^2 + b^2 + c^2}\cos(\theta). \qquad (3.9)$$

Expanding the binomials and simplifying, equation (3.9) is transformed to

$$-2ax - 2by - 2cz = -2\sqrt{x^2 + y^2 + z^2}\sqrt{a^2 + b^2 + c^2}\cos(\theta),$$

and from this we have

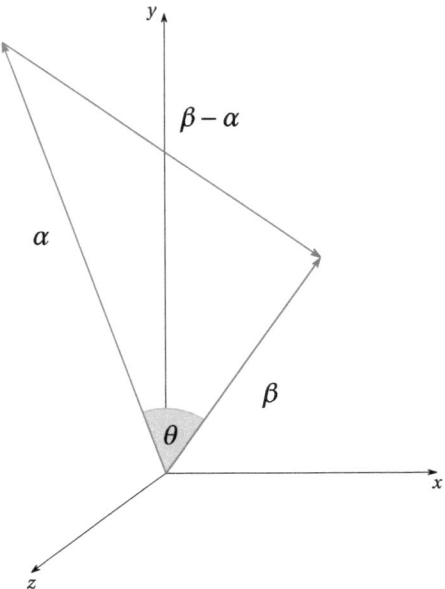

Figure 3.11: Angle between two vectors in $\mathbb{R}^3$.

$$\cos(\theta) = \frac{ax + by + cz}{\sqrt{x^2 + y^2 + z^2}\sqrt{a^2 + b^2 + c^2}}. \tag{3.10}$$

Analogously to $\mathbb{R}^2$, vectors $\alpha = (x, y, z)$ and $\beta = (a, b, c)$ are perpendicular if and only if $\cos(\theta) = 0$. Using equation (3.10), this happens if and only if $ax + by + cz = 0$.

As we have done in $\mathbb{R}^2$, we define the inner product of α and β by $\langle \alpha, \beta \rangle := xa + yb + zc$, obtaining that α is perpendicular to β if and only if $\langle \alpha, \beta \rangle = 0$.

With the above notation, the angle between two vectors in $\mathbb{R}^3$ is obtained from equation (3.11) by taking cosine inverse,

$$\cos(\theta) = \frac{\langle \alpha, \beta \rangle}{\|\alpha\|\|\beta\|}. \tag{3.11}$$

Exercise 3.3.5. In this exercise, compute the requested angle.
1. Compute the angles between the given pairs of vectors $(1, 2)$ and $(-1, 3)$, as well as $(1, 2, 3)$ and $(-1, 3, 4)$.
2. Determine the angle between the lines whose equations are $2x + y = 1$ and $x + y = 3$.

Exercise 3.3.6 (Lagrange's identity). Part of the exercise consists in asking you to provide a precise definition of $\mathbb{R}^n$.
1. Define an inner product in $\mathbb{R}^n$ extending that defined in $\mathbb{R}^2$ and $\mathbb{R}^3$.
2. Given $u = (a_1, a_2, \ldots, a_n)$ and $v = (b_1, b_2, \ldots, b_n)$ in $\mathbb{R}^n$, prove:

(a) $\langle u, u \rangle \langle v, v \rangle = \sum_{i=1}^{n} a_i^2 b_i^2 + \sum_{j=1}^{n} \left(\sum_{i \neq j} a_i^2 b_j^2 \right)$.

(b) $\langle u, v \rangle^2 = \sum_{i=1}^{n} a_i^2 b_i^2 + 2 \sum_{1 \leq i < j \leq n} a_i b_i a_j b_j$.

(c) Using the above results, prove that

$$\langle u, u \rangle \langle v, v \rangle - \langle u, v \rangle^2 = \sum_{1 \leq i < j \leq n} (a_i b_j - a_j b_i)^2 \quad \text{(Lagrange's identity)}.$$

3. Using Lagrange's identity, obtain a proof of the following important inequality:

$$|x_1 y_1 + x_2 y_2 + \cdots + x_n y_n| \leq \sqrt{x_1^2 + x_2^2 + \cdots + x_n^2} \sqrt{y_1^2 + y_2^2 + \cdots + y_n^2},$$

called Cauchy–Schwarz inequality.

4. Prove that the inner product in $\mathbb{R}^n$ satisfies:

(a) (symmetry) for every $u, v \in \mathbb{R}^n$, $\langle u, v \rangle = \langle v, u \rangle$,

(b) (positiveness) for every $u \in \mathbb{R}^n$, $\langle u, u \rangle \geq 0$ and $\langle u, u \rangle = 0$ if and only if $u = 0$,

(c) (linearity on the left) for every $u, v, w \in \mathbb{R}^n$ and for every $\lambda, \mu \in \mathbb{R}$, $\langle \lambda u + \mu v, w \rangle = \lambda \langle u, w \rangle + \mu \langle v, w \rangle$.

Remark 3.3.1. Notice that from the first and third properties of part 4 above, we have $\langle u, \lambda v + \mu w \rangle = \lambda \langle u, v \rangle + \mu \langle u, w \rangle$, for every $u, v, w \in \mathbb{R}^n$ and for every $\lambda, \mu \in \mathbb{R}$ (linearity on the right).

3.3.2 Orthogonal projection of a vector along another one

In several geometric or physical problems, it is important to determine the orthogonal projection of a vector along another one. For instance, when computing the distance from a point to a line, we draw a perpendicular segment from the point to the line and measure its length. The same idea can be used to compute the distance from one point to a plane. Given two nonzero vectors, u and v, we can decompose u as the sum of two vectors: one orthogonal to v, denoted x, and another parallel to v, which is of the form λv. See Figure 3.12.

The conditions above can be described as $u = x + \lambda v$. From this, one has $x = u - \lambda v$. Now, from the assumptions on x and v, we have $0 = \langle x, v \rangle = \langle u - \lambda v, v \rangle = \langle u, v \rangle - \lambda \langle v, v \rangle$. From the latter equation and taking into account that $v \neq 0$, we obtain that $\lambda = \frac{\langle u, v \rangle}{\langle v, v \rangle} = \frac{\langle u, v \rangle}{\|v\|^2}$.

Definition 3.3.4. With the same assumptions as above, the vector $\frac{\langle u, v \rangle}{\|v\|^2} v$ is called the orthogonal projection of u along the vector v and it is denoted by $\text{Proj}_v u = \frac{\langle u, v \rangle}{\|v\|^2} v$.

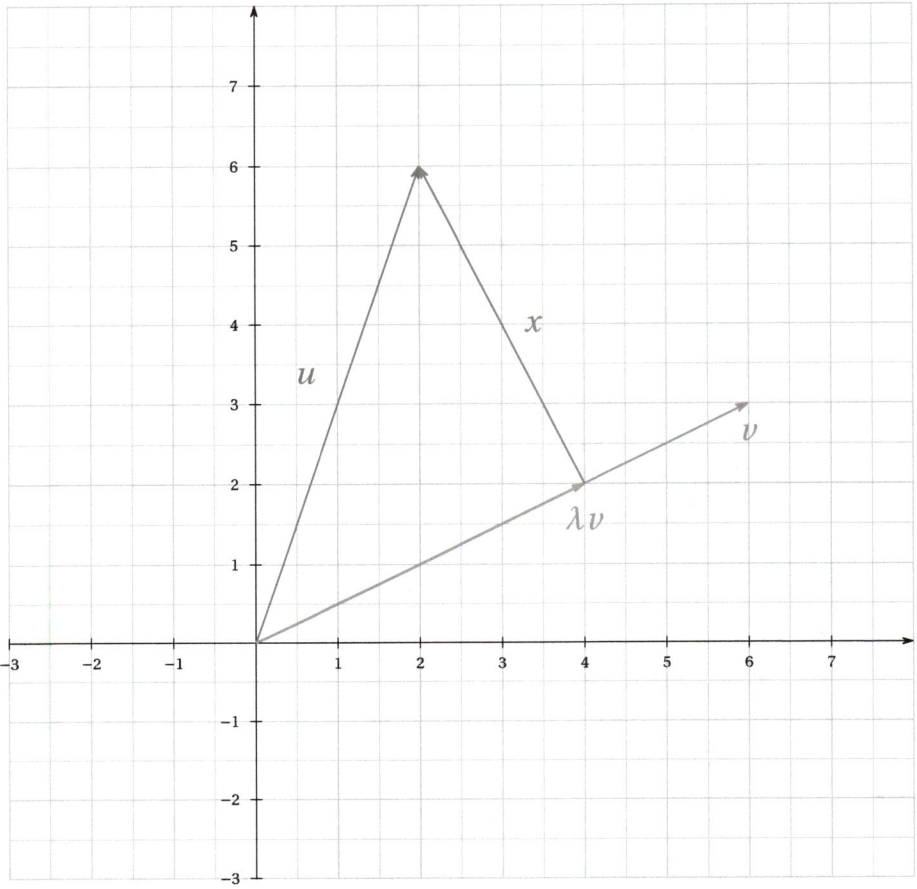

Figure 3.12: Orthogonal projection of u along v.

Remark 3.3.2. Notice that the vector $\frac{v}{\|v\|}$ has norm one, hence the vector $\text{Proj}_v\, u$ has length $\frac{\langle u,v \rangle}{\|v\|}$ in the v direction. In some texts, the orthogonal projection of u along v is defined as the scalar $\frac{\langle u,v \rangle}{\|v\|}$. Particularly, u is parallel to v if and only if $\|u\| = \frac{|\langle v,u \rangle|}{\|v\|}$.

Example 3.3.1. Find the orthogonal projection of $u = (1,2,3)$ along $v = (2,0,1)$. Sketch it geometrically.

Discussion. The orthogonal projection of u along v is obtained by applying Definition 3.3.4, that is, $\text{Proj}_v\, u = \frac{\langle (1,2,3),(2,0,1) \rangle}{\langle (2,0,1),(2,0,1) \rangle}(2,0,1) = \frac{2+3}{4+1}(2,0,1) = (2,0,1)$. Notice that the projection of u along v is v itself.

Exercise 3.3.7. In each of the following cases, express u as the sum of its orthogonal projection along v and another vector orthogonal to v:

1. $u = (3, 8), v = (1, 0)$;
2. $u = (1, 2, 3), v = (1, 1, 0)$;
3. $u = (2, 1, 1), v = (1, 2, 0)$.

An application to analytic geometry

Given a straight line L whose equation is $Ax + By + C = 0$, and a point $P = (x_0, y_0)$ that is not on L, determine the distance from P to L.

Without loss of generality, we can assume that $AB \neq 0$, since, when A or B is zero, the line is vertical or horizontal and the distance to P is easily calculated.

Figure 3.13 shows the geometric considerations that guide the discussion.

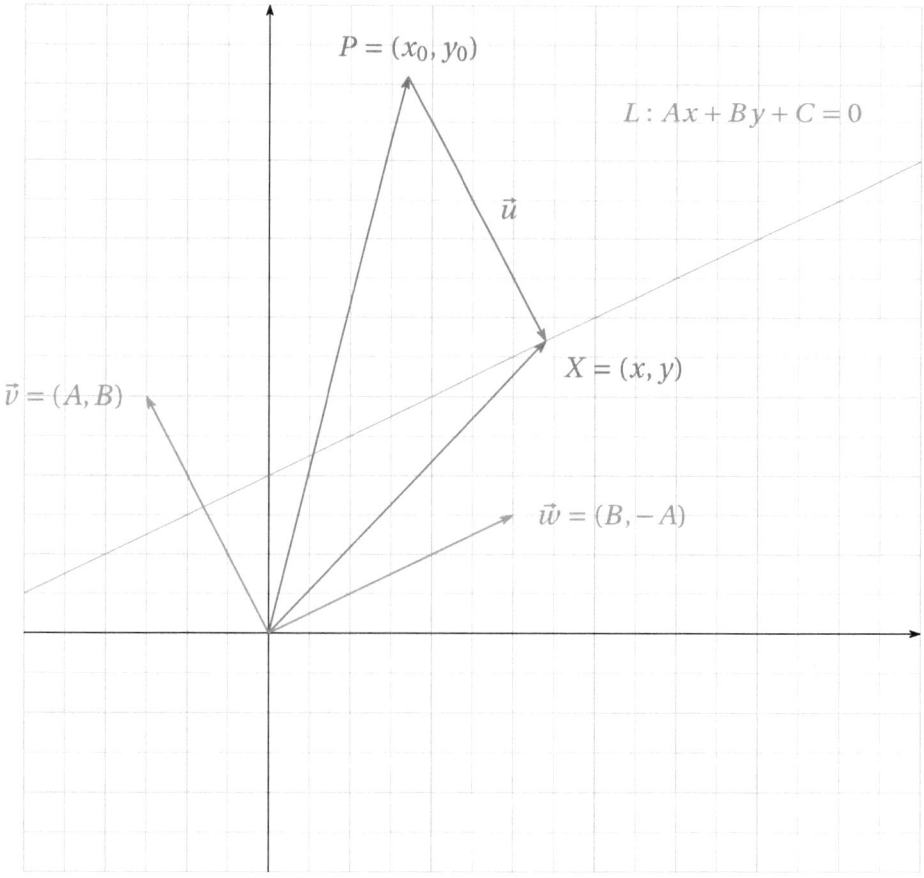

Figure 3.13: Distance from point $P = (x_0, y_0)$ to line $L : Ax + By + C = 0$.

We should notice that the distance from P to L is the norm of $u = P - X = (x_0, y_0) - (x, y)$. We also have that the vector $v = (A, B)$ is perpendicular to L, since the vector $w = (B, -A)$ is parallel to L, hence u is parallel to v. From Remark 3.3.2, we have

$$
\begin{aligned}
\|u\| &= \frac{|\langle v, u \rangle|}{\|v\|} \\
&= \frac{|A(x_0 - x) + B(y_0 - y)|}{\sqrt{A^2 + B^2}} \\
&= \frac{|Ax_0 - Ax + By_0 - By|}{\sqrt{A^2 + B^2}} \\
&= \frac{|Ax_0 + By_0 + C|}{\sqrt{A^2 + B^2}},
\end{aligned}
\tag{3.12}
$$

since the point (x, y) lies on L, and then $-Ax - By = C$. From this we have the well-known formula

$$
d(P, L) = \frac{|Ax_0 + By_0 + C|}{\sqrt{A^2 + B^2}},
\tag{3.13}
$$

to compute the distance from point $P = (x_0, y_0)$ to line L whose equation is $Ax + By + C = 0$.

3.3.3 Vectors orthogonal to a given one

Given a vector $u \in \mathbb{R}^2$, we are interested in finding all those vectors that are orthogonal to u. If $u = (0, 0)$, then all vectors are orthogonal to u. Otherwise, if $u = (a, b) \neq (0, 0)$ then we need to find those vectors $X = (x, y)$ such that $\langle u, X \rangle = ax + by = 0$, in other words, we are searching for vectors which belong to the line whose equation is $ax + by = 0$.

If v is a nonzero vector in L, for instance, $v = (-b, a)$, then we are interested in vectors of the form λv, with $\lambda \in \mathbb{R}$. See Figure 3.14.

If $u = (a, b, c) \in \mathbb{R}^3$, then the vectors $X = (x, y, z)$ that are orthogonal to $u = (a, b, c)$ are those that satisfy $\langle u, X \rangle = ax + by + cz = 0$ and, as in the previous case, there are two possibilities.

If $u = (0, 0, 0)$, then every element of $\mathbb{R}^3$ satisfies the condition, otherwise, the vectors X must satisfy the equation $\langle u, X \rangle = ax + by + cz = 0$. We will prove that the collection of those vectors lies in a plane that passes through the origin, Figure 3.15.

In fact, let us consider the equation $ax + by + cz = 0$, and assume, for instance, that $a \neq 0$, then we can write x in the form $x = -\frac{b}{a}y - \frac{c}{a}z$. Setting $y = 0$ and $z = -a$, we have the vector $v_1 = (c, 0, -a)$. Interchanging roles between y and z in the above case, we have another vector $v_2 = (b, -a, 0)$.

Claim 1. Vectors v_1 and v_2 are linearly independent.

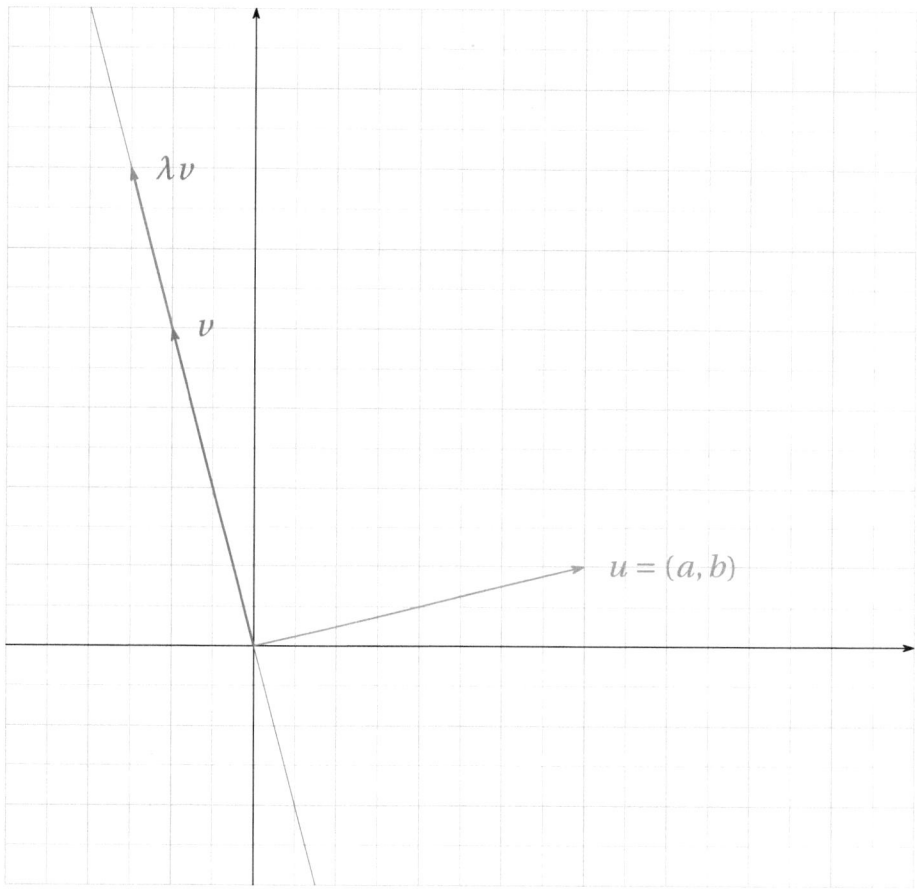

Figure 3.14: Orthogonal vectors to $u = (a, b)$.

If $xv_1 + yv_2 = (xc + by, -ax, -ay) = (0, 0, 0)$, then the assumption on a implies that $x = y = 0$, proving the claim.

Claim 2. The sets $\{(x, y, z) : ax + by + cz = 0\}$ and $\{\mu v_1 + \lambda v_2 : \mu, \lambda \in \mathbb{R}\}$ are equal.

The justification of Claim 2 is left as an exercise.

From the above discussion, we have that $\{(x, y, z) : ax + by + cz = 0\}$ is a plane that passes through the origin and is orthogonal to the vector $N = (a, b, c)$.

Example 3.3.2. Find the equation of the plane that is orthogonal to $(1, -1, 2)$ and passes through the origin. Also, find the vectors that generate it.

Discussion. The plane that is orthogonal to $N = (1, -1, 2)$ and passes through the origin is the set of points $P = \{(x, y, z) \in \mathbb{R}^3 \mid \langle N, X \rangle = \langle (1, -1, 2), (x, y, z) \rangle = x - y + 2z = 0\}$. To find

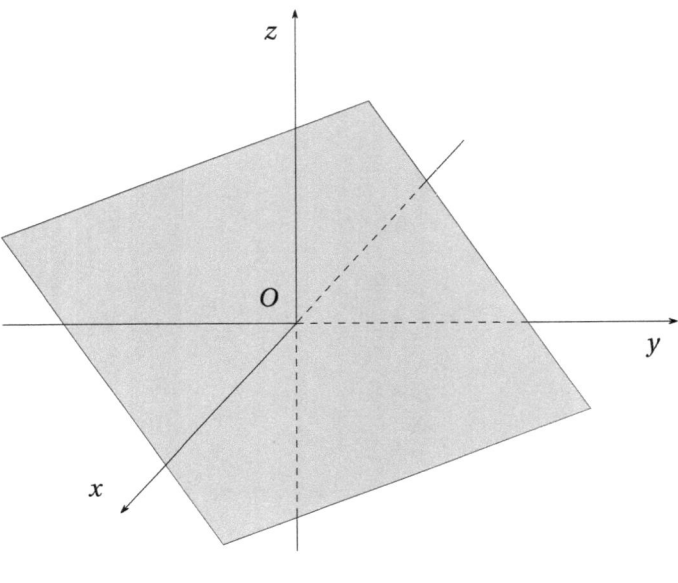

$$ax + by + cz = 0$$

Figure 3.15: A plane passing through the origin has equation $ax + by + cz = 0$.

vectors that generate the plane, we need to find two vectors $\alpha, \beta \in P$ that are linearly independent. From the equation that defines the plane P, one has that $\alpha = (1, 1, 0)$ and $\beta = (2, 0, -1)$ belong to P and are linearly independent, hence they generate P.

3.3.4 Cross product in $\mathbb{R}^3$

Let u and v be two linearly independent vectors. What are the vectors $X = (x, y, z)$ that satisfy $\langle u, X \rangle = \langle v, X \rangle = 0$?

Assume that $u = (a, b, c)$ and $v = (a_1, b_1, c_1)$ are linearly independent. The orthogonality condition on u and v with X leads to the system of linear equations

$$ax + by + cz = 0, \tag{3.14}$$

$$a_1 x + b_1 y + c_1 z = 0. \tag{3.15}$$

Multiplying equation (3.14) by a_1 and equation (3.15) by $-a$, adding them up, and factorizing, we have

$$(a_1 b - a b_1) y + (a_1 c - a c_1) z = 0. \tag{3.16}$$

By exploration one can obtain solutions of (3.16). For instance, $z = -(a_1b - ab_1)$ and $y = (a_1c - ac_1)$ is a solution. Substituting into (3.14), we obtain $ax + b(a_1c - ac_1) - c(a_1b - ab_1) = 0$, which simplifies to $ax = a(bc_1 - b_1c)$. Hence, a solution of (3.14) and (3.15) is the vector $X = (bc_1 - b_1c, a_1c - ac_1, ab_1 - a_1b)$, that is, X is orthogonal to u and v.

Exercise 3.3.8. Let $u = (a, b, c)$ and $v = (a_1, b_1, c_1)$ be vectors in $\mathbb{R}^3$. Then the vector $X = (bc_1 - b_1c, a_1c - ac_1, ab_1 - a_1b)$ is not zero if and only if the vectors u and v are linearly independent.

Definition 3.3.5. Given $u = (a, b, c)$ and $v = (a_1, b_1, c_1)$ in $\mathbb{R}^3$, the *cross product* or *vector product* of u and v is defined and denoted by $u \times v := (bc_1 - b_1c, a_1c - ac_1, ab_1 - a_1b)$.

As a mnemonic rule, we use the following representation:

$$u \times v := \begin{vmatrix} i & j & k \\ a & b & c \\ a_1 & b_1 & c_1 \end{vmatrix} := (bc_1 - b_1c)i + (a_1c - ac_1)j + (ab_1 - a_1b)k,$$

where $i = (1, 0, 0)$, $j = (0, 1, 0)$, and $k = (0, 0, 1)$.

Example 3.3.3. Given $(1, 2, 0)$ and $(1, 3, 0)$, compute $(1, 2, 0) \times (1, 3, 0)$.

Discussion. According to Definition 3.3.5, we have

$$(1, 2, 0) \times (1, 3, 0) = \begin{vmatrix} i & j & k \\ 1 & 2 & 0 \\ 1 & 3 & 0 \end{vmatrix} = (2 \cdot 0 - 3 \cdot 0)i + (1 \cdot 0 - 1 \cdot 0)j + (3 - 2)k = (0, 0, 1).$$

Geometric meaning of $\|u \times v\|$

We have that the angle θ between nonzero vectors u and v is given by

$$\cos(\theta) = \frac{\langle u, v \rangle}{\|u\|\|v\|}. \tag{3.17}$$

On the other hand, we have the well-known identity $\cos^2(\theta) + \sin^2(\theta) = 1$, from where we obtain

$$\sin^2(\theta) = 1 - \cos^2(\theta) = \frac{\langle u, u \rangle \langle v, v \rangle - \langle u, v \rangle^2}{\langle u, u \rangle \langle v, v \rangle}. \tag{3.18}$$

Assigning coordinates to $u = (a_1, a_2, a_3)$ and $v = (b_1, b_2, b_3)$, and using Lagrange's identity Exercise 3.3.6, 2c, p. 82, one has

$$\langle u, u \rangle \langle v, v \rangle - \langle u, v \rangle^2 = (a_1b_2 - a_2b_1)^2 + (a_1b_3 - a_3b_1)^2 + (a_2b_3 - a_3b_2)^2$$
$$= \|u \times v\|^2.$$

From this and equation (3.18), we obtain $\|u \times v\|^2 = \|u\|^2 \|v\|^2 \sin^2(\theta)$.
Since the angle θ between two vectors satisfies $0 \le \theta \le \pi$, we get $\sin(\theta) \ge 0$, hence

$$\|u \times v\| = \|u\| \|v\| \sin(\theta). \tag{3.19}$$

The expression on the right-hand side of (3.19) is the area of a parallelogram whose sides are u and v, as you can convince yourself by drawing a picture.

Remark 3.3.3. Equation (3.19) gives a formula to compute the area of a parallelogram determined by the vectors u and v (explain). It also gives an alternative method to compute the angle between vectors. One disadvantage of this method is that the formula cannot be extended to $\mathbb{R}^n$, while equation (3.18) can.

Example 3.3.4. Compute the area of the parallelogram determined by the vectors $u = (2, 1, 0)$ and $v = (1, 3, 0)$.

Discussion. We have that $u \times v = \begin{vmatrix} i & j & k \\ 2 & 1 & 0 \\ 1 & 3 & 0 \end{vmatrix} = (0, 0, 5)$. From this, the area is $\|(0, 0, 5)\| = 5$.

3.3.5 Cartesian equation of a plane

Recall that the parametric equations of a plane, Definition 3.3.1, are given by

$$x = x_0 + at + a_1 s, \tag{3.20}$$
$$y = y_0 + bt + b_1 s, \tag{3.21}$$
$$z = z_0 + ct + c_1 s, \tag{3.22}$$

where $P = (x_0, y_0, z_0)$, $\alpha = (a, b, c)$, $\beta = (a_1, b_1, c_1)$, and α and β are linearly independent.
   Multiplying equation (3.20) by $-b$ and equation 3.21 by a and adding them up, one has

$$ay - bx = ay_0 - bx_0 + (ab_1 - ba_1)s. \tag{3.23}$$

Multiplying equations (3.20) and (3.22) by $-c$ and a, respectively, and adding them up, one has

$$az - cx = az_0 - cx_0 + (ac_1 - ca_1)s. \tag{3.24}$$

It is straightforward to verify that α and β are linearly independent if and only if at least one of $(ab_1 - ba_1)$ and $(ac_1 - ca_1)$ is not zero. Hence we may assume that $(ab_1 - ba_1) \neq 0$. From (3.23), solving for s and substituting into (3.24), one has

$$az - cx = az_0 - cx_0 + \frac{(ac_1 - ca_1)}{ab_1 - ba_1}[ay - bx + bx_0 - ay_0]. \tag{3.25}$$

Simplifying and moving terms to one side, (3.25) becomes

$$(bc_1 - cb_1)(x - x_0) + (ca_1 - ac_1)(y - y_0) + (ab_1 - a_1b)(z - z_0) = 0. \qquad (3.26)$$

We recognize that equation (3.26) can be written in the form

$$\langle a \times \beta, (x - x_0, y - y_0, z - z_0) \rangle = 0. \qquad (3.27)$$

If $a \times \beta = (A, B, C)$, then equation (3.27) can be written as

$$A(x - x_0) + B(y - y_0) + C(z - z_0) = 0, \qquad (3.28)$$

or in the form

$$Ax + By + Cz + D = 0, \qquad (3.29)$$

where $D = -(Ax_0 + By_0 + Cz_0)$. Equation (3.29) is called the cartesian equation of a plane.

We know, from euclidean geometry, that a plane is completely determined by three noncollinear points. If those points are in $\mathbb{R}^3$ and are denoted by P, Q, and R, we want to find the equation of the plane that passes through them. The procedure that we will use is based on the following. Given a vector $N \neq 0$, there is a unique plane orthogonal to N that passes through the origin, from this we have that given a vector $N \neq 0$ and one point $S \in \mathbb{R}^3$, there is a unique plane orthogonal to N which passes through S.

If P, Q, and R are noncollinear points in $\mathbb{R}^3$, vectors $a = Q - P$ and $\beta = R - P$ are linearly independent (you are invited to prove it), from where $N = a \times \beta = (A, B, C)$ is nonzero and orthogonal to a and β. If $X = (x, y, z)$ is a vector in the plane that passes through P, then N is orthogonal to $X - P$. Suppose that the coordinates of P are $P = (a_1, b_1, c_1)$, then the orthogonality condition on N and $X - P$ leads to

$$\begin{aligned}
\langle (A, B, C), (x - a_1, y - b_1, z - c_1) \rangle &= A(x - a_1) + B(y - b_1) + C(z - c_1) \\
&= Ax + By + Cz + D \\
&= 0,
\end{aligned}$$

where $D = -(Aa_1 + Bb_1 + Cc_1)$.

Summarizing, the equation of the plane that passes through the points $P = (a_1, b_1, c_1)$, $Q = (a_2, b_2, c_2)$, and $R = (a_3, b_3, c_3)$ is given by

$$Ax + By + Cz + D = 0, \qquad (3.30)$$

where $D = -(Aa_1 + Bb_1 + Cc_1)$ and $(A, B, C) = (Q - P) \times (R - P)$.

The previous discussion helps us to find the equation of a plane that is normal to a nonzero vector $N = (A, B, C)$ and passes through a point $P = (x_0, y_0, z_0)$. More precisely, such an equation is of the form

$$Ax + By + Cz + D = 0, \quad \text{where } D = -\langle P, N \rangle. \tag{3.31}$$

Conversely, an equation of the form (3.31), with at least one of A, B, and C nonzero, represents a plane that has $N = (A, B, C)$ as a normal vector. This holds since at least one of the coefficients of the variables is nonzero, hence there is a point $P = (x_0, y_0, z_0)$ whose coordinates satisfy $Ax_0 + By_0 + Cz_0 + D = 0$. This equation can be written as $D = -Ax_0 - By_0 - Cz_0$. Substituting and rearranging terms in the original equation, we have $A(x - x_0) + B(y - y_0) + C(z - z_0) = 0$, which is interpreted as follows: the set of points $X = (x, y, z) \in \mathbb{R}^3$ that satisfy $\langle X - P, N \rangle = 0$, where $N = (A, B, C)$, is a plane orthogonal to N.

Definition 3.3.6. The cartesian equation of a plane is $Ax + By + Cz + D = 0$, where A, B, C, and D are real constants and at least one of A, B, and C is not zero.

Example 3.3.5. Find the equation of the plane that passes through the points $(1, 2, 0)$, $(0, 1, 1)$, and $(1, 2, 3)$.

Discussion. Let us take $P = (1, 2, 0)$, then, according to the above notation, one has $\alpha = (0, 1, 1) - (1, 2, 0) = (-1, -1, 1)$ and $\beta = (1, 2, 3) - (1, 2, 0) = (0, 0, 3)$. From this we have that the cross product of α and β is given by

$$N = \begin{vmatrix} i & j & k \\ -1 & -1 & 1 \\ 0 & 0 & 3 \end{vmatrix} = -3i + 3j = (-3, 3, 0).$$

Therefore, $\langle (-3, 3, 0), (x - 1, y - 2, z) \rangle = -3(x - 1) + 3(y - 2) = -3x + 3y - 3$, hence the equation of the plane is $x - y + 1 = 0$, and one verifies that the coordinates of $(1, 2, 0)$, $(0, 1, 1)$, and $(1, 2, 3)$ satisfy this equation.

3.4 Exercises

1. Consider a circle and three points on it, P, Q, and R, so that exactly two of them are on a diameter, say P and Q. Prove that PRQ is a right triangle at R.
2. Consider three vectors A, B, and C in $\mathbb{R}^3$. Under which conditions are $A \times B$ and $B \times C$ linearly independent?
3. Let $U = (a, b, c) \in \mathbb{R}^3$ so that $abc \neq 0$. Show that the collection of points (x, y, z) lying on the line that passes through $P = (x_0, y_0, z_0)$ in the direction U are exactly those that satisfy the equations $\frac{x - x_0}{a} = \frac{y - y_0}{b} = \frac{z - z_0}{c}$.
4. Let L and L_1 be two parallel lines whose equations are $Ax + By + C = 0$ and $Ax + By + D = 0$, respectively. Prove that the distance from L to L_1 is given by $d = \frac{|C - D|}{\sqrt{A^2 + B^2}}$.
5. Assume that the equation of a sphere is given by

$$x^2 + y^2 + z^2 + Ax + By + Cz + D = 0. \tag{3.32}$$

Let $P = (x_1, y_1, z_1)$, $Q = (x_2, y_2, z_2)$, $R = (x_3, y_3, z_3)$, $S = (x_4, y_4, z_4)$ be four points in $\mathbb{R}^3$. Find necessary and sufficient conditions on P, Q, R, and S in order that they determine the coefficients A, B, C, and D in (3.32).

6. Find the equation of the plane that passes through the point $(1, 1, 1)$ and is perpendicular to the line generated by the vector $(1, -1, 0)$.

7. Given the planes P_1 and P_2 with normal vectors N_1 and N_2, respectively, show that the angle between P_1 and P_2 is given by the angle between N_1 and N_2.

8. Does the line passing through $(1, 2, 3)$ and $(1, 0, 3)$ intersect the xy-plane? If so, give the intersection point.

9. Find the set of points where the planes given by the equations $x - y - 3z + 3 = 0$ and $x + y - z + 2 = 0$ intersect.

10. In Figure 3.16, an ellipse E is shown, whose equation is $\frac{x^2}{a^2} + \frac{y^2}{b^2} = 1$ and its foci are F_1 and F_2. The line T is tangent to E at the point $Q = (x_0, y_0)$ and segments $F_1 P_1$ and $F_2 P_2$ are orthogonal to L.

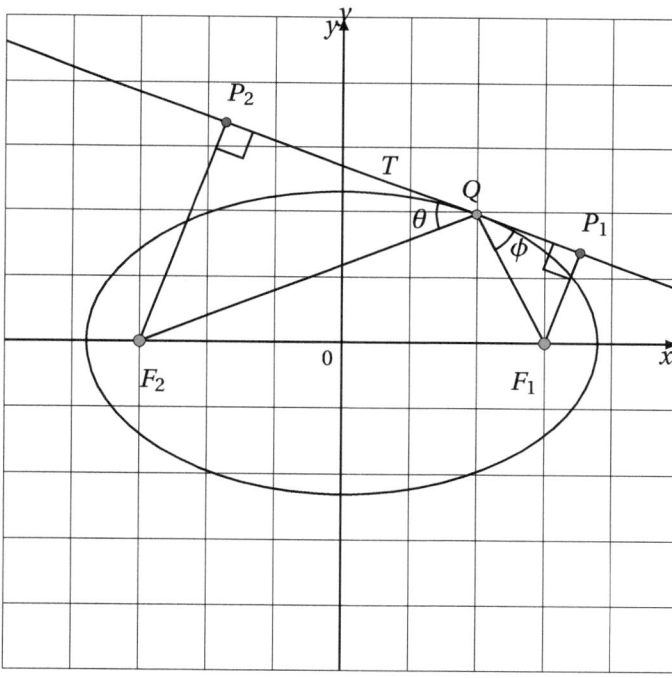

Figure 3.16: Equation of the tangent: $\frac{x_0}{a^2} x + \frac{y_0}{b^2} y = 1$.

(a) Show that the equation of L is $\frac{x_0}{a^2} x + \frac{y_0}{b^2} y = 1$.

(b) Let d_1 and d_2 denote the length of segments $F_1 P_1$ and $F_2 P_2$, respectively. Prove that $d_1 d_2 = b^2$.

(c) Prove that triangles F_2P_2Q and F_1P_1Q are similar, hence $\theta = \phi$. This property of an ellipse has applications in acoustics. One of these applications is in medicine (lithotripsy).

11. Given vectors u, v, and w in $\mathbb{R}^3$, prove that $(u \times v) \times w = \langle u, w \rangle v - \langle v, w \rangle u$ (*triple vector product*). Use this identity to prove that $(u \times v) \times w = u \times (v \times w)$ does not hold in general. Under which conditions does it hold?

12. Given vectors u, v, and w in $\mathbb{R}^3$, prove the identities:
    (a) $\langle (u+v), (u \times w) \times (u+v) \rangle = 0$.
    (b) $u \times (u \times (u \times v)) = \langle u, u \rangle (v \times u)$.
    (c) $(u \times v) \times (u \times w) = \langle u, v \times w \rangle u$.
    (d) $u \times (v \times w) + v \times (w \times u) + w \times (u \times v) = 0$.

13. Prove that the volume of the parallelepiped determined by the vectors u, v, and w is given by $|\langle u, v \times w \rangle|$. This number is called the *triple scalar product*.

14. Calculate the volume of the parallelepiped determined by the vectors $u = (1, 2, 0)$, $v = (4, 1, 0)$, and $w = (5, 3, 4)$.

15. A regular tetrahedron is a solid with four faces, each being an equilateral triangle. Let θ be the angle between any two faces. Prove that $\cos(\theta) = \frac{1}{3}$.

16. Show that the volume of a regular tetrahedron with edge length a is given by $V = \frac{a^3}{6\sqrt{2}}$.

17. A regular dodecahedron D is a solid with 12 faces, each being a regular pentagon. Prove that if the edge length of D is a, then its volume is $V = \frac{(15+7\sqrt{5})a^3}{4}$.

18. Find the distance from the line whose equation is $x + 3y = 6$ to the point $(3, 5)$.

19. Prove that vectors u, v, and w are linearly dependent if and only if $\langle u, v \times w \rangle = 0$. Give a geometric meaning of this result using Exercise 13.

20. Let u be a nonzero vector in $\mathbb{R}^3$, and suppose α, β, and γ are the angles between u and the axes x, y, and z, respectively. These angles are called *director angles* of u. Prove that $\cos^2(\alpha) + \cos^2(\beta) + \cos^2(\gamma) = 1$.
    (*Hint.*) Start by considering the case where u lies on one of the coordinate planes.

21. Determine if the points $(3, 7, -2)$, $(5, 5, 1)$, $(4, 0, -1)$, and $(6, -2, 2)$ are vertices of a parallelogram.

22. Compute the volume of the parallelepiped determined by the vectors $u = (1, 1, 1)$, $v = (1, 2, 3)$, and $w = (0, 1, 0)$.

23. In each of the following cases, find the equation of a plane P determined by the given conditions:
    (a) The plane P contains the points $(1, 0, -1)$, $(0, 1, 2)$, and $(1, 0, -2)$.
    (b) The plane P is normal to $(1, 2, 4)$ and contains the point $(1, 5, 9)$.
    (c) The plane P contains the points $(a, 0, 0)$, $(0, b, 0)$ and $(0, 0, c)$. If $abc \neq 0$, prove that the equation of the plane can be written as $\frac{x}{a} + \frac{y}{b} + \frac{z}{c} = 1$.

24. Find a formula to compute the distance from a point to a plane.
    (*Hint.*) Review the procedure to compute the distance from one point to a line.

25. By using vectors, prove that the heights of a triangle are concurrent.

26. In this exercise we invite the reader to search the web (click ellipsoid) for the terms that appear here and have not been defined. What is the equation of an ellipsoid centered at the origin? What is the equation of a paraboloid with vertex at the origin? Explore all possibilities considering, as a reference in the plane, the case of a parabola whose vertex is at the origin of coordinates.

27. Using equation (3.19), p. 90, prove the famous law of sines: for a triangle of angles A, B, and C with opposite sides a, b, and c, respectively, the equality $\frac{a}{\sin(A)} = \frac{b}{\sin(B)} = \frac{c}{\sin(C)}$ holds.

4 Vector spaces

It should be mentioned that several important mathematical problems can be stated and studied in $\mathbb{R}^n$, as a vector space. However, many more problems need to be discussed in more general settings, that is why it is necessary to consider general vector spaces.

We start the discussion by presenting some basic properties of $\mathbb{R}^n$ as a vector space and, having settled some properties of $\mathbb{R}^n$, the general definition of a vector space is presented. Also, basic results concerning finitely generated vector spaces are discussed.

4.1 The vector space $\mathbb{R}^n$

After discussing basic properties of $\mathbb{R}^2$ and $\mathbb{R}^3$ as vector spaces, it is natural to extend the discussion to $\mathbb{R}^n$, for $n > 3$. By doing this, we will be in a position to establish important connections between systems of linear equations and the concept of linear dependence, among others. To that end, we will start by discussing properties of $\mathbb{R}^n$ as a vector space.

Definition 4.1.1. For each positive integer n, define

$$\mathbb{R}^n = \{(x_1, x_2, \ldots, x_n) \mid x_i \in \mathbb{R} \text{ for every } i = 1, 2, \ldots, n\}.$$

The sum and product with scalars in $\mathbb{R}^n$ are defined as follows:
- Given $(x_1, x_2, \ldots, x_n)$ and $(y_1, y_2, \ldots, y_n) \in \mathbb{R}^n$, the sum is defined as

$$(x_1, x_2, \ldots, x_n) + (y_1, y_2, \ldots, y_n) = (x_1 + y_1, x_2 + y_2, \ldots, x_n + y_n).$$

- Given $(x_1, x_2, \ldots, x_n) \in \mathbb{R}^n$ and $c \in \mathbb{R}$, the product with the scalar c is defined as

$$c(x_1, x_2, \ldots, x_n) = (cx_1, cx_2, \ldots, cx_n).$$

Note that the sum and product with a scalar are analogous to those defined in $\mathbb{R}^2$ and $\mathbb{R}^3$. Theorem 4.1.1 summarizes the main properties that are derived from the definition of the sum and product with scalars in $\mathbb{R}^n$.

Theorem 4.1.1. *With the sum and product with scalars, $\mathbb{R}^n$ satisfies the following properties:*

1. *Properties of the sum.*
   (Commutativity) *For every α and $\beta \in \mathbb{R}^n$, $\alpha + \beta = \beta + \alpha$.*
   (Associativity) *For every α, β and γ in $\mathbb{R}^n$, one has $(\alpha + \beta) + \gamma = \alpha + (\beta + \gamma)$.*
   (Existence of additive identity) *There is an element in $\mathbb{R}^n$, called zero and denoted by 0, such that $0 + \alpha = \alpha$, for every $\alpha \in \mathbb{R}^n$.*
   (Existence of additive inverse) *For every $\alpha \in \mathbb{R}^n$, there is $\alpha' \in \mathbb{R}^n$ such that $\alpha + \alpha' = 0$.*

https://doi.org/10.1515/9783111135915-004

2. *Properties of the product by scalar.*
   (Unital property) *For every $\alpha \in \mathbb{R}^n$, $1\alpha = \alpha$, with $1 \in \mathbb{R}$.*
   (Associativity of scalar multiplication) *For every $\alpha \in \mathbb{R}^n$, and for every r and $s \in \mathbb{R}$, one has $r(s\alpha) = (rs)\alpha$.*
   (Distributivity of scalar multiplication) *The product with a scalar is distributive, that is, $(r + s)\alpha = r\alpha + s\alpha$; $r(\alpha + \beta) = r\alpha + r\beta$, for every $r, s \in \mathbb{R}$ and $\alpha, \beta \in \mathbb{R}^n$.*

Proof. We will prove only the commutativity property of the sum and the associativity of the scalar multiplication. The others are left to the reader as an exercise. Let $\alpha = (a_1, a_2, \ldots, a_n)$ and $\beta = (b_1, b_2, \ldots, b_n)$ be elements of $\mathbb{R}^n$, then

$$
\begin{aligned}
\alpha + \beta &= (a_1, a_2, \ldots, a_n) + (b_1, b_2, \ldots, b_n) \\
&= (a_1 + b_1, a_2 + b_2, \ldots, a_n + b_n) \\
&= (b_1 + a_1, b_2 + a_2, \ldots, b_n + a_n) \\
&= \beta + \alpha,
\end{aligned}
$$

proving the commutativity property of the sum.

The proof of the associativity of the scalar multiplication goes as follows. Let $r, s \in \mathbb{R}$ and consider $\alpha \in \mathbb{R}^n$, then

$$
\begin{aligned}
r(s\alpha) &= r(s(a_1, a_2, \ldots, a_n)) \\
&= r(sa_1, sa_2, \ldots, sa_n) \\
&= (rsa_1, rsa_2, \ldots, rsa_n) \\
&= (rs)(a_1, a_2, \ldots, a_n) \\
&= (rs)\alpha,
\end{aligned}
$$

proving the associativity of the scalar multiplication. $\qquad\square$

As it was done in $\mathbb{R}^2$ and $\mathbb{R}^3$, with the sum and product with a scalar in $\mathbb{R}^n$, a system of linear equations can be represented with a single equation. In order to show this, we present the following definition which is one of the most important concepts in linear algebra.

Definition 4.1.2. Let $A_1, A_2, \ldots, A_k$ be elements of $\mathbb{R}^n$ and $x_1, x_2, \ldots, x_k$ real numbers. The expression $x_1 A_1 + x_2 A_2 + \cdots + x_k A_k$ is called a linear combination of $A_1, A_2, \ldots, A_k$.

Using the properties of the sum and product with scalars in $\mathbb{R}^n$, a system of linear equations can be represented in a form that allows us to give another interpretation to the solutions. First of all, if A is an $m \times n$ matrix, then the columns of A can be thought as elements of $\mathbb{R}^m$ and its rows can be consider as elements of $\mathbb{R}^n$. With this interpretation, a system of linear equations

$$a_{11}x_1 + a_{12}x_2 + \cdots + a_{1n}x_n = b_1,$$
$$a_{21}x_1 + a_{22}x_2 + \cdots + a_{2n}x_n = b_2,$$
$$\vdots \qquad\qquad\qquad\qquad\qquad (4.1)$$
$$a_{m1}x_1 + a_{m1}x_1 + \cdots + a_{mn}x_n = b_m$$

can be represented as a single equation in terms of the columns of the coefficients' matrix A. More precisely, the system (4.1) can be written in the form

$$x_1 A_1 + x_2 A_2 + \cdots + x_n A_n = B, \qquad\qquad (4.2)$$

where $A_j = (a_{1j}, a_{2j}, \ldots, a_{mj})$ is the jth column of A and $B = (b_1, b_2, \ldots, b_m)$.

With this representation of (4.1), we have:

Remark 4.1.1. The system $AX = B$ has a solution if and only if B is a linear combination of $A_1, A_2, \ldots, A_n$, which are the columns of A.

When $B = (0, 0, \ldots, 0)$, the system (4.1) always has a solution, $x_1 = x_2 = \cdots = x_n = 0$. It might happen that, when $B = 0$, equation (4.1) has more than one solution, in this case it is said that the columns of A are linearly dependent. Otherwise, we say that the columns of A are linearly independent.

The linear dependence or linear independence of elements of $\mathbb{R}^n$ is important when solving homogeneous systems of linear equations. In the following section, we will discuss aspects related to them.

4.2 Linear dependence and dimension

Terms such as linear combination and linearly dependent vectors were presented in Section 3.2, p. 73. However, those terms are so important that we will discuss them here in a more general setting. We start with

Definition 4.2.1. Let $A_1, A_2, \ldots, A_k$ be elements in $\mathbb{R}^m$. We say that those elements are linearly independent if the equation $x_1 A_1 + x_2 A_2 + \cdots + x_k A_k = 0$ has only one solution $x_1 = x_2 = \cdots = x_k = 0$. If the equation has more than one solution, we say that the vectors are linearly dependent.

Remark 4.2.1. We consider that Definition 4.2.1 is one of the most important in the whole discussion concerning ideas from linear algebra. We recommend the reader to review it over and over by proposing a wide variety of examples.

The next example shows what to do when we need to decide if a collection of vectors, in $\mathbb{R}^4$, are linearly independent.

Example 4.2.1. Let $A_1 = (1, 2, -1, 0)$, $A_2 = (-1, 0, 1, 2)$, $A_3 = (1, 2, 3, 4)$, $A_4 = (1, -1, 2, 0)$ be elements of $\mathbb{R}^4$. Are these elements linearly independent?

Discussion. According to Definition 4.2.1, the given elements are linearly independent if the equation

$$xA_1 + yA_2 + zA_3 + wA_4 = 0 \tag{4.3}$$

has only the zero solution. We will see if this is the case.

Notice that equation (4.3) is equivalent to the system of linear equations

$$\begin{aligned} x - y + z + w &= 0, \\ 2x + 2z - w &= 0, \\ -x + y + 3z + 2w &= 0, \\ 2y + 4z &= 0. \end{aligned} \tag{4.4}$$

To decide whether the given vectors are linearly independent or not, we need to solve the system (4.4). We use the Gauss–Jordan reduction method to solve it. The augmented matrix of the system is $A = \begin{bmatrix} 1 & -1 & 1 & 1 & | & 0 \\ 2 & 0 & 2 & -1 & | & 0 \\ -1 & 1 & 3 & 2 & | & 0 \\ 0 & 2 & 4 & 0 & | & 0 \end{bmatrix}$. Applying elementary operations on the rows of A, we obtain its row reduced form $R = \begin{bmatrix} 1 & 0 & 0 & -\frac{5}{4} & | & 0 \\ 0 & 1 & 0 & -\frac{3}{2} & | & 0 \\ 0 & 0 & 1 & \frac{3}{4} & | & 0 \\ 0 & 0 & 0 & 0 & | & 0 \end{bmatrix}$, from where we immediately see that the system (4.4) has more than one solution, thus the given vectors are linearly dependent.

Remark 4.2.2. To decide if the collection of vectors $u_1, u_2, \ldots, u_k$ in $\mathbb{R}^n$ are linearly independent, we form the matrix A whose jth column is u_j and examine the system of linear equations $AX = 0$. The elements $u_1, u_2, \ldots, u_k$ are linearly independent if and only if the system of linear equations $AX = 0$ has only one solution. In particular, if one of the vectors is zero, then they are linearly dependent.

The next result is a useful characterization to decide when a collection of vectors are linearly dependent.

Theorem 4.2.1. *The vectors $u_1, u_2, \ldots, u_p$, with $p \geq 2$, are linearly dependent if and only if one of them is a linear combination of the others. Moreover, if $u_1 \neq 0$ then the vectors are linearly dependent if and only if one of them is a linear combination of the previous.*

Proof. It is clear that if one of the vectors is a linear combination of the others, then the vectors are linearly dependent.

Conversely, if the vectors are linearly dependent, then the equation $x_1 u_1 + x_2 u_2 + \cdots + x_p u_p = 0$ has a solution with at least one $x_i \neq 0$. From this condition, we obtain $u_i = -\frac{1}{x_i} \sum_{k \neq i} x_k u_k$, proving that u_i is a linear combination of the others.

If the vectors $u_1, u_2, \ldots, u_p$ are linearly dependent and $u_1 \neq 0$, let k be the greatest index so that $x_k \neq 0$, and $x_j = 0$ for every $j > k$. The assumption $u_1 \neq 0$ implies that $k \geq 2$, then the equation $x_1 u_1 + x_2 u_2 + \cdots + x_p u_p = 0$ can be transformed to $u_k = -\frac{1}{x_k} \sum_{i=1}^{k-1} x_i u_i$, that is, u_k is a linear combination of the previous vectors. $\qquad\square$

After proving Theorem 4.2.1, several question might arise. What is the maximum number of linearly independent elements that we can have in $\mathbb{R}^n$? In this regard, the following remark gives a partial answer to the posed question.

Remark 4.2.3. In $\mathbb{R}^m$ any collection of $l \geq m + 1$ vectors is linearly dependent.

Proof. The conclusion follows directly from Theorem 1.3.3, p. 23. $\qquad\square$

To push the discussion further, concerning the maximum number of linearly independent elements in $\mathbb{R}^n$, we need to introduce another important term. First of all, given any element $(x_1, x_2, \ldots, x_n) \in \mathbb{R}^n$, it can be written as $(x_1, x_2, \ldots, x_n) = x_1 e_1 + x_2 e_2 + \cdots + x_n e_n$, where $e_i = (0, 0, \ldots, 0, \underset{i}{1}, 0, \ldots, 0)$, for every $i \in \{1, 2, \ldots, n\}$, that is, any element in $\mathbb{R}^n$ is a linear combination of the elements $e_1, e_2, \ldots, e_n$. This property is captured in the following definition.

Definition 4.2.2. A collection of elements $a_1, a_2, \ldots, a_m$ of $\mathbb{R}^n$ are called generators if, for any $a \in \mathbb{R}^n$, there are scalars $x_1, x_2, \ldots, x_m$ such that $a = x_1 a_1 + x_2 a_2 + \cdots + x_m a_m$.

Before deciding what is the maximum number of elements that are linearly independent in $\mathbb{R}^n$, Theorem 4.2.2 compares the cardinality (number of elements) of a set that generates $\mathbb{R}^n$ with the cardinality of a set that is linearly independent. In order to provide a proof of the mentioned theorem, we need

Lemma 4.2.1. *Let $S = \{a_1, a_2, \ldots, a_k\}$ and $S_1 = \{\beta_1, \beta_2, \ldots, \beta_k\}$ be subsets of $\mathbb{R}^n$. Assume that S is linearly independent and that S_1 generates $\mathbb{R}^n$. Then S generates $\mathbb{R}^n$ and S_1 is linearly independent.*

Proof. We will prove that S generates $\mathbb{R}^n$ by showing that any element of S_1 is a linear combination of elements from S. The assumption that S_1 generates $\mathbb{R}^n$ implies that, for every $j \in \{1, 2, \ldots, k\}$, there are scalars $a_{1j}, a_{2j}, \ldots, a_{kj}$ so that

$$a_j = a_{1j}\beta_1 + a_{2j}\beta_2 + \cdots + a_{kj}\beta_k. \tag{4.5}$$

With the scalars from (4.5), define the matrix A whose entries are the elements a_{ij} for $i, j \in \{1, 2, \ldots, k\}$. We claim that A is invertible. From Theorem 2.2.3, p. 51, it is enough to show that the equation $AX = 0$ has a unique solution. We will write the former equation explicitly:

$$
\begin{aligned}
a_{11}x_1 + a_{12}x_2 + \cdots + a_{1k}x_k &= 0, \\
a_{21}x_1 + a_{22}x_2 + \cdots + a_{2k}x_k &= 0, \\
&\ \ \vdots \\
a_{k1}x_1 + a_{k2}x_2 + \cdots + a_{kk}x_k &= 0.
\end{aligned}
\tag{4.6}
$$

Multiplying each equation in (4.6) by the corresponding β_i, we obtain

$$
(a_{11}x_1 + a_{12}x_2 + \cdots + a_{1k}x_k)\beta_1 = 0,
$$
$$
(a_{21}x_1 + a_{22}x_2 + \cdots + a_{2k}x_k)\beta_2 = 0,
$$
$$
\vdots \tag{4.7}
$$
$$
(a_{k1}x_1 + a_{k2}x_2 + \cdots + a_{kk}x_k)\beta_k = 0.
$$

Adding equations from (4.7), using (4.5), and rewriting, we obtain $x_1 a_1 + x_2 a_2 + \cdots + x_k a_k = 0$. Now, the linear independence assumption on S implies that $x_1 = x_2 = \cdots = x_k = 0$, proving that A is invertible.

The collection of equations (4.5), $j \in \{1, 2, \ldots, k\}$, can be written in matrix form as

$$
\begin{bmatrix} a_1 \\ a_2 \\ \vdots \\ a_k \end{bmatrix} = A^t \begin{bmatrix} \beta_1 \\ \beta_2 \\ \vdots \\ \beta_k \end{bmatrix}. \tag{4.8}
$$

We also have that if A has an inverse, then so does A^t and $(A^t)^{-1} = (A^{-1})^t$. From this, multiplying on the left of (4.8) by $(A^t)^{-1}$, we have

$$
(A^t)^{-1} \begin{bmatrix} a_1 \\ a_2 \\ \vdots \\ a_k \end{bmatrix} = \begin{bmatrix} \beta_1 \\ \beta_2 \\ \vdots \\ \beta_k \end{bmatrix}, \tag{4.9}
$$

proving that each element of S_1 is a linear combination of the elements of S. Since S_1 generates, then so does S. The proof that S_1 is linearly independent is left as an exercise to the reader, Exercise 23, p. 115. □

Theorem 4.2.2. *If $a_1, a_2, \ldots, a_l$ are generators of $\mathbb{R}^n$, and $\beta_1, \beta_2, \ldots, \beta_k \in \mathbb{R}^n$ are linearly independent, then $k \le l$.*

Proof. We will use induction on k. If $k = 1$, then β_1 is the only element that we are considering and it can be represented as a linear combination of the elements $a_1, a_2, \ldots, a_l$, thus $l \ge 1 = k$, proving the base case. Assume that $k > 1$ and the result to be true for collections of $k - 1$ linearly independent elements. Since $a_1, a_2, \ldots, a_l$ are generators of $\mathbb{R}^n$, so are $\beta_k, a_1, a_2, \ldots, a_l$ and the enlarged collection is linearly dependent, hence, by Theorem 4.2.1, some a_r is a linear combination of the others. Then, the set $\{\beta_k, a_1, a_2, \ldots, a_l\} \setminus \{a_r\}$ also generates $\mathbb{R}^n$ and has l elements. On the other hand, the set $\{\beta_1, \beta_2, \ldots, \beta_{k-1}\}$ is linearly independent. By the induction assumption, $k - 1 \le l$. If $k - 1 = l$, then from Lemma 4.2.1, the set $\{\beta_1, \beta_2, \ldots, \beta_{k-1}\}$ also generates $\mathbb{R}^n$, contradicting that

$$\{\beta_1, \beta_2, \ldots, \beta_{k-1}, \beta_k\}$$

is linearly independent. From this we have $k - 1 < l$, thus $k \leq l$, finishing the proof. □

With Theorem 4.2.2, we are ready to improve the conclusion of Remark 4.2.3, but we need an important definition.

Definition 4.2.3. A set $\mathcal{B} \subseteq \mathbb{R}^n$ that generates $\mathbb{R}^n$ and is linearly independent is called a basis of $\mathbb{R}^n$.

Corollary 4.2.1. *Any two bases have the same cardinality.*

Proof. Let $\mathcal{B}$ and $\mathcal{B}_1$ be two bases. Since the set $\mathcal{B}_1$ spans $\mathbb{R}^n$ and the set $\mathcal{B}$ is linearly independent, Theorem 4.2.2 implies $|\mathcal{B}| \leq |\mathcal{B}_1|$. The conditions on the bases allow us to switch roles between them, which implies $|\mathcal{B}_1| \leq |\mathcal{B}|$. From this we obtain $|\mathcal{B}_1| = |\mathcal{B}|$, finishing the proof of the corollary. □

Definition 4.2.4. The cardinality of any basis $\mathcal{B} \subseteq \mathbb{R}^n$ is the dimension of $\mathbb{R}^n$, denoted by $\dim(\mathbb{R}^n)$.

Remark 4.2.4. It is readily verified that the elements $e_i = (0, 0, \ldots, 0, \underset{i}{1}, 0, \ldots, 0)$, $i \in \{1, 2, \ldots, n\}$, constitute a basis of $\mathbb{R}^n$, called the canonical basis, hence $\dim(\mathbb{R}^n) = n$.

One important problem in linear algebra is to construct a basis that satisfies specified conditions. The next result provides sufficient and necessary conditions for a set of n elements to be a basis of $\mathbb{R}^n$.

Theorem 4.2.3. *Let $\{\beta_1, \beta_2, \ldots, \beta_n\}$ be a subset of $\mathbb{R}^n$, then the following statements are equivalent:*
1. *The set $\{\beta_1, \beta_2, \ldots, \beta_n\}$ is a basis.*
2. *The set $\{\beta_1, \beta_2, \ldots, \beta_n\}$ is linearly independent.*
3. *The set $\{\beta_1, \beta_2, \ldots, \beta_n\}$ spans $\mathbb{R}^n$.*

Proof. $(1 \Rightarrow 2)$ By the definition of a basis, the set $\{\beta_1, \beta_2, \ldots, \beta_n\}$ is linearly independent.

$(2 \Rightarrow 3)$ If $\{\beta_1, \beta_2, \ldots, \beta_n\}$ does not span $\mathbb{R}^n$, then there is γ which is not a linear combination of the elements $\beta_1, \beta_2, \ldots, \beta_n$, hence $\{\beta_1, \beta_2, \ldots, \beta_n, \gamma\}$ is linearly independent, and from this we conclude that there is a subset of $\mathbb{R}^n$ with $n + 1$ elements which is linearly independent, contradicting Theorem 4.2.2.

$(3 \Rightarrow 1)$ We need to show only that $\{\beta_1, \beta_2, \ldots, \beta_n\}$ is linearly independent. If this were not the case, then one of the elements, say β_i, would be a linear combination of the others, hence one would have that $\mathbb{R}^n$ can be spanned by $n - 1$ elements, contradicting that the minimum number of elements that span $\mathbb{R}^n$ is n. □

The following example shows how we can use Theorem 4.2.3 to provide a basis with a specified condition.

Example 4.2.2. Find a basis of $\mathbb{R}^3$ that does not include the elements $(1, 0, 0)$, $(0, 1, 0)$, $(0, 0, 1)$.

Discussion. From Theorem 4.2.3, we need to propose a set with three elements which are linearly independent. We will start by considering the set $\{(1, 1, 0)\}$, which is linearly independent. Also $(1, 0, 1)$ is not a multiple of $(1, 1, 0)$, hence $\{(1, 1, 0), (1, 0, 1)\}$ is a linearly independent set. We will verify that the set $\{(1, 1, 0), (1, 0, 1), (0, 1, 1)\}$ is linearly independent. To see that, let us consider the equation $x(1, 1, 0) + y(1, 0, 1) + z(0, 1, 1) = (x + y, x + z, y + z) = (0, 0, 0)$, which is equivalent to the system

$$
\begin{aligned}
x + y &= 0, \\
x + z &= 0, \\
y + z &= 0.
\end{aligned}
\tag{4.10}
$$

Solving the system (4.10), one finds that $(0, 0, 0)$ is the only solution, thus showing that the set $\{(1, 1, 0), (1, 0, 1), (0, 1, 1)\}$ is a basis of $\mathbb{R}^3$ and satisfies the required condition.

When checking if a subset of $\mathbb{R}^n$ is a basis, there are some possibilities to consider. The set might be linearly independent, it might span $\mathbb{R}^n$, or neither of those. In this situation we can ask, if a set is linearly independent, can it be enlarged to become a basis? If a set generates $\mathbb{R}^n$, can we extract a basis from it? The answer to these questions is yes.

Theorem 4.2.4.
1. *Let $S = \{\beta_1, \beta_2, \ldots, \beta_k\}$ be a set of linearly independent vectors in $\mathbb{R}^n$, then S can be enlarged to become a basis.*
2. *Let $S_1 = \{\alpha_1, \alpha_2, \ldots, \alpha_m\}$ be a set of nonzero vectors which span $\mathbb{R}^n$, then S_1 contains a basis.*

Proof. 1. Let $S = \{\beta_1, \beta_2, \ldots, \beta_k\}$ be a subset of $\mathbb{R}^n$ which is linearly independent. If $k = n$, then Theorem 4.2.3 guarantees that S is basis and we have finished. If $k < n$, then there is β_{k+1} that is not a linear combination of the elements of S. This implies that $\{\beta_1, \beta_2, \ldots, \beta_k, \beta_{k+1}\}$ is linearly independent and has cardinality $k + 1$. We check again, if $n = k + 1$, we have finished, otherwise there is β_{k+2} which is not a linear combination of the elements of $\{\beta_1, \beta_2, \ldots, \beta_k, \beta_{k+1}\}$. This new set has cardinality $k + 2$ and is linearly independent. Now, we notice that in $n - k$ steps we have constructed a set $\mathcal{B}$ which is linearly independent, contains S, and has cardinality n. Applying Theorem 4.2.3, $\mathcal{B}$ is a basis.

2. If $m = n$, Theorem 4.2.3 implies that S_1 is a basis. If $m > n$, then there is $i \in \{1, 2, \ldots, m\}$ so that α_i is a linear combination of $\{\alpha_1, \alpha_2, \ldots, \alpha_{i-1}\}$ and the set $\{\alpha_1, \alpha_2, \ldots, \alpha_m\} \setminus \{\alpha_i\}$ generates. In $m - n$ steps we construct a set $\mathcal{B}_1$ which generates $\mathbb{R}^n$, is contained in S_1, and has cardinality n, hence Theorem 4.2.3 implies that it is a basis, finishing the proof of the theorem. $\qquad\square$

4.3 Subspaces of $\mathbb{R}^n$

There are several interesting subsets of $\mathbb{R}^3$ that satisfy the properties of Theorem 4.1.1. For instance, $W = \{(x, y, 0) \mid x, y \in \mathbb{R}\}$ satisfies the property that when adding two elements from W, their sum is in W, and when multiplying an element of W by a scalar the result belongs to W. From this, it is straightforward to verify that W satisfies all properties stated in Theorem 4.1.1.

In a more general setting, the set of points $(x, y, z) \in \mathbb{R}^3$ whose coordinates satisfy the equation $ax + by + cz = 0$, with at least one of a, b, and c not zero, also satisfies the properties with respect to the sum and scalar multiplication stated in Theorem 4.1.1.

Indeed, if (x, y, z) and (x_1, y_1, z_1) satisfy $ax + by + cz = 0$ and $ax_1 + by_1 + cz_1 = 0$, then the coordinates of $(x, y, z) + (x_1, y_1, z_1) = (x + x_1, y + y_1, z + z_1)$ also satisfy $a(x + x_1) + b(y + y_1) + c(z + z_1) = 0$, which is obtained by adding the equations $ax + by + cz = 0$ and $ax_1 + by_1 + cz_1 = 0$. If λ is a real number then the coordinates of $(\lambda x, \lambda y, \lambda z)$ satisfy $\lambda ax + \lambda by + \lambda cz = 0$. Having proved these two basic properties, we readily verify all the claims of Theorem 4.1.1.

There are many more interesting examples of subsets of $\mathbb{R}^3$ or, more generally, subsets of $\mathbb{R}^n$ that share the same properties, concerning the sum and product with a scalar, as the whole space. Those sets have a name.

Definition 4.3.1. Let W be a nonempty subset of $\mathbb{R}^n$. We say that W is a subspace if for every $\alpha, \beta \in W$ and $r \in \mathbb{R}^n$, $\alpha + \beta \in W$ and $r\alpha \in W$.

The conditions in Definition 4.3.1 establish that W is closed under the sum and product with scalars. It can be verified that W satisfies all the properties listed in Theorem 4.1.1. A reformulation of Definition 4.3.1 is given in Remark 4.3.1

Remark 4.3.1. Let W be nonempty subset of $\mathbb{R}^n$, then W is a subspace if and only if, for every $\alpha, \beta \in W$ and $r, s \in \mathbb{R}$, we have that $r\alpha + s\beta \in W$.

Proof. Assume that W is a subspace, that is, assume that for every $\alpha, \beta \in W$ and $r \in \mathbb{R}$, we have $\alpha + \beta \in W$ and $r\alpha \in W$. We need to prove that for every $\alpha, \beta \in W$ and $r, s \in \mathbb{R}^n$, $r\alpha + s\beta \in W$. The assumption on W implies $r\alpha, s\beta \in W$, from this and the assumption that the sum of any two elements from W is again in W, we have that $r\alpha + s\beta \in W$, as needed.

Conversely, assume that $r\alpha + s\beta \in W$, whenever $\alpha, \beta \in W$ and $r, s \in \mathbb{R}$. By taking $r = s = 1$, we have $1\alpha + 1\beta = \alpha + \beta \in W$. If $r \in \mathbb{R}$ and $s = 0$ then $r\alpha + 0\beta = r\alpha \in W$, proving the statement. $\square$

Example 4.3.1. Two important subspaces are $\{0\}$ and $\mathbb{R}^n$.

Example 4.3.2. We will determine which of the following subsets are subspaces.
1. Set $W = \{(x, y) \in \mathbb{R}^2 \mid x, y \in \mathbb{Z}\}$. Notice that the sum of two elements of W is again an element of W, since the coordinates are integers and the sum of integers is an

integer. If $\lambda = \sqrt{2}$ and $\alpha = (1, 2) \in W$, then $\lambda\alpha = (\sqrt{2}, 2\sqrt{2}) \notin W$, that is, W is not a subspace.

2. Set $W = \{(x, y) \in \mathbb{R}^2 \mid 0 \le x \le 1, 0 \le y \le 1\}$. The elements $(1, 1)$, $(1, 0.5)$ belong to W; however, $(1, 1) + (1, 0.5) = (2, 1.5) \notin W$, therefore W is not a subspace.

3. Set $W = \{(x, y) \in \mathbb{R}^2 \mid y = mx,$ with m a fixed real number$\}$. Given (x, y) and (v, w) in W, one has $y = mx$ and $w = mv$. Adding these equations, one has $y + w = mx + mv = m(x + v)$, that is, $(x, y) + (v, w) \in W$. Also, if $\lambda \in \mathbb{R}$ then $\lambda(x, y) = (\lambda x, \lambda y) \in W$, since $y = mx$ implies $\lambda y = m\lambda y$, we conclude that W is a subspace.

Exercise 4.3.1. Determine which of the following subsets are subspaces:

1. $W = \{(x, y) \in \mathbb{R}^2 \mid y = 2x\} \cup \{(x, y) \in \mathbb{R}^2 \mid y = 3x\}$;
2. $W = \{(x, y) \in \mathbb{R}^2 \mid y = 2x\} \cap \{(x, y) \in \mathbb{R}^2 \mid y = 3x\}$;
3. $W = \{(x, y) \in \mathbb{R}^2 \mid y = x^2\}$;
4. $W = \{(x_1, x_2, \ldots, x_n) \in \mathbb{R}^n \mid x_i \in \mathbb{Z}$ for every $i = 1, 2, \ldots, n\}$;
5. $W = \{(x_1, x_2, \ldots, x_n) \in \mathbb{R}^n \mid x_i \in \mathbb{Q}$ for every $i = 1, 2, \ldots, n\}$.

From the above examples, we have that there are subsets which are not subspaces. Then it is natural to ask how can we generate a subspace starting from a subset S? The generated subspace must contain finite sums of elements from S and scalar multiples of elements from S. More precisely,

Definition 4.3.2. Given a nonempty subset S in $\mathbb{R}^n$, the subspace generated or spanned by S, $\mathcal{L}(S)$, is defined by

$$\mathcal{L}(S) := \left\{ \sum_{i=1}^{k} a_i \alpha_i \mid a_i \in \mathbb{R}, \alpha_i \in S \text{ and } k \in \{1, 2, 3, \ldots\} \right\}.$$

When $\mathcal{L}(S) = \mathbb{R}^n$, S is called a generating set for $\mathbb{R}^n$.

In Exercise 3, p. 113, the reader is asked to provide a characterization of the subspace $\mathcal{L}(S)$ as the smallest subspace that contains S.

Given $W \ne \{0\}$, a subspace of $\mathbb{R}^n$, is it true that W has a basis?

Since W is not the zero subspace, W contains an element $\alpha_1 \ne 0$ and the subspace spanned by α_1, that is, $\{x\alpha_1 \mid x \in \mathbb{R}\} \subseteq W$. If $W = \{x\alpha_1 \mid x \in \mathbb{R}\}$, the set $\{\alpha_1\}$ is a basis of W (why?). If $W \ne \{x\alpha_1 \mid x \in \mathbb{R}\}$, then there exists $\alpha_2 \in W$ which is not a multiple of α_1, thus $\{\alpha_1, \alpha_2\}$ is linearly independent. If $W = \{x\alpha_1 + y\alpha_2 \mid x, y \in \mathbb{R}\}$, then $\{\alpha_1, \alpha_2\}$ is a basis of W, since it generates W and is linearly independent. If $W \ne \{x\alpha_1 + y\alpha_2 \mid x, y \in \mathbb{R}\}$, then there is $\alpha_3 \in W$ and it is not a linear combination of α_1, α_2, therefore $\{\alpha_1, \alpha_2, \alpha_3\}$ is linearly independent. This process must end in at most n steps, since there are no more than n linearly independent elements in $\mathbb{R}^n$. With this we have proved that W admits a basis with at most n elements. The above discussion can be summarized in:

Theorem 4.3.1. *Every nonzero subspace of* $\mathbb{R}^n$ *admits a basis with* $k \leq n$ *elements. Hence* $\dim(W) \leq n$.

In what follows, we will establish a close relationship between the subspace generated by the rows (row space) and the subspace generated by the columns (column space) of an $m \times n$ matrix. More precisely, we will prove that the dimension of the row space and the dimension of the column space of a matrix are equal. This number is called the *rank* of A.

Definition 4.3.3. Given an $m \times n$ matrix A, the subspace spanned by its columns is called the column space of A, and its dimension is called the column rank. Likewise, the space spanned by the rows of A is called the row space, and its dimension is called the row rank of A.

Theorem 4.3.2. *Let* A *be a nonzero* $m \times n$ *matrix. Then the row rank of* A *equals the column rank of* A.

Proof. From Exercise 24, p. 115, the row rank of A is equal to the number of nonzero rows of its row reduced form, R.

Assume that R has r nonzero rows, and that the main entry of row i of R is located in column t_i. From Exercise 25, p. 115, we have that R has column rank equal to r, which is equal to the row rank of R.

We will show that the column rank of A is equal to the column rank of R. For this we observe that the solution sets for $AX = 0$ and $RX = 0$ are equal. This is equivalent to saying that if $A_1, A_2, \ldots, A_n$ are the columns of A and $R_1, R_2, \ldots, R_n$ are the columns of R then for scalars $x_1, x_2, \ldots, x_n$, one has $x_1 A_1 + x_2 A_2 + \cdots + x_n A_n = 0$ if and only if $x_1 R_1 + x_2 R_2 + \cdots + x_n R_n = 0$. Let t be the column rank of A. Without loss of generality, we may assume that $A_1, A_2, \ldots, A_t$ are a basis for the column space of A. We will show that $t = r$.

Let $x_1, \ldots, x_t$ be scalars such that $x_1 R_1 + x_2 R_2 + \cdots + x_t R_t = 0$, then $x_1 A_1 + x_2 A_2 + \cdots + x_t A_t = 0$, hence $x_1 = x_2 = \cdots = x_t = 0$, that is, $R_1, R_2, \ldots, R_t$ are linearly independent, thus $t \leq r$.

If $x_1, \ldots, x_r$ satisfy $x_1 A_1 + x_2 A_2 + \cdots + x_r A_r = 0$, then $x_1 R_1 + x_2 R_2 + \cdots + x_r R_r = 0$, hence $x_1 = x_2 = \cdots = x_r = 0$, that is, $A_1, A_2, \ldots, A_r$ are linearly independent, thus $r \leq t$, finishing the proof of the theorem. $\square$

Definition 4.3.4. Given a matrix A, the rank of A is defined as the row rank or the column rank of A.

Remark 4.3.2. From the proof of Theorem 4.3.2, one can observe that the row rank of A is the number of nonzero rows of the row reduced form of A. If A is an $m \times n$ matrix, its rank is at most $\min\{m, n\}$.

Example 4.3.3. Let $A = \begin{bmatrix} 1 & 1 & -1 & 1 \\ 2 & -1 & 3 & 0 \\ 1 & 0 & -1 & 2 \end{bmatrix}$. Compute the rank of A.

Discussion. From Remark 4.3.2, the rank of A is the number of nonzero rows of the row reduced form of A.

Applying row operations on the rows of A, one has

$$\begin{bmatrix} 1 & 1 & -1 & 1 \\ 2 & -1 & 3 & 0 \\ 1 & 0 & -1 & 2 \end{bmatrix} \sim \begin{bmatrix} 1 & 1 & -1 & 1 \\ 0 & -3 & 5 & -2 \\ 0 & -1 & 0 & 1 \end{bmatrix} \sim \begin{bmatrix} 1 & 0 & -1 & 2 \\ 0 & 0 & 5 & -5 \\ 0 & -1 & 0 & 1 \end{bmatrix} \sim \begin{bmatrix} 1 & 0 & -1 & 2 \\ 0 & 1 & 0 & -1 \\ 0 & 0 & 1 & -1 \end{bmatrix}$$

$$\sim \begin{bmatrix} 1 & 0 & 0 & 1 \\ 0 & 1 & 0 & -1 \\ 0 & 0 & 1 & -1 \end{bmatrix}.$$

The last matrix is the row reduced form of A, so the rank of A is three.

According to Theorem 4.2.3, a set of n elements of $\mathbb{R}^n$ is a basis if and only if it is linearly independent. Thus, it is a good idea to have computational criteria to decide if a set of n elements of $\mathbb{R}^n$ is linearly independent. This and other conditions provide a refinement of Theorem 2.2.3, p. 51, which are established in Theorem 4.3.3. As we advance in the discussion of the contents of the book, new equivalent conditions will be incorporated to the theorem.

Theorem 4.3.3. *Let A be an $n \times n$ matrix. Then the following statements are equivalent:*
1. *The matrix A has an inverse.*
2. *The matrix A is row equivalent to I_n.*
3. *The rows of A are linearly independent.*
4. *The matrix A is a product of elementary matrices.*
5. *The system $AX = 0$ has a unique solution.*
6. *The columns of A are linearly independent.*
7. *For every $B = \begin{bmatrix} b_1 \\ b_2 \\ \vdots \\ b_n \end{bmatrix}$, the system $AX = B$ has a unique solution.*
8. *The columns of A span $\mathbb{R}^n$.*
9. *The rank of A is n.*

Proof. The proof is obtained, essentially, by reviewing the method to find the inverse of a matrix, using a row reduction process together with Theorem 4.3.2. □

One way to use Theorem 4.3.3 is by constructing a matrix starting with a set of n elements from $\mathbb{R}^n$ and deciding if such a matrix has an inverse.

We consider the vectors of Example 4.2.2 to illustrate the use of Theorem 4.3.3. Construct the matrix whose columns are the vectors $(1, 1, 0)$, $(1, 0, 1)$, $(0, 1, 1)$, that is, set matrix $A = \begin{pmatrix} 1 & 1 & 0 \\ 1 & 0 & 1 \\ 0 & 1 & 1 \end{pmatrix}$. Applying row operations on the rows of A to obtain the row reduced form of A, one has

$$A = \begin{pmatrix} 1 & 1 & 0 \\ 1 & 0 & 1 \\ 0 & 1 & 1 \end{pmatrix} \sim \begin{pmatrix} 1 & 1 & 0 \\ 0 & -1 & 1 \\ 0 & 1 & 1 \end{pmatrix} \sim \begin{pmatrix} 1 & 1 & 0 \\ 0 & -1 & 1 \\ 0 & 0 & 2 \end{pmatrix} \sim \begin{pmatrix} 1 & 1 & 0 \\ 0 & -1 & 1 \\ 0 & 0 & 1 \end{pmatrix} \sim \begin{pmatrix} 1 & 1 & 0 \\ 0 & -1 & 0 \\ 0 & 0 & 1 \end{pmatrix}.$$

From the last matrix, one concludes that the rows of A are linearly independent, hence the vectors $(1, 1, 0)$, $(1, 0, 1)$, $(0, 1, 1)$ are linearly independent.

4.3.1 Operations with subspaces

Given two subspaces W and U, we can define two new sets $W \cap U$ and $W \cup U$ and ask which of these sets, if any, is a subspace?

If $\alpha, \beta \in W \cap U$, then $\alpha, \beta \in W$ and $\alpha, \beta \in U$, since W and U are subspaces then $\alpha + \beta \in W$ and $\alpha + \beta \in U$, that is, $\alpha + \beta \in W \cap U$.

If $\alpha \in W \cap U$ and $x \in \mathbb{R}$, then $\alpha \in W$ and $\alpha \in U$, so again, using that W and U are subspaces, one has $x\alpha \in W$ and $x\alpha \in U$, that is, $x\alpha \in W \cap U$. From this we have that $W \cap U$ is a subspace. Actually, the same result holds for any collection of subspaces, that is, if $\{W_i\}_{i \in I}$ is any collection of subspaces, then $\bigcap_{i \in I} W_i$ is a subspace.

Exercise 4.3.2. Find necessary and sufficient conditions under which $W \cup U$ is a subspace. More generally, given $W_1, W_2, \ldots, W_k$, subspaces of $\mathbb{R}^n$, find necessary and sufficient conditions under which $W_1 \cup W_2 \cup \cdots \cup W_k$ is a subspace.

Assume that W and U are subspaces of $\mathbb{R}^n$. Define the sum of W and U as $W + U :=$ $\{w + u \mid w \in W, u \in U\}$. Then $W + U$ is a subspace, called the sum of W and U.

In fact, if $\alpha, \beta \in W + U$, then $\alpha = w_1 + u_1$ and $\beta = w_2 + u_2$, with $w_1, w_2 \in W$ and $u_1, u_2 \in U$, hence $\alpha + \beta = (w_1 + u_1) + (w_2 + u_2) = (w_1 + w_2) + (u_1 + u_2) \in W + U$.

If $c \in \mathbb{R}$, then $c\alpha = c(w_1 + u_1) = cw_1 + cu_1 \in W + U$, as claimed.

Exercise 4.3.3. Determine if the following subsets are subspaces and, if so, find a basis and compute their dimension:

1. $W = \{(x, y) \in \mathbb{R}^2 \mid 2x + 3y = 0\}$;
2. $W = \{(x, y, z) \in \mathbb{R}^3 \mid 2x + 3y - z = 0\}$;
3. $W = \{(x, y, z, w) \in \mathbb{R}^4 \mid 2x + 3y - z + w + 1 = 0\}$;
4. $W = \{(x, y, z, w) \in \mathbb{R}^4 \mid 2x + 3y - z + w = 0\}$;
5. $W = \{(x_1, x_2, \ldots, x_{n-1}, x_n) \in \mathbb{R}^n \mid x_1 + 2x_2 + 3x_3 + \cdots + nx_n = 0\}$;
6. $W = \{(x_1, x_2, \ldots, x_{n-1}, x_n) \in \mathbb{R}^n \mid x_1 x_2 x_3 \cdots x_n = 0\}$.

As we saw before, when starting with two subspaces W and U, we can form two new subspaces $W + U$ and $W \cap U$. We also have that $W, U \subseteq W + U$. How are the dimensions of these subspaces related? To obtain an idea about the answer, let us start by assuming that W and U are subspaces of $\mathbb{R}^3$ of dimension two. The following possibilities arise:

(a) If $U = W$, then $U + W = U$ and $U \cap W = U$. From these conditions, we have $\dim(W + U) = \dim(W) + \dim(U) - \dim(W \cap U)$.

(b) If $U \neq W$, then neither is contained in the other, since otherwise they must be equal, as both have dimension 2. From this condition, we can find $\alpha \in U \setminus W$. If β and γ form a basis of W, then the choice of α implies that β, γ, and α are linearly independent. We also have that $\{\alpha, \beta, \gamma\} \subseteq U + W \subseteq \mathbb{R}^3$, hence $U + W$ has dimension at least three. On the other hand, Theorem 4.3.1 implies that $U + W$ has dimension at most three, thus $U + W$ has dimension three.

The condition $U \neq W$, implies that $U \cap W \neq \{0\}$. We will argue by contradiction. If $U \cap W = \{0\}$, then $U + W$ would have dimension at least four, which is not possible. In fact, let $\{\alpha, \beta\}$ and $\{\gamma, \rho\}$ be bases of U and W, respectively. Considering the linear combination $a\alpha + b\beta + c\gamma + d\rho = 0$, from this equation we have $a\alpha + b\beta = -c\gamma - d\rho \in U \cap W = \{0\}$, thus $a\alpha + b\beta = 0 = -c\gamma - d\rho$. Now using that $\{\alpha, \beta\}$ and $\{\gamma, \rho\}$ are linearly independent, we obtain that $a = b = -c = -d = 0$, contradicting that in $\mathbb{R}^3$ there are at most three linearly independent elements.

From the above arguments, one has $\dim(W + U) = 3$ and $\dim(W \cap U) = 1$. In cases (a) and (b), we obtain that $\dim(W + U) = \dim(W) + \dim(U) - \dim(W \cap U)$ holds.

This particular case suggests that the previous equation might hold in general.

Theorem 4.3.4. *Let W and U be subspaces of $\mathbb{R}^n$. Then*

$$\dim(W + U) = \dim(W) + \dim(U) - \dim(W \cap U). \tag{4.11}$$

Proof. First, consider the case $W \cap U = \{0\}$. If one of W and U is zero, the result is clear. We may assume that both are nonzero. Let $\{\alpha_1, \ldots, \alpha_l\}$ and $\{\beta_1, \ldots, \beta_r\}$ be basis of W and U respectively.

Claim. $\{\alpha_1, \ldots, \alpha_l, \beta_1, \ldots, \beta_r\}$ is a basis of $W + U$. In fact, if $a_1\alpha_1 + \cdots + a_l\alpha_l + b_1\beta_1 + \cdots + b_r\beta_r = 0$, then $a_1\alpha_1 + \cdots + a_l\alpha_l = -(b_1\beta_1 + \cdots + b_r\beta_r)$ belongs to the intersection, therefore $a_1\alpha_1 + \cdots + a_l\alpha_l = -(b_1\beta_1 + \cdots + b_r\beta_r) = 0$.

Now from the assumption on $\{\alpha_1, \ldots, \alpha_l\}$ and $\{\beta_1, \ldots, \beta_r\}$ being basis, one has that these sets are linearly independent, hence $a_1 = a_2 = \cdots = a_l = b_1 = b_2 = \cdots = b_r = 0$, that is, the set $\{\alpha_1, \ldots, \alpha_l, \beta_1, \ldots, \beta_r\}$ is linearly independent.

Given $\alpha + \beta \in W + U$, with $\alpha \in W$ and $\beta \in U$, we have $\alpha = a_1\alpha_1 + \cdots + a_l\alpha_l$ and $\beta = b_1\beta_1 + \cdots + b_r\beta_r$, hence $\alpha + \beta = a_1\alpha_1 + \cdots + a_l\alpha_l + b_1\beta_1 + \cdots + b_r\beta_r$, proving that the set $\{\alpha_1, \ldots, \alpha_l, \beta_1, \ldots, \beta_r\}$ spans $W + U$.

The second case is $W \cap U \neq \{0\}$. Let $\{\gamma_1, \ldots, \gamma_s\}$ be a basis of $W \cap U$, then this is a linearly independent subset of W and U. Applying Theorem 4.2.4(1), the set $\{\gamma_1, \ldots, \gamma_s\}$ can be extended to bases of W and U, respectively. This bases will be $\{\gamma_1, \ldots, \gamma_s, \alpha_1, \ldots, \alpha_t\}$ and $\{\gamma_1, \ldots, \gamma_s, \beta_1, \ldots, \beta_m\}$, respectively.

Claim. The set $B = \{\gamma_1, \ldots, \gamma_s, \alpha_1, \ldots, \alpha_t, \beta_1, \ldots, \beta_m\}$ is a basis of $W + U$.

Indeed, if

$$a_1\gamma_1 + \cdots + a_s\gamma_s + b_1\alpha_1 + \cdots + b_t\alpha_t + c_1\beta_1 + \cdots + c_m\beta_m = 0, \tag{4.12}$$

then $a_1\gamma_1 + \cdots + a_s\gamma_s + b_1\alpha_1 + \cdots + b_t\alpha_t = -(c_1\beta_1 + \cdots + c_m\beta_m)$ is an element of $W \cap U$ (why?). From this, $c_1\beta_1 + \cdots + c_m\beta_m = d_1\gamma_1 + \cdots + d_s\gamma_s$, for some scalars $d_1, d_2, \ldots, d_s$. This equation implies $c_1\beta_1 + \cdots + c_m\beta_m - d_1\gamma_1 - \cdots - d_s\gamma_s = 0$. Since we have that $\{\gamma_1, \ldots, \gamma_s, \beta_1, \ldots, \beta_m\}$ is a basis, it is linearly independent, therefore $c_1 = c_2 = \cdots = c_m = d_1 = d_2 = \cdots = d_s = 0$. Using this condition in (4.12), we have that $a_1\gamma_1 + \cdots + a_s\gamma_s + b_1\alpha_1 + \cdots + b_t\alpha_t = 0$. Since $\{\gamma_1, \ldots, \gamma_s, \alpha_1, \ldots, \alpha_t\}$ is a basis of W, one concludes that $a_1 = a_2 = \cdots = a_s = b_1 = b_2 = \cdots = b_t = 0$, proving that $\{\gamma_1, \ldots, \gamma_s, \alpha_1, \ldots, \alpha_t, \beta_1, \ldots, \beta_m\}$ is linearly independent.

In Exercise 5, p. 113, we invite the reader to show that B spans $W + U$. □

4.4 General vector spaces

An approach to solve many problems in mathematics and other sciences uses the concept of an abstract vector space. One way to explain this might be the fact that when dealing with various mathematical objects, such as functions, matrices, among others, there is a need to add or multiply by scalars. The idea of an abstract vector space points to "operating" with mathematical objects that can help represent mathematical ideas.

In this section we present the axioms of an abstract vector space, starting with a generalization of what we have done in $\mathbb{R}^n$. More precisely, in $\mathbb{R}^n$ there is a sum and a product with scalars which satisfy properties stated in Theorem 4.1.1.

Those properties are fundamental and the needed ones that must hold in a general vector space. We make this precise in the following definition.

Definition 4.4.1. A vector space over the real numbers is a nonempty set V, with a binary operation, called the sum and a product with a scalar. They satisfy:
1. Properties of the sum.
    (a) (*Commutativity of the sum*) For every α and $\beta \in V$, $\alpha + \beta = \beta + \alpha$.
    (b) (*Associativity of the sum*) For every α, β and γ in V, $(\alpha + \beta) + \gamma = \alpha + (\beta + \gamma)$.
    (c) (*Existence of additive identity*) There is an element in V, called zero and denoted by 0, such that $0 + \alpha = \alpha$, for every $\alpha \in V$.
    (d) (*Existence of additive inverse*) For every $\alpha \in V$, there is an element $\alpha' \in V$ such that $\alpha + \alpha' = 0$.
2. Properties of the product with a scalar.
    (a) For every $\alpha \in V$, $1\alpha = \alpha$, with $1 \in \mathbb{R}$.
    (b) For every $\alpha \in V$, and for every λ and μ in $\mathbb{R}$, $\lambda(\mu\alpha) = (\lambda\mu)\alpha$.
    (c) The product with a scalar is distributive with respect to the sum, that is, for every λ, μ in $\mathbb{R}$ and for every $\alpha, \beta \in V$, $(\lambda + \mu)\alpha = \lambda\alpha + \mu\alpha$ and $\lambda(\alpha + \beta) = \lambda\alpha + \lambda\beta$.

If V is a vector space, its elements are called *vectors*.

Before we give examples such as those presented in linear algebra textbooks, we will present an example to illustrate the abstractness of the concept of a vector space.

Example 4.4.1. Consider your favorite item or pet, for instance, your mobile, and define the set $V = \{\text{mobile}\}$. We will define a sum in V. To make the notation easy, the unique element in V will be denoted by m.

Since V has only one element, m, the only way to add elements of V is by taking m and adding to itself, and the result must be m, thus there is only one way to define such a sum, that is, $m + m = m$ (notice that we are using the usual sign for the sum).

We need to define a product with scalars, that is, if $x \in \mathbb{R}$ then we need to declare the meaning of $x \cdot m$. As we did before, the only possibility is $x \cdot m = m$.

Is V a vector space with the defined sum and product by scalars?

1. Properties of the sum.
   (a) Since m is the only element in V, the definition of sum, $m + m = m$, guarantees commutativity.
   (b) Associativity also holds, since one has $(m + m) + m = m + (m + m) = m$.
   (c) There is an element in V, called zero and denoted by 0, so that $0 + m = m$. We take $0 = m$.
   (d) For every $a \in V$, there is a' so that $a + a' = 0$. Since V has a unique element, take $a' = m$.
2. Properties of the product with a scalar. We leave them to the reader as an exercise.

With the previous example, it is natural to ask if we can construct a real vector space with two, three, or more (finitely many) elements. The answer is no and we ask you to justify the answer.

Exercise 4.4.1. For each of the following sets and operations, determine if the corresponding set is a vector space.

1. Let $V = \mathbb{R}^n$, with the sum and product with a scalar already defined.
2. Let $V = \mathcal{M}_{m \times n}(\mathbb{R}) := \{A \mid A \text{ is an } m \times n \text{ matrix with real entries}\}$ be the set of $m \times n$ matrices. Define in V the usual sum of matrices and define the product with a scalar multiplying the scalar by each entry in the matrix.
3. Let V be the set of all functions from $\mathbb{R}$ to $\mathbb{R}$ with the usual sum of functions, and the usual product of a scalar and a function.
4. Let V be the set of polynomials with real coefficients. Define in V the usual sum of polynomials and the usual product of a polynomial and a scalar.
5. Given a positive integer n, let V be the set of polynomials of degree $\leq n$. Define in V the usual sum of polynomials and the usual product with a scalar.
6. Let V be the set of functions from $\mathbb{R}$ to $\mathbb{R}$ that are zero, except on a finite set of points.
7. Let $V = \mathbb{Z}^n := \{(x_1, x_2, \ldots, x_n) \mid x_i \text{ is an integer for each } i = 1, 2, \ldots, n\}$. Define the sum entry by entry and the product with a scalar, multiplying the scalar by each entry.

As it was done in $\mathbb{R}^n$, we will define some fundamental concepts for a general vector space.

Definition 4.4.2. Let V be a vector space, $S = \{a_1, a_2, \ldots, a_n\} \subseteq V$. It is said that S is *linearly independent* if the only scalars that satisfy the equation $x_1 a_1 + x_2 a_2 + \cdots + x_n a_n = 0$ are $x_1 = x_2 = \cdots = x_n = 0$. Otherwise, we say that S is linearly dependent.

Definition 4.4.3. A vector space V is said to be *finitely generated* if there is a finite subset $S = \{a_1, a_2, \ldots, a_k\} \subseteq V$ so that every element $a \in V$ is of the form $a = x_1 a_1 + x_2 a_2 + \cdots + x_k a_k$, for some scalars $x_1, x_2, \ldots, x_k \in \mathbb{R}$. The set S is called a generator set.

In what follows, we will consider basically finitely generated vector spaces. However, in a few examples we will consider vector spaces that are not finitely generated.

We have shown that $\mathbb{R}^n$ contains a basis and proved that any two bases have n elements. It is important to notice that the proof of Theorem 4.2.2, p. 101, and its corollary apply to any finitely generated vector space.

Remark 4.4.1. Let V be a vector space and let $a_1, a_2, \ldots, a_n$ be elements of V. If $\beta = a_1 a_1 + \cdots + a_n a_n$ and some a_i is a linear combination of the other vectors, $a_1, \ldots, a_{i-1}, a_{i+1}, \ldots, a_n$, then β is a linear combination of $a_1, \ldots, a_{i-1}, a_{i+1}, \ldots, a_n$.

Proof. Since $a_i = \sum_{j \neq i} x_j a_j$ and $\beta = \sum_{k=1}^{n} a_k a_k$, substituting the first equation into the second, one obtains $\beta = \sum_{k \neq i} (a_k + a_i x_k) a_k$. □

Given that the concept of a basis is fundamental, we will state it here in its general form.

Definition 4.4.4. Let V be a vector space and $\mathcal{B} \subseteq V$. We say that $\mathcal{B}$ is a basis if it spans V and is linearly independent.

Theorem 4.4.1. *Let $V \neq \{0\}$ be a finitely generated vector space with $S = \{a_1, a_2, \ldots, a_m\}$ a spanning set of V. Then we can extract a basis of V from S.*

Proof. If S is linearly independent we have finished. Hence we may assume that S is linearly dependent. If $0 \in S$, we take it way and S still generates. Since S is linearly dependent, there is $a_i \in S$ which is a linear combination of the other elements of S. Applying Remark 4.4.1, the set $S_1 = S \setminus \{a_i\}$ also spans V. Continuing the process as many times as needed, we will arrive at a set $\mathcal{B} \subseteq S$ which spans V and is linearly independent. □

Theorem 4.4.2. *Let V be a finitely generated vector space, and let $S = \{a_1, a_2, \ldots, a_m\} \subseteq V$ be a set of generators of V. Furthermore, assume that $\{\beta_1, \beta_2, \ldots, \beta_r\} \subseteq V$ is linearly independent, then $r \leq m$.*

The proof of Theorem 4.4.2 follows the same lines as the proof of Theorem 4.2.2, p. 101. We invite the reader to review that proof.

Corollary 4.4.1. *In a finitely generated vector space, any two bases have the same number of elements.*

For the proof of Corollary 4.4.1, see the proof of Corollary 4.2.1, p. 102.

Definition 4.4.5. Let V be a finitely generated vector space, and let $\mathcal{B}$ be a basis of V. The cardinality of $\mathcal{B}$ is called the dimension of V and denoted by $\dim(V)$.

Theorem 4.4.3. *Let V be a finitely generated vector space, and let $S = \{\beta_1, \beta_2, \ldots, \beta_r\}$ be a linearly independent subset of V. Then there is a basis of V which contains S.*

Proof. Let $\{a_1, a_2, \ldots, a_m\}$ be a set of generators of V which does not include the zero vector, then

$$S_1 = \{\beta_1, \beta_2, \ldots, \beta_r, a_1, a_2, \ldots, a_m\}$$

also generates V. Following the process as in the proof of Theorem 4.4.1, p. 112, we can extract a basis of V from S_1 which includes the elements of S. □

Definition 4.4.6. Let V be a vector space, and let U and W be subspaces of V. We say that V is the direct sum of U and W if $V = U + W$ and $U \cap W = \{0\}$. In this case we use the notation $V = U \oplus W$.

For instance, $\mathbb{R}^2$ is the direct sum of any two different one-dimensional subspaces; $\mathbb{R}^3$ is the direct sum of a one-dimensional subspace W and a two-dimensional subspace U that does not contain W.

Remark 4.4.2. We consider important to point out that all terms and results stated in Section 4.3 can be formulated for subspaces in the general case. The proofs of the results follow the same lines as those in the cited section. It would be a good idea that the reader take a close look at those results and try to prove them.

4.5 Exercises

1. Review all theorems stated for $\mathbb{R}^n$ and state them for a finitely generated vector space V.
2. Give examples of subsets in $\mathbb{R}^2$ that are closed under addition and are not closed under product with a scalar.
3. Let V be a vector space, and S be a nonempty subset of V. The subspace spanned by S is denoted and defined by $\mathcal{L}(S) := \left\{ \sum_{j=1}^n x_j a_j : x_j \in \mathbb{R}, a_j \in S, n \in \mathbb{N} \right\}$. Prove that $\mathcal{L}(S) = \bigcap W$, where the intersection is taken over all subspaces that contain S. What is the subspace generated by the empty set? With this in mind, what is the dimension of the zero subspace? Does any subspace have a basis?
4. Give the definition of a direct sum for $k > 2$ subspaces.
5. Finish the proof of Theorem 4.3.4, p. 109.
6. Prove that any three elements in $\mathbb{R}^2$ are linearly dependent. What is the condition on k so that any set containing k elements is linearly dependent in $\mathbb{R}^n$?
7. Use SageMath or any other computational system to decide if the proposed sets are linearly independent. First of all, identify the vector space that contains them.

(a) $S = \{(1,2,1), (1,2,-1), (1,2,1)\}$;

(b) $S = \{(1,2,3,3), (13,0,0,1), (1,-1,0,1), (1,2,3,30)\}$;

(c) $S = \{(1,2,3,3,1), (1,2,0,0,1), (1,-1,-1,0,1), (1,2,3,3,-1), (1,1,1,1,10)\}$;

(d) $S = \{(1,2,3,3,0,0), (1,2,-1,0,2,1), (1,3,3,-1,0,1), (1,2,3,3,1,0), (1,2,3,3,-1,1),$ $(1,2,-1,0,-2,-1)\}$;

(e) $S = \{(1,2,3,3,0,0,1), (1,2,-1,3,0,0,1), (1,2,3,3,-1,0,1)\}$.

8. Prove that $\mathbb{R}^n$ is the direct sum of two subspaces, a subspace of dimension 1 and the other of dimension $n-1$. Go further, prove that a vector space of dimension n is the direct sum of n subspaces of dimension 1.

9. Let V be a vector space of positive dimension, and let S be a finite subset with more than one element. Prove that S is not a subspace.

10. Consider the following equation: $x(1,-2,3) + y(2,-4,0) + z(0,1,-1) = (1,0,0)$. Are there real numbers x, y, and z that satisfy it?
Is the set $\{(1,-2,3), (2,-4,0), (0,1,-1)\}$ linearly independent?

11. Find several sets with three elements that generate $\mathbb{R}^3$. Explain why those sets are linearly independent.

12. Let n be a positive integer, and let V be the set of polynomials with real coefficients of degree $\leq n$. Show that V is a vector space and determine its dimension.

13. Set $W_1 = \{(x,y,z) \in \mathbb{R}^3 \mid x + y - z = 0\}$ and $W_2 = \{(x,y,z) \in \mathbb{R}^3 \mid 2x - y + z = 0\}$. Prove that W_1 and W_2 are subspaces of $\mathbb{R}^3$. Determine $W_1 \cap W_2$.

14. Find subspaces W_1, W_2, and W_3 of $\mathbb{R}^n$ to show that the equation $W_1 \cap (W_2 + W_3) = (W_1 \cap W_2) + (W_1 \cap W_3)$ does not hold in general.

15. Let V be a vector space of dimension $n > 2$, and let W_1 and W_2 be subspaces of V with $\dim(W_1) = n - 2$. Determine the minimum value of $\dim(W_2)$ so that $W_1 \cap W_2 \neq \{0\}$.

16. Let V be the vector space of all functions from $\mathbb{R}$ to $\mathbb{R}$,

$$P = \{f \in V : f \text{ is even, that is, } f(x) = f(-x)\},$$
$$I = \{f \in V : f \text{ is odd, that is, } f(-x) = -f(x)\}.$$

Prove that I and P are subspaces and $V = P \oplus I$.

17. Let V be a vector space, $S \subseteq V$. Prove that S is linearly dependent if and only if S contains a proper subset T so that S and T span the same subspace.

18. Let α_1, α_2, α_3 be linearly independent elements of the vector space V. Determine which of the following sets are linearly independent:

(a) $\{\alpha_1 + \alpha_2, \alpha_2, \alpha_2 - \alpha_3\}$.

(b) $\{\alpha_1 + \alpha_2 + \alpha_3, \alpha_1 - \alpha_3\}$.

(c) $\{2\alpha_1 + \alpha_2, \alpha_2, \alpha_1 - 2\alpha_2\}$.

19. Let W_1 and W_2 be subspaces of V so that $V = W_1 \oplus W_2$. Prove that every $\alpha \in V$ has a unique representation in the form $\alpha = \alpha_1 + \alpha_2$, with $\alpha_1 \in W_1$ and $\alpha_2 \in W_2$.

20. Given $S = \{(1,-3,0), (-1,3,-2), (0,0,-2)\} \subseteq \mathbb{R}^3$, is there a linearly independent subset of S which spans the same subspace as S?

21. Consider $S = \{(1,2,3,4,5), (1,0,-1,2,-2), (-1,0,1,0,-1), (-2,0,-1,2,-1)\} \subseteq \mathbb{R}^5$.

(a) Determine if S is linearly independent.

(b) Let W be the subspace spanned by S. Determine $\dim(W)$.

(c) Let W be as in (b). Determine if the following elements belong to W: (a) $\alpha = (1,1,1,1,1)$, (b) $\alpha = (0,-1,1,-1,0)$, and (c) $\alpha = (-1,1,-1,1,-1)$.

22. Assume that $X_1, \ldots, X_n \in \mathbb{R}^n$, where $X_j = (a_{1j}, \ldots, a_{nj})$ and $j \in \{1, \ldots, n\}$. Let A be the matrix whose columns are the given X_j's. Prove that $\{X_1, \ldots, X_n\}$ is linearly independent if and only if A has an inverse.

23. Finish the proof of Lemma 4.2.1, that is, prove that the set S_1 declared there is linearly independent.

24. Let A be an $m \times n$ matrix and let R be its row reduced form. Prove that A and R have the same row space, hence they have the same row rank, which is the number of nonzero rows of R.

25. Let A and R be as in Exercise 24, and let $t_1 < t_2 < \cdots < t_r$ be the columns of R which contain the main entries of R. Prove that these columns generate the column space of R and are linearly independent. Conclude that R has rank r.

26. Let A be an $n \times n$ matrix and let $X_1, X_2, \ldots, X_n$ be linearly independent elements of $\mathbb{R}^n$. Prove that A is the zero matrix if and only if $AX_j = 0$ for every $j \in \{1, 2, \ldots, n\}$. (*Hint.*) Let E_j be the jth canonical element of $\mathbb{R}^n$. Prove that if $AE_j = 0$ for every $j \in \{1, 2, \ldots, n\}$, then $A = 0$. Now argue that any E_j is a linear combination of $X_1, X_2, \ldots, X_n$ and use that $AX_j = 0$.

27. Let A be a $p \times q$ matrix. Furthermore, assume that the rows and columns of A are linearly independent. Prove that $p = q$.

28. Let A and B be matrices of sizes $m \times n$ and $n \times p$, respectively. Assume that AB has rank p. Prove that the columns of B are linearly independent and that $p \leq n$.

29. Let A be an $n \times n$ matrix so that $A^k = 0$, for some k. Prove that $I_n - A$ has an inverse.

30. We say that a square matrix A is nilpotent if $A^r = 0$ for some $r \geq 1$. Let A and B be nilpotent matrices of the same size and assume that $AB = BA$. Prove that AB and $A + B$ are nilpotent.

31. Given the system

$$2x_1 + 3x_2 + x_3 + 4x_4 - 9x_5 = 0,$$
$$x_1 + x_2 + x_3 + x_4 - 3x_5 = 0,$$
$$x_1 + x_2 + x_3 + x_4 - 5x_5 = 0,$$
$$2x_1 + 2x_2 + 2x_3 + 3x_4 - 8x_5 = 0,$$

determine the solution space and compute its dimension.

32. Let W_1 and W_2 be subspaces of V such that $\dim(W_1) = \dim(W_2)$. Prove that $W_1 = W_2$ if and only if $W_1 \subseteq W_2$ or $W_2 \subseteq W_1$.

33. For this exercise the reader might need to consult Appendix B, Section A.3. The complex numbers can be considered as a real vector space. What is its dimension?

34. Let n be a positive integer, and let $A = (a_{ij})$ be an $n \times n$ matrix. Assume that $a_{i1} + a_{i2} + \cdots + a_{in} = 0$ for every $i = 1, 2, \ldots, n$. Prove that A is singular, that is, A has no inverse. (*Hint.*) The rows of A satisfy the equation $x_1 + x_2 + \cdots + x_n = 0$.

35. Let V be a vector space of dimension $n \geq 2$. Prove that V cannot be represented as a finite union of proper subspaces.
    (*Hint.*) Use induction to prove that the union of subspaces $W_1, W_2, \ldots, W_k$ is a subspace if and only there is $j \leq k$, so that $W_l \subseteq W_j$, for every $l \in \{1, 2, \ldots, k\}$.

36. Consider the vector space $V = \mathcal{M}_{n \times n}(\mathbb{R})$ (see Exercise 4.4.1, 2, p. 111). Prove that $\mathcal{S}_{n \times n}(\mathbb{R}) := \{A \in \mathcal{M}_{n \times n}(\mathbb{R}) \mid A^t = A\}$ and $\mathcal{A}_{n \times n}(\mathbb{R}) := \{A \in \mathcal{M}_{n \times n}(\mathbb{R}) \mid A^t = -A\}$ are subspaces of V and $V = \mathcal{S}_{n \times n}(\mathbb{R}) \oplus \mathcal{A}_{n \times n}(\mathbb{R})$. What is the dimension of each one?
    (*Hint.*) To compute the dimension of $\mathcal{S}_{n \times n}(\mathbb{R})$, consider the set of nonzero symmetric matrices with as many zeros as possible. Now count them and show that this set is a basis for the set of symmetric matrices. Start with $n = 2, 3, 4$ to get an idea of how to proceed.

5 Linear transformations and matrices

The concept of a function is one of the most important in all mathematics. By using this concept, many relationships among mathematical objects can be formulated. For instance, in calculus, fundamental problems related with the concepts of limit and continuity are stated in the language of functions. We will see in the discussion that the main problems in linear algebra, such as solving equations of the form $AX = B$ or $AX = \lambda X$, can be deeply and widely understood by approaching them via the use of linear transformations, which ultimately are functions.

5.1 Definitions and basic results

In the following discussion, we will recall Example 2.1.4, p. 39.

Example 5.1.1. Consider a consortium with n companies that produce m different consumer products. Assume that all companies produce the same consumer products. Under these assumptions, the production of each consumer product is described below.

For each million dollars that are invested in company 1, its production is:

$$a_{11} \text{ dollars of consumer product 1,}$$
$$a_{21} \text{ dollars of consumer product 2,}$$
$$\vdots$$
$$a_{m1} \text{ dollars of consumer product } m.$$

In general, for each million dollars invested in the jth company, it produces:

$$a_{1j} \text{ dollars of consumer product 1,}$$
$$a_{2j} \text{ dollars of consumer product 2,}$$
$$\vdots$$
$$a_{mj} \text{ dollars of consumer product } m.$$

If we denote by x_j the amount of money invested in company j, then the investment in all companies can be represented by an element $(x_1, x_2, x_3, \ldots, x_n)$ of $\mathbb{R}^n$, and the total value of consumer product i is

$$a_{i1}x_1 + a_{i2}x_2 + \cdots + a_{in}x_n = c_i, \quad \text{for } i = 1, 2, \ldots, m.$$

This data can be represented in functional notation, that is,

$$T(x_1, x_2, \ldots, x_n) = (c_1, c_2, \ldots, c_m). \tag{5.1}$$

https://doi.org/10.1515/9783111135915-005

The vector $C = (c_1, c_2, \ldots, c_m)$ could be called the production vector associated to the investment vector $X = (x_1, x_2, x_3, \ldots, x_n)$.

Assume that there is another investment vector $Y = (y_1, y_2, \ldots, y_n)$ with production vector $D = (d_1, d_2, \ldots, d_m)$, then

$$T(X + Y) = T(X) + T(Y) = C + D. \qquad (5.2)$$

If the investment vector X is multiplied by a scalar λ, then the corresponding production vector C is also multiplied by λ, that is,

$$T(\lambda X) = \lambda C = \lambda T(X). \qquad (5.3)$$

There are many applied problems that, when formulated in the language of functions, satisfy (5.2) and (5.3). These properties are called linearity properties of T. The basic idea of linearity is established in the following definition.

Definition 5.1.1. Let V and W be two vector spaces. A linear transformation $T : V \to W$ is a function that satisfies:
1. For every $\alpha, \beta \in V$, $T(\alpha + \beta) = T(\alpha) + T(\beta)$.
2. For every scalar $r \in \mathbb{R}$ and for every $\alpha \in V$, $T(r\alpha) = rT(\alpha)$.

When $V = W$ and T is a linear transformation, we say that T is a linear operator or just an operator.

If $V = W = \mathbb{R}$, the function $T(x) = x$ is a linear transformation, since $T(x + y) = x + y = T(x) + T(y)$ and $T(rx) = rx = rT(x)$, for all real numbers x, y, r.

A more general case occurs when considering a function like $f(x) = ax$, for some fixed real number a. The reader is invited to verify that this function is a linear transformation from $\mathbb{R}$ to $\mathbb{R}$.

An example of a function that is not a linear transformation is $f(x) = x + 1$, since $f(x + y) = x + y + 1 \neq f(x) + f(y) = (x + 1) + (y + 1)$.

The examples above motivate the following questions. The first two are left to the reader, and we encourage you to try to find the answer.
1. Is there a function $f : \mathbb{R} \to \mathbb{R}$ which satisfies $f(x + y) = f(x) + f(y)$ for every $x, y \in \mathbb{R}$ and is not a linear transformation?
2. Is there a function $f : \mathbb{R} \to \mathbb{R}$ which satisfies $f(rx) = rf(x)$ for every $r, x \in \mathbb{R}$ and is not a linear transformation?
3. Let $T : \mathbb{R} \to \mathbb{R}$ be a linear transformation. What is the algebraic representation of $T(x)$ for every x?

Concerning the last question, we have that if $T : \mathbb{R} \to \mathbb{R}$ is a linear transformation, then $T(x) = T(x1) = xT(1)$. Let $a = T(1)$, from which we conclude that $T(x) = ax$, for every $x \in \mathbb{R}$. Conversely, if $T(x) = ax$ for some fixed $a \in \mathbb{R}$, then T is a linear transformation.

Summarizing, a function $T : \mathbb{R} \to \mathbb{R}$ is a linear transformation, if and only if there is a fixed $a \in \mathbb{R}$ so that $T(x) = ax$ for every $x \in \mathbb{R}$. From this discussion we have that the linear transformations from $\mathbb{R}$ to $\mathbb{R}$ are exactly the linear functions whose graphs pass through the origin. This, translated to algebraic language, means that if $T : \mathbb{R} \to \mathbb{R}$ is a linear transformation, then $T(0) = 0$.

The latter property of a linear transformation from $\mathbb{R} \to \mathbb{R}$ holds in general, as the following remark shows.

Remark 5.1.1. If $T : V \to W$ is a linear transformation, then $T(0) = 0$. This holds since $T(0) = T(0 + 0) = T(0) + T(0)$. Simplifying one obtains $T(0) = 0$, as claimed.

What is the algebraic representation of a linear transformation from $\mathbb{R}^2$ to $\mathbb{R}$?

To answer this question, we recall that an element (x, y) of $\mathbb{R}^2$ can be represented in terms of the canonical basis, that is, $(x, y) = xe_1 + ye_2$. If $T : \mathbb{R}^2 \to \mathbb{R}$ is a linear transformation, then, using the representation of (x, y) and the properties of T, one concludes that $T(x, y) = T(xe_1 + ye_2) = xT(e_1) + yT(e_2)$. Defining real numbers $a_1 = T(e_1)$ and $a_2 = T(e_2)$, which only depend on T, we have $T(x, y) = a_1 x + a_2 y$. The latter equation represents the fact that T is completely determined by its action on a basis, in this case, the canonical basis. We should also notice that the action of T is a linear combination of the components of the vector (x, y).

We invite the reader to verify that the converse also holds. That is, if there are constants $a_1, a_2 \in \mathbb{R}$ so that $T(x, y) = a_1 x + a_2 y$, then T is a linear transformation.

In the discussion above, we describe how the linear transformations T from $\mathbb{R}$ to $\mathbb{R}$ and from $\mathbb{R}^2$ to $\mathbb{R}$ look like. To achieve this, we use a representation of the elements of $\mathbb{R}^2$ as a linear combination of the canonical basis.

If $T : \mathbb{R}^2 \to \mathbb{R}^2$ is a linear transformation, then $T(x, y)$ is a linear combination of the canonical elements, that is, there are real numbers a_1 and a_2, which depend on T and on (x, y), so that $T(x, y) = a_1 e_1 + a_2 e_2$. To emphasize that a_1 and a_2 depend on (x, y), we use the notation $a_1 = T_1(x, y)$ and $a_2 = T_2(x, y)$, where T_1 and T_2 are functions from $\mathbb{R}^2$ to $\mathbb{R}$. With this, the scalars a_1 and a_2 are the values of the functions T_1 and T_2 at (x, y).

Using this terminology and notation, we can say that T is determined by T_1 and T_2 which are functions from $\mathbb{R}^2 \to \mathbb{R}$. This can be written as $T = (T_1, T_2)$ and it can be interpreted as $T(x, y) = (T_1(x, y), T_2(x, y))$.

If T is given by $T = (T_1, T_2)$, the functions T_1 and T_2 are called the coordinate functions of T.

From above, we have that a linear transformation $T : \mathbb{R}^2 \to \mathbb{R}^2$ is completely determined by its coordinate functions $T_1, T_2 : \mathbb{R}^2 \to \mathbb{R}$. The next result establishes conditions under which $T = (T_1, T_2)$ is a linear transformation in terms of its coordinate functions. Actually, the result holds in general for functions $T : \mathbb{R}^n \to \mathbb{R}^m$.

Theorem 5.1.1. *The function $T : \mathbb{R}^n \to \mathbb{R}^m$, represented by $T = (T_1, \ldots, T_m)$, is a linear transformation if and only if its coordinate functions $T_i : \mathbb{R}^n \to \mathbb{R}$, $i \in \{1, 2, \ldots, m\}$ are linear transformations.*

Proof. Assume that T is a linear transformation, then

$$T(X + Y) = T(X) + T(Y)$$
$$= (T_1(X), T_2(X), \ldots, T_m(X)) + (T_1(Y), T_2(Y), \ldots, T_m(Y))$$
$$= (T_1(X) + T_1(Y), T_2(X) + T_2(Y), \ldots, T_m(X) + T_m(Y)). \tag{5.4}$$

On the other hand,

$$T(X + Y) = (T_1(X + Y), T_2(X + Y), \ldots, T_m(X + Y)). \tag{5.5}$$

From (5.4) and (5.5), one obtains that $T_i(X + Y) = T_i(X) + T_i(Y)$ for every $i \in \{1, 2, \ldots, m\}$.

Conversely, if each T_i is linear, then $T_i(X+Y) = T_i(X)+T_i(Y)$ for every $i \in \{1, 2, \ldots, m\}$, and so

$$T(X + Y) = (T_1(X + Y), T_2(X + Y), \ldots, T_m(X + Y))$$
$$= (T_1(X) + T_1(Y), T_2(X) + T_2(Y), \ldots, T_m(X) + T_m(Y))$$
$$= (T_1(X), T_2(X), \ldots, T_m(X)) + (T_1(Y), T_2(Y), \ldots, T_m(Y))$$
$$= T(X) + T(Y). \tag{5.6}$$

Along the same line of reasoning, one proves that $T(aX) = aT(X)$ if and only if $T_i(aX) = aT_i(X)$ for every $i \in \{1, 2, \ldots, m\}$, finishing the proof. $\qquad\square$

Using Theorem 5.1.1, it is not difficult to construct linear transformations from $\mathbb{R}^2$ to $\mathbb{R}^2$. For instance, the function $T(x, y) = (x - y, 3x + 4y)$ is a linear transformation.

Exercise 5.1.1. Decide which of the given functions are linear transformations:

1. $T(x, y) = (x, y)$;
2. $T(x, y) = (x + 1, y - 1)$;
3. $T(x, y) = (x + y, y - x)$;
4. $T(x, y) = (x + \cos(y), y)$;
5. $T(x, y) = (ax + by, cx + dy)$, with a, b, c, and d being constants.

In what follows, we are interested in exploring the image of a set under a linear transformation.

Example 5.1.2. Let $T : \mathbb{R}^2 \to \mathbb{R}^2$ be given by $T(x, y) = (2x + y, x + y)$. What is the image of the line, whose equation is $y = x$, under the transformation T? More generally, what is the image of a line whose equation is $y = mx$, with m a fixed nonzero real number?

To answer the first question, we will consider points on the line whose equation is $y = x$, then we need to evaluate T at points of the form (x, x), that is, $T(x, x) = (2x + x, x + x) = (3x, 2x) = x(3, 2)$. When x varies in the real numbers, $x(3, 2)$ describes the line whose equation is $y = \frac{2}{3}x$. Then we have that T transforms the line $y = x$ into the line $y = \frac{2}{3}x$, as shown in Figure 5.1.

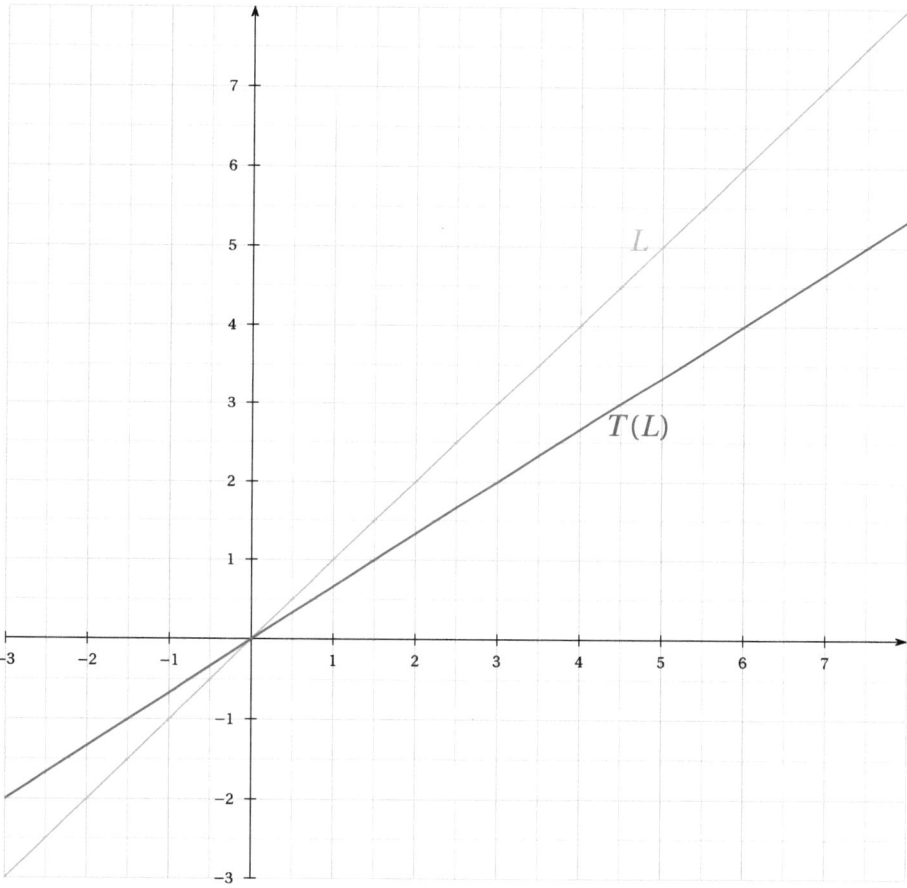

Figure 5.1: Image of L under the linear transformation $T(x,y) = (2x + y, x + y)$.

The second question is addressed analogously, that is, we evaluate T at points of the form (x, mx), then $T(x, mx) = (2x + mx, x + mx) = x(2 + m, 1 + m)$. From this, one has that T transforms the line $y = mx$ into the line whose points are obtained by multiplying the vector $(2 + m, 1 + m)$ by a scalar x. Notice that the vector $(2 + m, 1 + m)$ is nonzero for every $m \in \mathbb{R}$. It is interesting to notice that the line whose slope is $m = -2$ is transformed into the y axis, while the line of slope $m = -1$ is transformed into the horizontal axis (explain).

The points of the unit square are by definition $\{(x, y) \mid 0 \le x \le 1, 0 \le y \le 1\}$. An interesting question is: What is the image of the unit square under T? One possible approach to answer the question could be to evaluate T, at the boundary of the square, that is, to obtain the vectors $T(x, 0)$, $T(1, y)$, $T(x, 1)$, and $T(0, y)$, where $0 \leq x \leq 1$, and $0 \leq y \leq 1$.

We have $T(x, 0) = (2x, x) = x(2, 1)$, $T(1, y) = (2 + y, 1 + y) = (2, 1) + y(1, 1)$, $T(x, 1) = (2x + 1, x + 1) = (2x, x) + (1, 1) = x(2, 1) + (1, 1)$, and $T(0, y) = (y, y) = y(1, 1)$. Since x and y are real numbers between zero and one, the images of the sides of the square are segments as shown on the right of Figure 5.2. From that information, one has that the image of the unit square, under T, is the parallelogram shown in the mentioned figure.

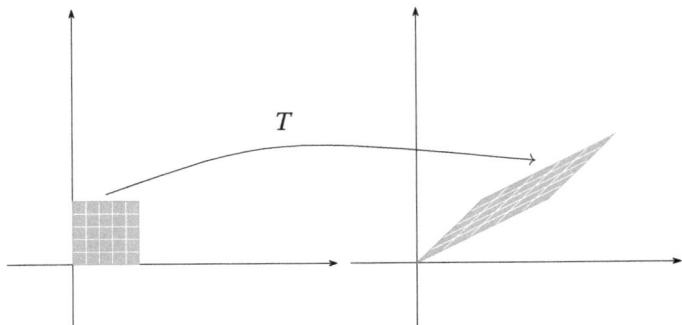

Figure 5.2: Image of the unit square under the transformation $T(x, y) = (2x + y, x + y)$.

Exercise 5.1.2. Use GeoGebra, or any other computational system, to answer the questions posed below. After obtaining the images by using a computer system, obtain an algebraic representation to verify that the answer given by the computer is correct.
1. If $T : \mathbb{R}^2 \to \mathbb{R}^2$ is defined by $T(x, y) = (2x - y, x + 2y)$, what is the image of the circle of radius 1 and center at the origin under T?
2. How should T be defined so that it transforms circles into circles?

In the previous discussion for linear transformations from $\mathbb{R}$ to $\mathbb{R}$ and from $\mathbb{R}^2$ to $\mathbb{R}$, it was noticed that the transformations were determined by their effect on a basis, hence it is natural to ask if this holds in general. The answer is given in the next result.

Theorem 5.1.2. *Let V and W be vector spaces. Suppose $\{a_1, a_2, \ldots, a_n\}$ is a basis of V and let $\beta_1, \beta_2, \ldots, \beta_n$ be any collection of n elements in W. Then there is a unique linear transformation $T : V \to W$ so that $T(a_i) = \beta_i$, for every $i \in \{1, 2, \ldots, n\}$.*

Proof. To define T, we will proceed as follows. We define T at each a_i, more precisely, set $T(a_i) := \beta_i$ for each $i = 1, 2, \ldots, n$. In general, if $a = x_1 a_1 + x_2 a_2 + \cdots + x_n a_n$, define $T(a) = x_1 T(a_1) + x_2 T(a_2) + \cdots + x_n T(a_n)$.

We have that each $a \in V$ can be represented, in a unique way, as a linear combination of the elements of $\{a_1, a_2, \ldots, a_n\}$, hence the definition of T is consistent. The rest of the proof is left to the reader as an exercise. $\qquad\square$

5.2 Geometric linear transformations in $\mathbb{R}^2$

In this section we discuss some special cases of linear transformations that are interesting from the geometric point of view. These linear transformations are rotations and translations. As you can imagine, these transformations have geometric properties such as preserving distances or angles. We start by considering:

(a) *Reflections.* The idea of a reflection is closely connected with observing the image of an object reflected through a mirror. In a plane, a reflection can be thought as reflecting an object through a line, which can be assumed to contain the origin of coordinates. With this assumption, if W is the line with equation $y = mx$, for a fixed m, the transformation leaves the points of W fixed and any vector u, orthogonal to W, is transformed to its negative. See Figure 5.3.

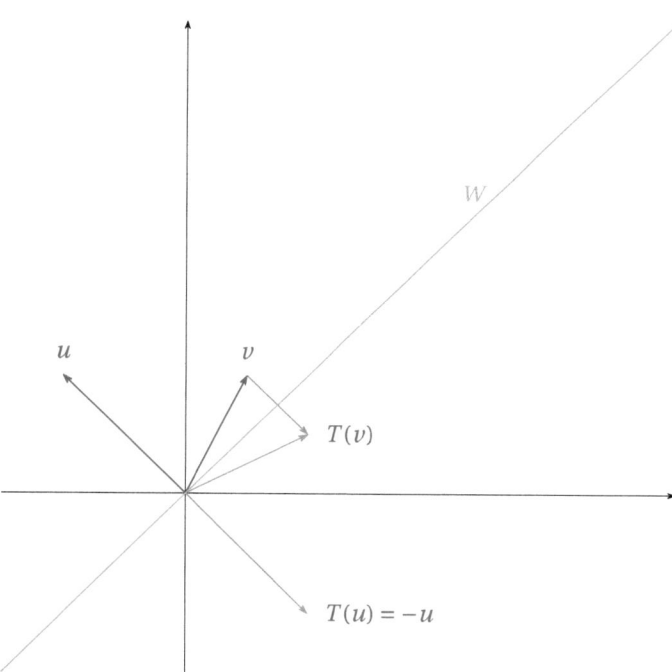

Figure 5.3: Reflection through the subspace W.

If w is a generator of W and v is any vector, we know from Definition 3.3.4, p. 83, that the orthogonal projection of v along w is given by $\frac{\langle v, w \rangle}{\langle w, w \rangle} w$ and $v = X + \frac{\langle v, w \rangle}{\langle w, w \rangle} w$, with X orthogonal to w. Based on this, using that $T(X) = -X$ and that T fixes the elements of W, we have

$$T(v) = T(X) + \frac{\langle v, w \rangle}{\langle w, w \rangle} w$$

$$= -X + \frac{\langle v, w \rangle}{\langle w, w \rangle} w$$

$$= -v + \frac{\langle v, w \rangle}{\langle w, w \rangle} w + \frac{\langle v, w \rangle}{\langle w, w \rangle} w$$

$$= 2\frac{\langle v, w \rangle}{\langle w, w \rangle} w - v, \tag{5.7}$$

or in a more concise form,

$$T(v) = 2\frac{\langle v, w \rangle}{\langle w, w \rangle} w - v, \tag{5.8}$$

where w is a generator of W, the line with respect to which the reflection takes place.

Example 5.2.1. Find an expression for the reflection through the line $y = 2x$.

Discussion. To apply equation (5.8), we need to find a vector w that generates the subspace $W = \{(x, y) \in \mathbb{R}^2 \mid y = 2x\}$. One such generator is, for instance, $w = (1, 2)$. For this w and any element $v = (x, y)$ in $\mathbb{R}^2$, we have $\langle w, w \rangle = 5$ and $\langle v, w \rangle = x + 2y$. Now using equation (5.8), we obtain

$$T(x, y) = \frac{2}{5} \langle (x, y), (1, 2) \rangle (1, 2) - (x, y)$$

$$= \frac{2}{5}(x + 2y)(1, 2) - (x, y)$$

$$= \frac{1}{5}(-3x + 4y, 4x + 3y). \tag{5.9}$$

Figure 5.4 shows the result of applying the linear transformation T to an ellipse whose foci are F_1 and F_2. Also, one can verify that T fixes the points of the line $y = 2x$, that is, if we apply T to points of the form $(x, 2x)$, then

$$T(x, 2x) = \frac{1}{5}(-3x + 4(2x), 4x + 3(2x)) = \frac{1}{5}(5x, 10x) = (x, 2x),$$

as it was expected.

Equation (5.7) shows that a reflection on the plane is determined by a nonzero vector w, or equivalently, by a line that passes through the origin. This idea can be extended to $\mathbb{R}^3$, or in general to $\mathbb{R}^n$, since, to define a reflection in $\mathbb{R}^n$, we need to define a "plane" that passes through the origin. This can be done by considering a nonzero vector $u = (a_1, a_2, \ldots, a_n) \in \mathbb{R}^n$, then the plane is the set of points $v = (x_1, x_2, \ldots, x_n) \in \mathbb{R}^n$ that satisfy

$$\langle v, u \rangle = a_1 x_1 + a_2 x_2 + \cdots + a_n x_n = 0. \tag{5.10}$$

Alternatively, the elements of W satisfy a homogeneous equation, hence it is a subspace of dimension $n - 1$, Exercise 13.

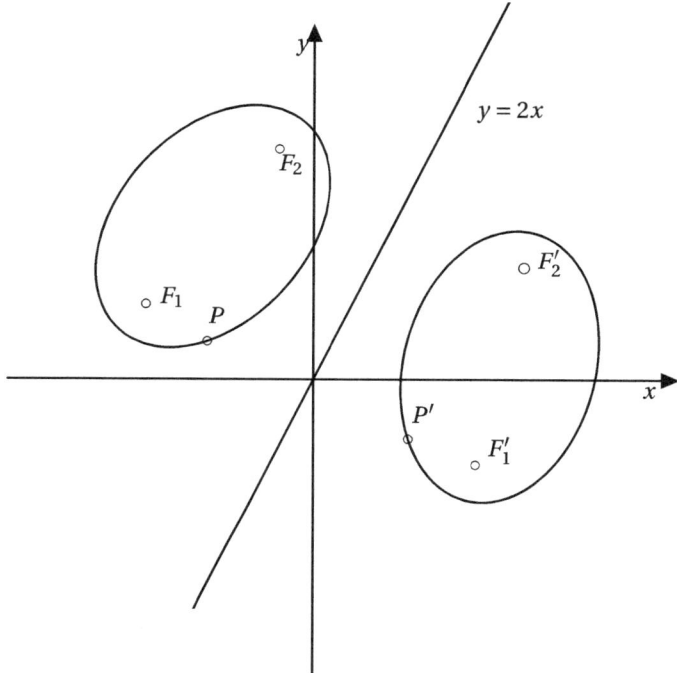

Figure 5.4: The reflection of the ellipse with foci F_1 and F_2 is the ellipse with foci F_1' and F_2'.

We notice that W is determined by u and the reflection that fixes W and sends multiples of u to their negatives is given by

$$T(v) = v - 2\frac{\langle v, u \rangle}{\langle u, u \rangle}u, \qquad (5.11)$$

as can be verified directly.

Remark 5.2.1. One should notice the relationship between equations (5.8) and (5.11). The former is obtained from a generator of W, while the latter comes from an orthogonal vector to W.

(b) *Rotations.* From equation (A.12), p. 269, given a vector $u \in \mathbb{R}^2$, it can be represented in polar form as $u = r(\cos(\theta), \sin(\theta))$, with r the norm of u, and θ the angle between u and the horizontal axis. If u is rotated by an angle ω and T_ω represents the function that rotates, then this can be written as

$$T_\omega(u) = r(\cos(\theta + \omega), \sin(\theta + \omega)) = u',$$

and its geometric representation is shown in Figure 5.5.

Recalling the addition formulas for sine and cosine, we have:

$$\sin(\theta + \omega) = \sin(\theta)\cos(\omega) + \sin(\omega)\cos(\theta),$$
$$\cos(\theta + \omega) = \cos(\theta)\cos(\omega) - \sin(\theta)\sin(\omega). \tag{5.12}$$

From the equations in (5.12), the expression for $T_\omega(u)$ is

$$T_\omega(u) = r\big(\cos(\theta + \omega), \sin(\theta + \omega)\big)$$
$$= r\big(\cos(\theta)\cos(\omega) - \sin(\theta)\sin(\omega), \sin(\theta)\cos(\omega) + \sin(\omega)\cos(\theta)\big). \tag{5.13}$$

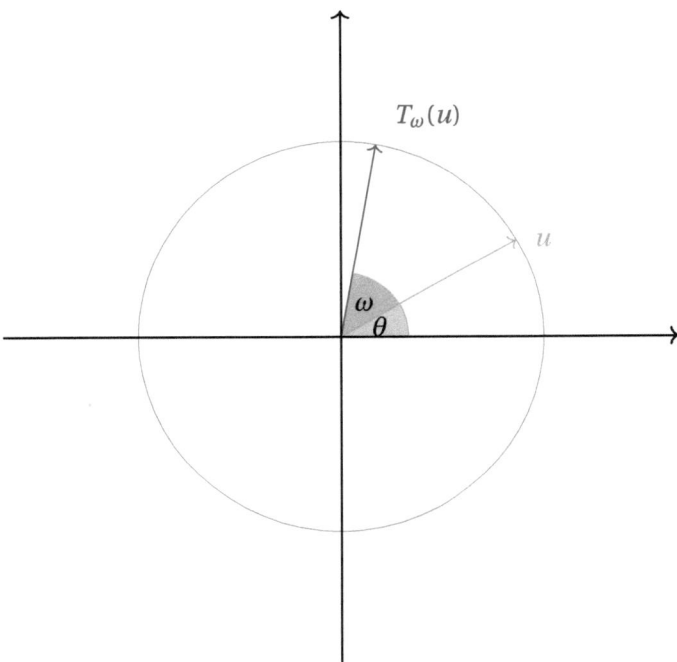

Figure 5.5: Rotation by an angle ω.

Representing the vector $T_\omega(u)$ as a column vector, we have

$$T_\omega(u) = \begin{bmatrix} \cos(\omega) & -\sin(\omega) \\ \sin(\omega) & \cos(\omega) \end{bmatrix} \begin{bmatrix} r\cos(\theta) \\ r\sin(\theta) \end{bmatrix}. \tag{5.14}$$

The matrix

$$R_\omega = \begin{bmatrix} \cos(\omega) & -\sin(\omega) \\ \sin(\omega) & \cos(\omega) \end{bmatrix} \tag{5.15}$$

is called a rotation matrix and represents T_ω in the canonical basis.

If the representation of u is given by $u = (x, y)$, then $T_\omega(x, y)$ is obtained from

$$T_\omega(x, y) = \begin{bmatrix} \cos(\omega) & -\sin(\omega) \\ \sin(\omega) & \cos(\omega) \end{bmatrix} \begin{bmatrix} x \\ y \end{bmatrix} = \begin{bmatrix} \cos(\omega)x & -\sin(\omega)y \\ \sin(\omega)x & +\cos(\omega)y \end{bmatrix}. \tag{5.16}$$

This result can be expressed in the form

$$T_\omega(x, y) = (\cos(\omega)x - \sin(\omega)y, \; \sin(\omega)x + \cos(\omega)y). \tag{5.17}$$

Example 5.2.2. Describe algebraically the linear transformation that rotates the plane by an angle of $\frac{\pi}{4}$ radians.

Discussion. According to equation (5.17), the transformation is given by

$$T_{\pi/4}(x, y) = (\cos(\pi/4)x - \sin(\pi/4)y, \; \sin(\pi/4)x + \cos(\pi/4)y) = \frac{1}{\sqrt{2}}(x - y, x + y).$$

Solve the following exercises.

Exercise 5.2.1.
1. Prove that rotations and reflections in $\mathbb{R}^2$ are linear transformations.
2. Describe the geometric result of composing a rotation and a reflection.
3. Define a translation. Is it a linear transformation?
4. An isometry of the plane is a function $f : \mathbb{R}^2 \to \mathbb{R}^2$ which preserves distance, that is, $\|f(\alpha) - f(\beta)\| = \|\alpha - \beta\|$ for every $\alpha, \beta \in \mathbb{R}^2$. Prove that rotations, reflections, and translations are isometries.
5. Consider a square C whose sides are parallel to the coordinate axes. Given $T(x, y) = (2x + y, x + y)$, find the image of C under T.

5.3 Range and kernel of a linear transformation

Given a linear transformation T, there are two subspaces associated with it, which are important to describe properties of T.

Definition 5.3.1. Let $T : V \to W$ be a linear transformation. The set $N_T = \{\alpha \in V \mid T(\alpha) = 0\}$ is called the kernel of T. The range or image of T, denoted by R_T, is defined by $R_T = \{\beta \in W \mid \text{there is } \alpha \in V \text{ such that } T(\alpha) = \beta\}$.

Exercise 5.3.1. Prove that the kernel and range of a linear transformation are subspaces.

Exercise 5.3.2. For each of the following linear transformations, find the range and kernel:
1. Let $T : \mathbb{R}^2 \to \mathbb{R}^2$ be defined by $T(x, y) = (x - y, 3x + 2y)$.

2. Let n be a positive integer, let V be the vector space of polynomials with real coefficients of degree at most n, and let T be the function given by $T(p(x)) = p'(x)$, with $p'(x)$ the derivative of $p(x)$.
3. Let $T : \mathbb{R}^n \to \mathbb{R}$ be defined by $T(x_1, x_2, \ldots, x_n) = x_1 + x_2 + x_3 + \cdots + x_n$.
4. Let V be the space of 2×2 matrices and $A = \left(\begin{smallmatrix} 1 & 2 \\ 0 & 1 \end{smallmatrix}\right)$. Define $T_A : V \to V$ by $T_A(X) = AX$, where $X = \left(\begin{smallmatrix} x & y \\ z & w \end{smallmatrix}\right)$.

Exercise 5.3.3. Review the definition of an injective and surjective function and prove the following concerning a linear transformation $T : V \to W$:
1. The linear transformation T is one-to-one if and only if $N_T = \{0\}$.
2. The linear transformation T is surjective if and only if $R_T = W$.

Definition 5.3.2. Let V and W be vector spaces, and let $T : V \to W$ be a linear transformation.
1. We say that T is nonsingular if it is one-to-one.
2. If T is bijective, it is called an isomorphism, and we say that V is isomorphic to W, written $V \cong W$.

The following results are among the most important in linear algebra for finite-dimensional vector spaces.

Theorem 5.3.1. *Let V be a vector space. Then $V \cong \mathbb{R}^n$ if and only if V has dimension n.*

Proof. Assume that V has dimension n and let $\{a_1, a_2, \ldots, a_n\}$ be a basis of V. We will prove that there is a bijective linear transformation $T : V \to \mathbb{R}^n$. From Theorem 5.1.2, there is a unique linear transformation T so that $T(a_i) = e_i$, where e_i is the ith canonical element of $\mathbb{R}^n$. If $a = x_1 a_1 + \cdots + x_n a_n \in V$ and $T(a) = x_1 e_1 + \cdots + x_n e_n = 0$, then, using that the canonical elements are linearly independent, we have $x_i = 0$, for every $i \in \{1, 2, \ldots, n\}$, that is, T is injective. Also, given $X = x_1 e_1 + \cdots + x_n e_n \in \mathbb{R}^n$ and defining $a = x_1 a_1 + \cdots + x_n a_n$, we have that $T(a) = X$, proving that T is surjective, therefore V is isomorphic to $\mathbb{R}^n$.

Conversely, assume that there is a bijective linear transformation $T : V \to \mathbb{R}^n$, and let $\{e_1, e_2, \ldots, e_n\}$ denote the canonical basis of $\mathbb{R}^n$. Since T is bijective, we get that there is $S = \{a_1, a_2, \ldots, a_n\} \subseteq V$ such that $T(a_j) = e_j$ for every $j \in \{1, 2, \ldots, n\}$. We claim that S is a basis of V. Assume that $x_1 a_1 + \cdots + x_n a_n = 0$, for some scalars. Then applying T to this equation, we have that $T(x_1 a_1 + \cdots + x_n a_n) = T(0) = 0$. Using the linearity of T and the choice of S, we obtain $x_1 e_1 + \cdots + x_n e_n = 0$, from where we conclude that $x_1 = x_2 = \cdots = x_n = 0$, that is, S is linearly independent. We have to prove that S spans V. Given $\beta \in V$, we can write

$$T(\beta) = y_1 e_1 + y_2 e_2 + \cdots + y_n e_n$$
$$= y_1 T(a_1) + y_2 T(a_2) + \cdots + y_n T(a_n)$$

$$= T(y_1 a_1 + y_2 a_2 + \cdots + y_n a_n), \tag{5.18}$$

for some scalars $y_1, y_2, \ldots, y_n$. From (5.18) and the assumption that T is bijective, hence injective, we conclude that $\beta = y_1 a_1 + y_2 a_2 + \cdots + y_n a_n$, proving that S spans V. □

An important corollary of Theorem 5.3.1 is

Corollary 5.3.1. *Given positive integers m and n, $\mathbb{R}^m \cong \mathbb{R}^n$ if and only if $m = n$.*

The reader is invited to provide the details of the proof of Corollary 5.3.1.

Theorem 5.3.2. *If V and W are vector spaces, with V finitely generated and $T : V \to W$ being a linear transformation, then*

$$\dim(V) = \dim(N_T) + \dim(R_T). \tag{5.19}$$

Proof. If T is nonsingular, then, according to Exercise 18, p. 140, T transforms linearly independent sets into linearly independent sets, thus the dimension of V is equal to the dimension of R_T, and also the kernel of T is zero, so that equation (5.19) holds. If T is singular, then $N_T \neq \{0\}$. Let $\{a_1, a_2, \ldots, a_r\}$ be a basis of N_T. From Theorem 4.4.3, we can extend this set to a basis of V. Let us say that $\{a_1, a_2, \ldots, a_r, a_{r+1}, \ldots, a_n\}$ is such a basis. We will prove that $\{T(a_{r+1}), \ldots, T(a_n)\}$ is a basis of R_T, from where equation (5.19) follows directly.

Assume $x_{r+1} T(a_{r+1}) + \cdots + x_n T(a_n) = 0$. Then, using the linearity of T, we obtain

$$T(x_{r+1} a_{r+1} + \cdots + x_n a_n) = 0,$$

that is, $x_{r+1} a_{r+1} + \cdots + x_n a_n \in N_T$, therefore there are real numbers $x_1, x_2, \ldots, x_r$ so that $x_1 a_1 + \cdots + x_r a_r = x_{r+1} a_{r+1} + \cdots + x_n a_n$. From this equation, we conclude that all scalars $x_{r+1}, \ldots, x_n$ are zero, proving that the set $\{T(a_{r+1}), \ldots, T(a_n)\}$ is linearly independent. If $\beta \in R_T$, then there is $\alpha = a_1 a_1 + \cdots + a_n a_n \in V$ such that $T(\alpha) = \beta$. On the other hand,

$$
\begin{aligned}
T(\alpha) &= T(a_1 a_1 + \cdots + a_n a_n) \\
&= a_1 T(a_1) + \cdots + a_r T(a_r) + a_{r+1} T(a_{r+1}) + \cdots + a_n T(a_n) \\
&= 0 + \cdots + 0 + a_{r+1} T(a_{r+1}) + \cdots + a_n T(a_n) \\
&= \beta,
\end{aligned}
$$

proving that $\{T(a_{r+1}), \ldots, T(a_n)\}$ generates R_T, finishing the proof of the theorem. □

Exercise 5.3.4. Let m and n be positive integers. Prove:
1. If $T : \mathbb{R}^n \to \mathbb{R}^m$ is a one-to-one linear transformation, then $n \leq m$.
2. If $T : \mathbb{R}^n \to \mathbb{R}^m$ is a surjective linear transformation, then $m \leq n$.

Exercise 5.3.5. Let V and W be two finitely generated vector spaces. Prove that $V \cong W$ if and only if $\dim(V) = \dim(W)$.

5.4 Matrices and linear transformations

In this section we will establish a close relationship between linear transformations and matrices. In this respect, it is appropriate to say that when performing calculations with linear transformations, using matrix representations is probably the right way to make good progress.

We have proved, see Theorem 5.1.2, p. 122, that a linear transformation is determined by its action on a basis. Hence, if $T : V \rightarrow W$ is a linear transformation and $\{\alpha_1, \alpha_2, \ldots, \alpha_n\}$ and $\{\beta_1, \beta_2, \ldots, \beta_m\}$ are bases of V and W, respectively, then for each $j \in \{1, \ldots, n\}$, $T(\alpha_j)$ is represented as a linear combination of $\{\beta_1, \beta_2, \ldots, \beta_m\}$, that is, there are unique scalars $a_{1j}, a_{2j}, \ldots, a_{mj}$ such that

$$T(\alpha_j) = a_{1j}\beta_1 + a_{2j}\beta_2 + \cdots + a_{mj}\beta_m. \tag{5.20}$$

Notice that the scalars a_{ij} depend only on T and the chosen bases. With these scalars, we form the matrix

$$A_{[\alpha,\beta]} = \begin{bmatrix} a_{11} & a_{12} & \cdots & a_{1n} \\ a_{21} & a_{22} & \cdots & a_{2n} \\ \vdots & \vdots & \ddots & \vdots \\ a_{m1} & a_{m2} & \cdots & a_{mn} \end{bmatrix}.$$

Definition 5.4.1. The matrix $A_{[\alpha,\beta]}$ is called the matrix associated to T with respect to the bases $\{\alpha_1, \alpha_2, \ldots, \alpha_n\}$ and $\{\beta_1, \beta_2, \ldots, \beta_m\}$. If there is no confusion, the matrix is written without indexes.

Example 5.4.1. Let $T : \mathbb{R}^2 \rightarrow \mathbb{R}^5$ be given by $T(x, y) = (2x + y, x - y, 4x - 3y, x, y)$. Find the matrix associated to T with respect to the canonical bases of $\mathbb{R}^2$ and $\mathbb{R}^5$.

Discussion. To find the matrix of T with respect to the bases, evaluate T at $e_1 = (1, 0)$ and at $e_2 = (0, 1)$, and represent those elements as linear combinations of the canonical basis in $\mathbb{R}^5$. The coefficients that appear in $T(e_1)$ are the entries of the first column, and the coefficients of $T(e_2)$ form the second column. We have:

$$T(1, 0) = (2, 1, 4, 1, 0)$$
$$= 2(1, 0, 0, 0, 0) + (0, 1, 0, 0, 0) + 4(0, 0, 1, 0, 0) + (0, 0, 0, 1, 0) + 0(0, 0, 0, 0, 1), \text{and}$$
$$T(0, 1) = (1, -1, -3, 0, 1)$$
$$= (1, 0, 0, 0, 0) - (0, 1, 0, 0, 0) - 3(0, 0, 1, 0, 0) + 0(0, 0, 0, 1, 0) + (0, 0, 0, 0, 1).$$

From this we have that the matrix associated to T is $A = \begin{pmatrix} 2 & 1 \\ 1 & -1 \\ 4 & -3 \\ 1 & 0 \\ 0 & 1 \end{pmatrix}$.

Example 5.4.2. It is straightforward to verify that $\alpha_1 = (1, 1)$ and $\alpha_2 = (3, -2)$ are linearly independent, hence they form a basis of $\mathbb{R}^2$. Given the vectors $(4, 5)$ and $(6, -1)$, according to Theorem 5.1.2, there is a unique linear transformation $T : \mathbb{R}^2 \to \mathbb{R}^2$ given by $T(a_1) = (4, 5)$ and $T(a_2) = (6, -1)$. Find the matrix associated to T with respect to the canonical basis of $\mathbb{R}^2$.

Discussion. To determine the matrix of T with respect to the canonical basis, we need to know $T(e_1) = (a, b)$ and $T(e_2) = (c, d)$. Using the equations $T(1, 1) = T(e_1) + T(e_2)$ and $T(3, -2) = 3T(e_1) - 2T(e_2)$, we obtain $(4, 5) = (a + c, b + d)$ and $(6, -1) = (3a - 2c, 3b - 2d)$, which are equivalent to

$$
\begin{aligned}
a + c &= 4, \\
b + d &= 5, \\
3a - 2c &= 6, \\
3b - 2d &= -1.
\end{aligned}
\tag{5.21}
$$

Solving this system leads to $T(e_1) = \left(\frac{14}{5}, \frac{9}{5}\right)$ and $T(e_2) = \left(\frac{6}{5}, \frac{16}{5}\right)$, from where the associated matrix to T is obtained as

$$
A = \begin{bmatrix} \frac{14}{5} & \frac{6}{5} \\ \frac{9}{5} & \frac{16}{5} \end{bmatrix}.
$$

From the information above, we can compute $T(x, y)$ by using A, that is,

$$
AX = \begin{bmatrix} \frac{14}{5} & \frac{6}{5} \\ \frac{9}{5} & \frac{16}{5} \end{bmatrix} \begin{bmatrix} x \\ y \end{bmatrix} = \begin{bmatrix} \frac{16}{5}x + \frac{6}{5}y \\ \frac{9}{5}x + \frac{16}{5}y \end{bmatrix}.
$$

From this equation, $T(x, y) = \left(\frac{16}{5}x + \frac{6}{5}y, \frac{9}{5}x + \frac{16}{5}y\right)$.

Exercise 5.4.1. Let P_n be the vector space of the real polynomials of degree at most n, and let D be the differential operator, that is, $D(p(x)) = p'(x)$. Decide if D is a linear transformation and, if so, find the matrix associated to D with respect to the basis $\{1, x, x^2, \dots, x^n\}$. What is the kernel and range of D?

Several mathematical problems can be stated by using the composition of functions, and the new function possesses the same properties as those of the original functions. For example, the composition of continuous functions is continuous. The same holds for differentiable functions. Then it is natural to ask if the composition of linear transformations a linear transformation.

Assuming that V, W, and U are vector spaces and there are linear transformations $T : V \to W$ and $T_1 : W \to U$, we can define the function $T_1 \circ T : V \to U$ which is a linear transformation.

In fact, if $\alpha, \beta \in V$ then

$$
(T_1 \circ T)(\alpha + \beta) = T_1(T(\alpha + \beta)) = T_1(T(\alpha) + T(\beta)) = T_1(T(\alpha)) + T_1(T(\beta)).
$$

Likewise, $(T_1 \circ T)(a\alpha) = a(T_1 \circ T)(\alpha)$. This proves the assertion.

Let V, W, and U be vector spaces with bases

$$\{\alpha_1, \alpha_2, \dots, \alpha_n\}, \quad \{\beta_1, \beta_2, \dots, \beta_m\}, \quad \text{and} \quad \{\gamma_1, \gamma_2, \dots, \gamma_r\},$$

respectively. Let $T : V \to W$ and $T_1 : W \to U$ be linear transformations with matrices $A_{[\alpha,\beta]}$ and $B_{[\beta,\gamma]}$ associated to T and T_1, respectively, with respect to the corresponding bases. A question is: What is the matrix associated to $T_1 \circ T$ with respect to the bases $\{\alpha_1, \alpha_2, \dots, \alpha_n\}$ and $\{\gamma_1, \gamma_2, \dots, \gamma_r\}$?

To determine this matrix, we need to evaluate $T_1 \circ T$ at each element of $\{\alpha_1, \alpha_2, \dots, \alpha_n\}$ and obtain the result as a linear combination of $\{\gamma_1, \gamma_2, \dots, \gamma_r\}$.

Suppose that the matrices $A_{[\alpha,\beta]}$ and $B_{[\beta,\gamma]}$ have entries a_{ij} and b_{ts}, respectively. Let us call C the matrix associated to $T_1 \circ T$ with respect to the bases $\{\alpha_1, \alpha_2, \dots, \alpha_n\}$ and $\{\gamma_1, \gamma_2, \dots, \gamma_r\}$. Notice that the matrix C has size $r \times n$ and its entries satisfy

$$(T_1 \circ T)(\alpha_j) = c_{1j}\gamma_1 + c_{2j}\gamma_2 + \cdots + c_{rj}\gamma_r. \tag{5.22}$$

Using the definition of the composition of functions and the definition of the matrix associated to a linear transformation, we have

$$
\begin{aligned}
(T_1 \circ T)(\alpha_j) &= T_1(T(\alpha_j)) \\
&= T_1(a_{1j}\beta_1 + a_{2j}\beta_2 + \cdots + a_{mj}\beta_m) \\
&= a_{1j}T_1(\beta_1) + a_{2j}T_1(\beta_2) + \cdots + a_{mj}T_1(\beta_m) \\
&= a_{1j}\left(\sum_{t=1}^{r} b_{t1}\gamma_t\right) + a_{2j}\left(\sum_{t=1}^{r} b_{t2}\gamma_t\right) + \cdots + a_{mj}\left(\sum_{t=1}^{r} b_{tm}\gamma_t\right) \\
&= \left(\sum_{k=1}^{m} b_{1k}a_{kj}\right)\gamma_1 + \left(\sum_{k=1}^{m} b_{2k}a_{kj}\right)\gamma_2 + \cdots + \left(\sum_{k=1}^{m} b_{rk}a_{kj}\right)\gamma_r. \tag{5.23}
\end{aligned}
$$

From equations (5.22), (5.23), and using that the representation of a vector in a basis is unique, we obtain

$$c_{ij} = b_{i1}a_{1j} + b_{i2}a_{2j} + \cdots + b_{im}a_{mj},$$

for every $i \in \{1, 2, \dots, r\}$ and $j \in \{1, 2, \dots, n\}$, i.e., the matrix associated to $T_1 \circ T$ is $B_{[\beta,\gamma]}A_{[\alpha,\beta]}$.

The previous discussion can be summarized in:

Theorem 5.4.1. *Let U, V, and W be vector spaces with bases $\{\alpha_1, \alpha_2, \dots, \alpha_n\}$, $\{\beta_1, \beta_2, \dots, \beta_m\}$ and $\{\gamma_1, \gamma_2, \dots, \gamma_r\}$, respectively. If $T : V \to W$ and $T_1 : W \to U$ are linear transformations with associated matrices $A_{[\alpha,\beta]}$ and $B_{[\beta,\gamma]}$ with respect to the given basis, then the matrix associated to $T_1 \circ T$ is $B_{[\beta,\gamma]}A_{[\alpha,\beta]}$.*

The reader is invited to review carefully the arguments provided in the discussion, prior to Theorem 5.4.1.

5.4.1 Change of basis matrix

When one chooses a basis in $\mathbb{R}^2$, a reference system is defined. For instance, when we choose the canonical basis and a vector is represented with respect to this basis, $a = (x, y)$, this means that to arrive at the point a, one has to travel from the origin x units in the direction of the vector $e_1 = (1, 0)$ and y units in the direction of $e_2 = (0, 1)$. However, to arrive at the point determined by a, we could choose a different reference system. For instance, the vectors $a_1 = (3, 2)$ and $a_2 = (1, \frac{3}{2})$ define another reference system.

Hence, to arrive at the point determined by a, we have to travel x_1 units in the direction given by a_1 and y_1 units in the direction of a_2. The pair of numbers x_1 and y_1 are called the *coordinates of a respect to the basis* $\{a_1 = (3, 2), a_2 = (1, 3/2)\}$.

Exercise 5.4.2. Draw a picture to illustrate the previous discussion.

An interesting question is: What is the relationship between the coordinates of a with respect to the canonical basis and with respect to the basis $\{a_1 = (3, 2), a_2 = (1, \frac{3}{2})\}$?

Before answering the posed question, we will discuss a more general idea that will lead us to provide an answer. This idea has to do with what is called the change of basis matrix.

Let V be a vector space with bases $\{a_1, a_2, \ldots, a_n\}$ and $\{a'_1, a'_2, \ldots, a'_n\}$, then each element a'_j can be expressed as a linear combination of $a_1, a_2, \ldots, a_n$, that is, there are scalars $p_{1j}, p_{2j}, \ldots, p_{nj}$ so that $a'_j = p_{1j}a_1 + p_{2j}a_2 + \cdots + p_{nj}a_n$. With these coefficients, we can form the matrix

$$
P = \begin{bmatrix} p_{11} & p_{12} & \cdots & p_{1n} \\ p_{21} & p_{22} & \cdots & p_{2n} \\ \vdots & \vdots & \ddots & \vdots \\ p_{n1} & p_{n2} & \cdots & p_{nn} \end{bmatrix}, \tag{5.24}
$$

that is, P is obtained by choosing the coefficients of a'_j to define the jth column.

Definition 5.4.2. With the notation as in (5.24), matrix P is called the matrix that changes basis $\{a_1, a_2, \ldots, a_n\}$ to $\{a'_1, a'_2, \ldots, a'_n\}$.

Remark 5.4.1. Given that $\{a'_1, a'_2, \ldots, a'_n\}$ is also a basis, each a_j can be represented as a linear combination of $\{a'_1, a'_2, \ldots, a'_n\}$ and a matrix Q is obtained, which turns out to be the inverse of P.

Exercise 5.4.3. Justify Remark 5.4.1.

To answer the posed question related to the coordinates of a with respect to the canonical basis and with respect to $\{a_1 = (3,2), a_2 = \left(1, \frac{3}{2}\right)\}$, we will discuss the general case and then apply it to the example.

Let V be a vector space, and let $B = \{a_1, a_2, \ldots, a_n\}$ be a basis of V. We know that for any $a \in V$ there are unique scalars $x_1, x_2, \ldots, x_n$ so that

$$a = x_1 a_1 + x_2 a_2 + \cdots + x_n a_n. \tag{5.25}$$

Definition 5.4.3. Let V be a vector space, $B = \{a_1, a_2, \ldots, a_n\}$ a basis of V, and $a \in V$ an element of V represented by equation (5.25). The scalars $x_1, x_2, \ldots, x_n$ are called the coordinates of a with respect to the basis $B = \{a_1, a_2, \ldots, a_n\}$, and $X = (x_1, x_2, \ldots, x_n)_B$ is called the coordinate vector of a with respect to the given basis.

Let $\{a_1, a_2, \ldots, a_n\}$ and $\{a'_1, a'_2, \ldots, a'_n\}$ be bases of V. Given $a \in V$, there are two representations $a = \sum_{i=1}^{n} x_i a_i$ and $a = \sum_{i=1}^{n} y_i a'_i$. If P is the change of basis matrix from $\{a_1, a_2, \ldots, a_n\}$ to $\{a'_1, a'_2, \ldots, a'_n\}$, then we will show that $X = PY$, where X is the coordinate vector of a with respect to the basis $\{a_1, a_2, \ldots, a_n\}$ and Y is the coordinate vector of a with respect to the basis $\{a'_1, a'_2, \ldots, a'_n\}$.

Just for notation, we will represent the relation between the elements of the bases by the "matrix" equation

$$\begin{bmatrix} p_{11} & p_{21} & \cdots & p_{n1} \\ p_{12} & p_{22} & \cdots & p_{n2} \\ \vdots & \vdots & \ddots & \vdots \\ p_{1n} & p_{2n} & \cdots & p_{nn} \end{bmatrix} \begin{bmatrix} a_1 \\ a_2 \\ \vdots \\ a_n \end{bmatrix} = \begin{bmatrix} a'_1 \\ a'_2 \\ \vdots \\ a'_n \end{bmatrix}. \tag{5.26}$$

Notice that the first factor in equation (5.26) is the transpose of P.

Likewise, we can represent the vector a using its coordinates in each basis as

$$a = \begin{bmatrix} x_1 & x_2 & \cdots & x_n \end{bmatrix} \begin{bmatrix} a_1 \\ a_2 \\ \vdots \\ a_n \end{bmatrix} = \begin{bmatrix} y_1 & y_2 & \cdots & y_n \end{bmatrix} \begin{bmatrix} a'_1 \\ a'_2 \\ \vdots \\ a'_n \end{bmatrix}.$$

From equation (5.26), one has

$$a = \begin{bmatrix} x_1 & x_2 & \cdots & x_n \end{bmatrix} \begin{bmatrix} a_1 \\ a_2 \\ \vdots \\ a_n \end{bmatrix} = \begin{bmatrix} y_1 & y_2 & \cdots & y_n \end{bmatrix} \begin{bmatrix} p_{11} & p_{21} & \cdots & p_{n1} \\ p_{12} & p_{22} & \cdots & p_{n2} \\ \vdots & \vdots & \ddots & \vdots \\ p_{1n} & p_{2n} & \cdots & p_{nn} \end{bmatrix} \begin{bmatrix} a_1 \\ a_2 \\ \vdots \\ a_n \end{bmatrix}.$$

It is well known that the representation of a vector with respect to a basis is unique, then one has that $X^t = Y^t P^t$ or, equivalently,

$$X = PY. \tag{5.27}$$

We know that if P has an inverse, then the equation $X = PY$ is equivalent to $P^{-1}X = Y$, that is, the coordinates of a with respect to the new basis $\{a_1', a_2', \ldots, a_n'\}$ are obtained from $Y = P^{-1}X$.

The question that originated the discussion can be addressed as follows. We need to find the inverse of the change of basis matrix from the canonical basis to $\{(3, 2), (1, 3/2)\}$. The change of basis matrix is $P = \begin{bmatrix} 3 & 1 \\ 2 & \frac{3}{2} \end{bmatrix}$. From this we find that $P^{-1} = \begin{bmatrix} \frac{3}{5} & -\frac{2}{5} \\ -\frac{4}{5} & \frac{6}{5} \end{bmatrix}$. If Y is the coordinate vector of a with respect to $\{(3, 2), (1, 3/2)\}$ and X is the coordinate vector of a with respect to the canonical basis, then $Y = P^{-1}X = \begin{bmatrix} \frac{3}{5} & -\frac{2}{5} \\ -\frac{4}{5} & \frac{6}{5} \end{bmatrix} X$ or, equivalently, $y_1 = \frac{3}{5}x_1 - \frac{2}{5}x_2$ and $y_2 = -\frac{4}{5}x_1 + \frac{6}{5}x_2$.

Example 5.4.3. Given the canonical basis $\{e_1, e_2, e_3\}$ of $\mathbb{R}^3$, find the coordinates of $a = e_1 - 2e_2 + 5e_3$ with respect to $\{(1, 2, 0), (1, -1, 0), (1, 1, 1)\}$.

Discussion. The change of basis matrix is $P = \begin{bmatrix} 1 & 1 & 1 \\ 2 & -1 & 1 \\ 0 & 0 & 1 \end{bmatrix}$. Computing the inverse of P, one finds that

$$P^{-1} = \begin{bmatrix} \frac{1}{3} & \frac{1}{3} & -\frac{2}{3} \\ \frac{2}{3} & -\frac{1}{3} & -\frac{1}{3} \\ 0 & 0 & 1 \end{bmatrix}.$$

From this, one has that the coordinates of a with respect to the new basis are given by

$$Y = \begin{bmatrix} y_1 \\ y_2 \\ y_3 \end{bmatrix} = \begin{bmatrix} \frac{1}{3} & \frac{1}{3} & -\frac{2}{3} \\ \frac{2}{3} & -\frac{1}{3} & -\frac{1}{3} \\ 0 & 0 & 1 \end{bmatrix} \begin{bmatrix} 1 \\ -2 \\ 5 \end{bmatrix} = \begin{bmatrix} -\frac{11}{3} \\ -\frac{1}{3} \\ 5 \end{bmatrix}.$$

The previous equation represents the coordinates of $(1, -2, 5)$ with respect to $\{(1, 2, 0), (1, -1, 0), (1, 1, 1)\}$, that is, $(1, -2, 5) = -\frac{11}{3}(1, 2, 0) - \frac{1}{3}(1, -1, 0) + 5(1, 1, 1)$.

In several applications, it is important to find a matrix that represents a linear transformation. One of the conditions on the matrix could be that such a matrix has as few nonzero entries as possible. This depends, of course, on the linear transformation. We know that, given bases of V and W and a linear transformation $T : V \to W$, there is an associated matrix with respect to the given bases. More precisely, if $\{a_1, a_2, \ldots, a_n\}$ and $\{\beta_1, \beta_2, \ldots, \beta_m\}$ are bases of V and W, respectively, the associated matrix of T is determined. If the given bases are changed to $\{a_1', a_2', \ldots, a_n'\}$ and $\{\beta_1', \beta_2', \ldots, \beta_m'\}$, in these new bases T has another associated matrix B. What is the relationship between A, B, and the change of basis matrices? The answer is given in Theorem 5.4.2.

Theorem 5.4.2 (Change of basis theorem). *Let $T : V \to W$ be a linear transformation. Assume that A is the matrix associated to T with respect to the given bases $\{a_1, a_2, \ldots, a_n\}$*

in V and $\{\beta_1, \beta_2, \ldots, \beta_m\}$ in W. If the given bases are changed to $\{\alpha'_1, \alpha'_2, \ldots, \alpha'_n\}$ and $\{\beta'_1, \beta'_2, \ldots, \beta'_m\}$, with matrices P and Q, respectively, and if B is the matrix associated to T in the new bases, then

$$B = Q^{-1}AP. \tag{5.28}$$

Proof. Assume that the entries of A, B, P, and Q are denoted by a_{ij}, b_{ij}, p_{ij}, and q_{ij}, respectively. From the definition of the matrix associated to T with respect to the bases $\{\alpha_1, \alpha_2, \ldots, \alpha_n\}$ and $\{\beta_1, \beta_2, \ldots, \beta_m\}$, we have

$$T(\alpha_j) = \sum_{i=1}^{m} a_{ij}\beta_i. \tag{5.29}$$

Likewise, for the bases $\{\alpha'_1, \alpha'_2, \ldots, \alpha'_n\}$ and $\{\beta'_1, \beta'_2, \ldots, \beta'_m\}$, we have

$$T(\alpha'_j) = \sum_{i=1}^{m} b_{ij}\beta'_i. \tag{5.30}$$

Also, from the definition of change of basis matrix, we have

$$\alpha'_j = \sum_{r=1}^{n} p_{rj}\alpha_r \quad \text{and} \quad \beta'_i = \sum_{k=1}^{m} q_{ki}\beta_k. \tag{5.31}$$

Substituting (5.31) into (5.30), using (5.29) and the linearity of T, we have

$$\begin{aligned}
T\left(\sum_{r=1}^{n} p_{rj}\alpha_r\right) &= \sum_{i=1}^{m} b_{ij} \sum_{k=1}^{m} q_{ki}\beta_k, \\
\sum_{r=1}^{n} p_{rj}T(\alpha_r) &= \sum_{i=1}^{m} b_{ij} \sum_{k=1}^{m} q_{ki}\beta_k, \\
\sum_{r=1}^{n} p_{rj}\left(\sum_{i=1}^{m} a_{ir}\beta_i\right) &= \sum_{i=1}^{m} b_{ij} \sum_{k=1}^{m} q_{ki}\beta_k, \\
\sum_{i=1}^{m}\left(\sum_{r=1}^{n} p_{rj}a_{ir}\right)\beta_i &= \sum_{k=1}^{m}\left(\sum_{i=1}^{m} b_{ij}q_{ki}\right)\beta_k.
\end{aligned} \tag{5.32}$$

Since the summation indices are dummy variables, from the last line of (5.32), we have

$$\sum_{r=1}^{n} p_{rj}a_{kr} = \sum_{i=1}^{m} b_{ij}q_{ki}. \tag{5.33}$$

Now, we notice that the left-hand side of (5.33) is the (k,j)th entry of AP and the right-hand side is the (k,j)th entry of QB, thus $QB = AP$. Using that Q has an inverse, the later equation is equivalent to $B = Q^{-1}AP$, as claimed. $\qquad\square$

Corollary 5.4.1. *Let* $T : V \to V$ *be a linear transformation. If* $\alpha_i = \beta_i$ *and* $\alpha'_i = \beta'_i$ *for every* $i \in \{1, 2, \ldots, n\}$, *then the matrix associated to* T *with respect to the new basis is* $P^{-1}AP$, *where* P *is the change of basis matrix.*

Proof. The proof follows directly from Theorem 5.4.2 by substituting Q with P in the equation $B = Q^{-1}AP$. $\qquad\qquad\square$

Definition 5.4.4. Let A and B be $n \times n$ matrices. It is said that B is similar to A if there is a nonsingular matrix P so that $B = P^{-1}AP$.

Exercise 5.4.4. Prove the following:
1. *(reflexivity)* Every matrix is similar to itself.
2. *(symmetry)* If B is similar to A, then A is similar to B.
3. *(transitivity)* If C is similar to B and B is similar a A, then C is similar to A.

These properties imply that the vector space of the $n \times n$ real matrices is partitioned into disjoint subsets. The elements of a given subset represent a linear transformation $T : V \to V$ (V has dimension n), hence two matrices A and B are similar if and only if they represent T in some bases of V.

Example 5.4.4. Let $T : \mathbb{R}^3 \to \mathbb{R}^3$ be given by $T(x, y, z) = (x + y - z, 2x - y + 3z, x - z)$. Find the associated matrix of T with respect to $\{(1, 2, 0), (1, -1, 0), (1, 1, 1)\}$.

Discussion. First of all, we need to find the matrix associated to T with respect to the canonical basis, which is obtained by evaluating T at the canonical vectors.

We have $T(1, 0, 0) = (1, 2, 1)$, $T(0, 1, 0) = (1, -1, 0)$, and $T(0, 0, 1) = (-1, 3, -1)$, then the associated matrix to T with respect to the canonical basis is

$$A = \begin{bmatrix} 1 & 1 & -1 \\ 2 & -1 & 3 \\ 1 & 0 & -1 \end{bmatrix}.$$

The change of basis matrix is $P = \begin{bmatrix} 1 & 1 & 1 \\ 2 & -1 & 1 \\ 0 & 0 & 1 \end{bmatrix}$, and, applying the row operations method to find the inverse of P, one has

$$\begin{bmatrix} 1 & 1 & 1 & 1 & 0 & 0 \\ 2 & -1 & 1 & 0 & 1 & 0 \\ 0 & 0 & 1 & 0 & 0 & 1 \end{bmatrix} \sim \begin{bmatrix} 1 & 1 & 1 & 1 & 0 & 0 \\ 0 & -3 & -1 & -2 & 1 & 0 \\ 0 & 0 & 1 & 0 & 0 & 1 \end{bmatrix} \sim \begin{bmatrix} 1 & 1 & 0 & 1 & 0 & -1 \\ 0 & -3 & 0 & -2 & 1 & 1 \\ 0 & 0 & 1 & 0 & 0 & 1 \end{bmatrix}$$

$$\sim \begin{bmatrix} 1 & 1 & 0 & 1 & 0 & -1 \\ 0 & 1 & 0 & 2/3 & -1/3 & -1/3 \\ 0 & 0 & 1 & 0 & 0 & 1 \end{bmatrix} \sim \begin{bmatrix} 1 & 0 & 0 & 1/3 & 1/3 & -2/3 \\ 0 & 1 & 0 & 2/3 & -1/3 & -1/3 \\ 0 & 0 & 1 & 0 & 0 & 1 \end{bmatrix}.$$

From above we have $P^{-1} = \frac{1}{3} \begin{bmatrix} 1 & 1 & -2 \\ 2 & -1 & -1 \\ 0 & 0 & 3 \end{bmatrix}$. Applying Theorem 5.4.2, we conclude that the associated matrix of T with respect to the basis $\{(1, 2, 0), (1, -1, 0), (1, 1, 1)\}$ is

$$B = \frac{1}{3} \begin{bmatrix} 1 & 1 & -2 \\ 2 & -1 & -1 \\ 0 & 0 & 3 \end{bmatrix} \begin{bmatrix} 1 & 1 & -1 \\ 2 & -1 & 3 \\ 1 & 0 & -1 \end{bmatrix} \begin{bmatrix} 1 & 1 & 1 \\ 2 & -1 & 1 \\ 0 & 0 & 1 \end{bmatrix} = \frac{1}{3} \begin{bmatrix} 1 & 1 & 5 \\ 5 & -4 & -2 \\ 3 & 3 & 0 \end{bmatrix}.$$

5.4.2 The vector space of linear transformations

In this subsection we present a discussion about the vector space of linear transformations and construct an explicit basis for it.

Let V and W be vector spaces of dimension n and m, respectively. The set of linear transformations from V to W, denoted by $\mathcal{L}(V; W)$, is a vector space with the usual sum of functions and the usual product of a function and a scalar. Explicitly,

- (*Sum*) Given $T, T_1 \in \mathcal{L}(V; W)$, define $(T + T_1)(a) := T(a) + T_1(a)$.
- (*Product with a scalar*) Given a scalar λ and $T \in \mathcal{L}(V; W)$, define $(\lambda T)(a) := \lambda T(a)$.

With the sum and product with a scalar, $\mathcal{L}(V; W)$ has the structure of a vector space. The proof is left to the reader.

Recall that given $\{a_1, a_2, \ldots, a_n\}$ and $\{\beta_1, \beta_2, \ldots, \beta_m\}$, bases of V and W, respectively, and $T \in \mathcal{L}(V; W)$, there is a matrix associated to T with respect to the given bases, hence when considering $T, T_1 \in \mathcal{L}(V; W)$, there are matrices A and B associated to T and T_1. What is the matrix associated to $T + T_1$? From the definition of the associated matrix, one has that the jth column of A is obtained from $T(a_j) = a_{1j}\beta_1 + a_{2j}\beta_2 + \cdots + a_{mj}\beta_m$ by considering the scalars $a_{1j}, \ldots, a_{mj}$ as the jth column of A. In a similar way, the jth column for B is obtained from $T_1(a_j) = b_{1j}\beta_1 + b_{2j}\beta_2 + \cdots + b_{mj}\beta_m$.

Now, using the definition of the sum of two functions, we have

$$\begin{aligned}
(T + T_1)(a_j) &= T(a_j) + T_1(a_j) \\
&= (a_{1j}\beta_1 + a_{2j}\beta_2 + \cdots + a_{mj}\beta_m) + (b_{1j}\beta_1 + b_{2j}\beta_2 + \cdots + b_{mj}\beta_m) \\
&= (a_{1j} + b_{1j})\beta_1 + (a_{2j} + b_{2j})\beta_2 + \cdots + (a_{mj} + b_{mj})\beta_m.
\end{aligned} \tag{5.34}$$

From equation (5.34), one has that the matrix associated to $T + T_1$ is $A + B$. A similar calculation as before shows that λT has associated matrix λA. From this small discussion, we have that there is a function $\varphi : \mathcal{L}(V; W) \to \mathcal{M}_{m \times n}(\mathbb{R})$ given by the rule $\varphi(T) := A$.

The properties of φ are stated in

Theorem 5.4.3. *Let V and W be vector spaces of dimension n and m, respectively. Then φ is an isomorphism, consequently, the dimension of $\mathcal{L}(V; W)$ is mn.*

Proof. The reader is invited to provide the details of the first part of the proof, however, the second part is proven below since we will have the opportunity to exhibit a basis for $\mathcal{L}(V; W)$. Let $\{a_1, a_2, \ldots, a_n\}$ and $\{\beta_1, \beta_2, \ldots, \beta_m\}$ be bases of V and W, respectively. For each $i \in \{1, 2, \ldots, m\}$ and $j \in \{1, 2, \ldots, n\}$, we will define linear transformations $T_{ij} : V \to W$ which form a basis of $\mathcal{L}(V; W)$. From Theorem 5.1.2, we know that a linear transformation is determined by its action on a basis, thus it is enough to define T_{ij} as follows:

$$T_{ij}(a_k) = \begin{cases} \beta_i & \text{if } j = k, \\ 0 & \text{otherwise.} \end{cases}$$

We will prove that $B = \{T_{ij} \mid 1 \le i \le m, 1 \le j \le n\}$ is a basis. First, we will prove that B is linearly independent. To achieve this, consider the equation $\sum_{T_{ij} \in B} a_{ij} T_{ij} = 0$. Given that this equation defines the zero function, $0 = \sum_{T_{ij} \in B} a_{ij} T_{ij}(a_k) = \sum_{i=1}^{m} a_{ik} \beta_i$, for every $k \in \{1, \ldots, n\}$. On the other hand, this is a linear combination of the basis of W, so the scalars a_{ik} must be zero for every $i \in \{1, 2, \ldots, m\}$, hence $a_{ik} = 0$ for every $i \in \{1, 2, \ldots, m\}$ and $k \in \{1, 2, \ldots, n\}$, proving what was claimed.

In order to show that B spans $\mathcal{L}(V; W)$, we consider $T \in \mathcal{L}(V; W)$ and prove that there are scalars b_{ij} so that $\sum_{T_{ij} \in B} b_{ij} T_{ij} = T$. We know T, hence we know $T(a_k)$, for every $k \in \{1, 2, \ldots, n\}$, that is, we know $T(a_k) = \sum_{i=1}^{m} a_{ik} \beta_i$, for every $k \in \{1, 2, \ldots, n\}$. Define $b_{ij} := a_{ij}$ and let $S = \sum_{T_{ij} \in B} b_{ij} T_{ij}$. According to the above calculations, $T(a_k) = \sum_{T_{ij} \in B} b_{ij} T_{ij}(a_k) = S(a_k)$, proving what is needed. The set B has nm elements, hence the dimension of $\mathcal{L}(V; W)$ is nm. □

5.5 Exercises

1. In the space of polynomials of degree at most three, find a basis that includes the elements x and $x + x^2$.
2. Find a basis for each of the following spaces:
   (a) $\mathcal{L}(\mathbb{R}^4; \mathcal{M}_{2 \times 2}(\mathbb{R}))$;
   (b) $\mathcal{L}(P_2; \mathcal{M}_{2 \times 3}(\mathbb{R}))$.
3. Let $T : \mathbb{R}^3 \to \mathcal{M}_{3 \times 3}(\mathbb{R})$ be given by $T(a_1, a_2, a_3) = \begin{bmatrix} 0 & -a_3 & a_2 \\ a_3 & 0 & -a_1 \\ -a_2 & a_1 & 0 \end{bmatrix}$. Prove:
   (a) Transformation T is injective and linear, whose image is the subspace of anti-symmetric matrices.
   (b) If $(b_1, b_2, b_3) \in \mathbb{R}^3$, then the cross product of (a_1, a_2, a_3) and (b_1, b_2, b_3) is given by $(a_1, a_2, a_3) \times (b_1, b_2, b_3) = \begin{bmatrix} 0 & -a_3 & a_2 \\ a_3 & 0 & -a_1 \\ -a_2 & a_1 & 0 \end{bmatrix} \begin{bmatrix} b_1 \\ b_2 \\ b_3 \end{bmatrix}$.
   (c) Use (a) and (b) to prove that the cross product, as a function from $\mathbb{R}^3 \times \mathbb{R}^3 \to \mathbb{R}^3$, is linear in each entry.
4. If W_1 and W_2 are subspaces of dimension three in $\mathbb{R}^4$, what are the possible dimensions of $W_1 \cap W_2$ and $W_1 + W_2$?
5. Prove that the composition of linear transformations is a linear transformation.
6. Assume that the line L passes through the origin and forms an angle θ with the horizontal axis. Prove that the rotation determined by L is given by $T_\theta(x, y) = (\cos(2\theta)x + \sin(2\theta)y, \sin(2\theta)x - \cos(2\theta)y)$. Represent T_θ with a matrix.
7. Let $T : \mathbb{R}^2 \to \mathbb{R}^2$ be given by $T(x, y) = \frac{1}{5}(4y - 3x, 4x + 3y)$. Prove that T^2 is the identity.

8. Let $T : \mathbb{R}^2 \to \mathbb{R}^2$ be given by $T(x,y) = (ay-bx, ax+by)$. Find necessary and sufficient conditions on a and b in order that $T^2 = I$, the identity. Under these conditions, is T a reflection? What is the geometric meaning of $T^2 = I$?

9. If $T : \mathbb{R}^2 \to \mathbb{R}^2$ is a linear transformation given by $T(1,0) = (a,b)$ and $T(0,1) = (c,d)$. Find the algebraic representation of $T(x,y)$ for every $(x,y) \in \mathbb{R}^2$.

10. Let $T : \mathbb{R}^3 \to \mathbb{R}^4$ be given by $T(x,y,z) = (x-y, y-z, z-x, x)$. Determine N_T and R_T.

11. Find an injective linear transformation from $\mathbb{R}^3$ to the space of polynomials of degree at most three.

12. Let V and W be two vector spaces and let $T : V \to W$ be a function. Prove that T is linear if and only if $T(x\alpha + y\beta) = xT(\alpha) + yT(\beta)$, for every $x,y \in \mathbb{R}$ and $\alpha, \beta \in V$.

13. Let $T : \mathbb{R}^n \to \mathbb{R}$ be a function. Prove that T is linear if and only if there are $a_1, a_2, \ldots, a_n \in \mathbb{R}$ so that $T(x_1, \ldots, x_n) = a_1 x_1 + a_2 x_2 + \cdots + a_n x_n$. Additionally, if $a_i \neq 0$ for some $i \in \{1, 2, \ldots, n\}$, prove that the set $W = \{(x_1 x_2, \ldots, x_n) \in \mathbb{R}^n \mid T(x_1, x_2, \ldots, x_n) = 0\}$ is a subspace of dimension $n - 1$.

14. Let $T : \mathbb{R}^n \to \mathbb{R}^m$ be a function with coordinate functions $T_1, T_2, \ldots, T_m$. Prove that T is linear if and only if each T_i is linear, $i \in \{1, 2, \ldots, m\}$.

15. Let $T : \mathbb{R} \to \mathbb{R}$ be a function so that $T(x+y) = T(x) + T(y)$ for every $x,y \in \mathbb{R}$. Prove that the following statements are equivalent:
    (a) The function T is a linear transformation.
    (b) The function T is continuous in $\mathbb{R}$.
    (c) The function T is continuous at $x_0 \in \mathbb{R}$, for some fixed x_0.

16. Assume that V is a vector space with a basis $\{a_1, a_2, \ldots, a_n\}$ and let $T : V \to V$ be a linear transformation whose action at the basic elements is $T(a_n) = 0$, and $T(a_i) = a_{i+1}$ for each $i \in \{1, 2, \ldots, n-1\}$.
    (a) Find the matrix A, associated to T with respect to the given basis.
    (b) Prove that $T^n = 0$, where T^n means T composed with itself n times.
    (c) From above conclude that $A^n = 0$.

17. Determine which of the following statements are true. Justify your answers.
    (a) Let V be a vector space, and let $S = \{a_1, a_2, a_3\} \subseteq V$. If any two elements of S are linearly independent, then S is linearly independent.
    (b) Any $n \times n$ matrix A can represent a linear transformation $T : \mathbb{R}^n \to \mathbb{R}^n$ with respect to a given basis.
    (c) If a linear transformation $T : V \to V$ is onto and V has finite dimension, then T is bijective.
    (d) If $W \neq \mathbb{R}^n$ is a subspace, then $\dim(W) < n$.
    (e) A linear transformation $T : \mathbb{R}^3 \to \mathbb{R}^3$ can transform a line onto a plane.
    (f) If the image of a linear transformation $T : \mathbb{R}^n \to \mathbb{R}^m$ is $\mathbb{R}^m$, then $m > n$.

18. Prove that a linear transformation T is nonsingular if and only if T transforms linearly independent sets into linearly independent sets.

19. Let V be a vector space of finite dimension, and let $T : V \to V$ be a linear transformation. Prove that T is nonsingular if and only if T is bijective, if and only if T is surjective.

20. Let V be a vector space and let $W \subseteq V$ be a subspace. Is W the kernel of a linear transformation $T : V \to V$? Is W the image of a linear transformation?

21. Let V be the vector space of real polynomials, and consider the linear transformation $T : V \to V$ defined by $T(f(x)) := a_0 x + \frac{a_1 x^2}{2} + \cdots + \frac{a_k x^{k+1}}{k+1}$, where $f(x) = a_0 + a_1 x + \cdots + a_k x^k$. Let $D : V \to V$ be the derivative operator, that is, $D(f(x)) = f'(x)$. Prove that $D \circ T = \mathrm{id}$, but neither D nor T is an isomorphism. Which of them is injective or surjective?

22. Let $T \in \mathcal{L}(V; V)$. Prove that $\dim (R_T)$ is equal to the number of nonzero rows of the row reduced form of any matrix associated to T.

23. Let $T \in \mathcal{L}(V; V)$ be an operator on V. Prove that $T^2 = 0$ if and only if $R_T \subseteq N_T$.

24. Find an example of a linear transformation $T : V \to V$ such that $R_T \cap N_T \neq \{0\}$.

25. Let $T \in \mathcal{L}(V, V)$ be an operator. Prove that there is a nonzero operator $S \in \mathcal{L}(V; V)$ so that $T \circ S = 0$ if and only if $N_T \neq \{0\}$.

26. Let V be a vector space of finite dimension, $S, T \in \mathcal{L}(V; V)$ so that $S \circ T = \mathrm{id}$. Prove that $T \circ S = \mathrm{id}$. (Exercise 21 shows that the conclusion might be false if V has infinite dimension).

27. Let V be a vector space of dimension n. Denote by V^* the vector space $\mathcal{L}(V; \mathbb{R})$. We know that this space has dimension n. Prove this again by the following argument. If $\{a_1, a_2, \ldots, a_n\}$ is a basis of V, then there exists a basis $\{f_1, f_2 \ldots, f_n\}$ of V^* so that $f_i(a_j) = 1$ if $i = j$, and $f_i(a_j) = 0$ otherwise. The basis $\{f_1, f_2, \ldots, f_n\}$ is called the dual basis of $\{a_1, a_2, \ldots, a_n\}$ and V^* is called the dual of V.

28. Let $A = (a_{ij})$ be a real $n \times n$ matrix which satisfies:
    (i) $a_{ij} < 0$ for $i \neq j$;
    (ii) $a_{i1} + a_{i2} + \cdots + a_{in} > 0$ for every $i = 1, \ldots, n$.
    Prove that A has an inverse.
    (*Hint.*) If A has no inverse, then the system $AX = 0$ has a nontrivial solution. Let
    $$X = \begin{bmatrix} x_1 \\ x_2 \\ \vdots \\ x_n \end{bmatrix}$$
    be such a solution. If x_k is a nonzero component of X which satisfies $|x_k| \geq x_i$ for every $i \in \{1, 2, \ldots, n\}$, then, multiplying by -1, we can assume that $x_k > 0$. The assumption on the entries of A implies $a_{kj} x_j \geq a_{kj} x_k$ for every $j \in \{1, 2, \ldots, n\} \setminus \{k\}$. Use this, (ii), and $a_{k1} x_1 + \cdots + a_{kk} x_k + \cdots + a_{kn} x_n = 0$ to arrive at a contradiction.

29. In this exercise we present a sketch of the proof that the row rank and column rank of a matrix are equal. The main ideas in the proof are borrowed from [17]. Let n be an integer ≥ 3 and for $k \in \{1, 2, \ldots, n\}$, define
    $$W_k = \left\{ (x_1, x_2, \ldots, x_n) \in \mathbb{R}^n \mid x_k = \sum_{j \neq k} c_j x_j, \ c_j \in \mathbb{R} \right\}.$$

    (a) Prove that W_k is a subspace of dimension $n - 1$.

(b) Define $T_k : \mathbb{R}^n \to \mathbb{R}^{n-1}$ by $T(x_1, x_2, \ldots, x_k, \ldots, x_n) = (x_1, x_2, \ldots, x_{k-1}, x_{k+1}, \ldots, x_n)$. Prove that T_k is a linear transformation and its kernel is generated by the canonical element e_k.

(c) Let A be an $m \times n$ matrix and denote the rows of A by $R_1, R_2, \ldots, R_m$ as elements of $\mathbb{R}^n$. Similarly, denote the columns of A by $C_1, C_2, \ldots, C_n$ as elements of $\mathbb{R}^m$. Assume that $R_k = \sum_{j \neq k} c_j R_j$ and let A' be the matrix obtained from A by deleting row R_k. It is clear that A and A' have the same row rank. Prove that A and A' have the same column rank. To achieve this, notice that $T_k(C_j) = C'_j$ is the jth column of A'; the columns of A are in W_k and T_k restricted to W_k is injective.

(d) Applying the previous argument to A^t, one sees that if the columns of A are linearly dependent, by dropping the column that is represented in terms of the others, the new matrix has the same row rank and column rank as A. Continuing this process, we end up with a matrix B which has row and column ranks equal to those of A. The big difference is that the columns and rows of B are linearly independent. We can assume that this new matrix B is of size $p \times q$. Then the linearly independent p rows are elements of $\mathbb{R}^q$, hence $p \leq q$. Similarly, the q columns of B are linearly independent elements of $\mathbb{R}^p$, thus $q \leq p$, therefore $p = q$, finishing the proof that the row and column ranks of A are equal.

6 Determinants

The theory of determinants is an important theoretical tool when studying systems of linear equations. Actually, it has several applications in geometry, physics, and other scientific disciplines. One instance of this will be seen in Section 6.3. The approach to discuss determinants that we are taking to explore basic properties is motivated by considering the area of a parallelogram or the volume of a parallelepiped. More precisely, given two linearly independent vectors in $\mathbb{R}^2$, a parallelogram is determined by those vectors, see Figure 6.1, and its area can be computed using the length and angle between its sides. The same idea can be extended to $\mathbb{R}^3$, when three linearly independent vectors are given, in this case a parallelepiped is determined and its volume can be calculated by using the length of the vectors.

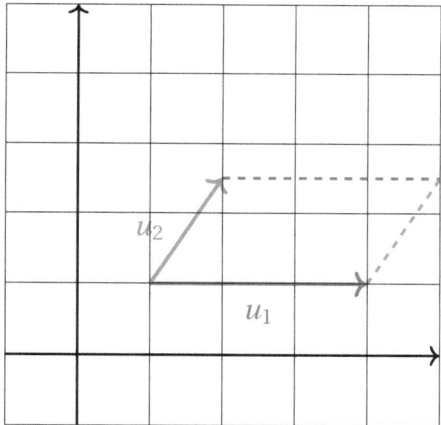

Figure 6.1: Parallelogram determined by vectors u_1 and u_2.

6.1 Determinants and volumes

In Chapter 3, equation (3.19), we found a formula to compute the area of a parallelogram whose sides are determined by vectors u_1 and u_2. Now, we want to examine properties of the function that assigns to each pair of vectors, the area of the parallelogram determined by them. Let us denote that function by $A(u_1, u_2)$ and observe some properties of A. An intuitive argument, based on a graphical representation of the parallelogram, shows that A satisfies

1. $A(e_1, e_2) = 1$ (area of the unit square).
2. If λ is a positive real number, then $A(\lambda u_1, u_2) = A(u_1, \lambda u_2) = \lambda A(u_1, u_2)$.
3. $A(u_1 + u_2, u_2) = A(u_1, u_1 + u_2) = A(u_1, u_2)$.
4. $A(u_1, u_2) \neq 0$ if and only if u_1 and u_2 are linearly independent.

https://doi.org/10.1515/9783111135915-006

Property 1 is clear, since the canonical vectors determine the unit square. Property 2 is obtained by noticing that when multiplying one of the vectors by λ, the resulting parallelogram has the same height and the base is λ times the original base.

Exercise 6.1.1. Justify properties 3 and 4.

According to Remark 3.3.3, p. 90, the area of a parallelogram determined by vectors $u = (a, b)$ and $v = (a_1, b_1)$ is given by

$$\|u \times v\| = \left\| \begin{bmatrix} i & j & k \\ a & b & 0 \\ a_1 & b_1 & 0 \end{bmatrix} \right\| := \|(ab_1 - a_1 b)k\| = |ab_1 - a_1 b|. \tag{6.1}$$

Using equation (6.1), we have a formula that expresses the area of a parallelogram in terms of the coordinates of the vectors, that is, with the notation for the function A, one has

$$A(u, v) = |ab_1 - a_1 b|.$$

The expression $ab_1 - a_1 b$ will be called the *determinant* of the vectors u and v, or the determinant of the matrix $B = \begin{bmatrix} a & b \\ a_1 & b_1 \end{bmatrix}$ obtained by placing the coordinates of the vectors as rows.

Exercise 6.1.2. Compute the area of the parallelogram determined by the pair of vectors:
1. $u = (2, 3), v = (4, 5)$;
2. $u = (0, -2), v = (2, \pi)$;
3. $u = (-1, 3), v = (0, \sqrt{2})$;
4. $u = (2, 3), v = (2, 6)$.

Starting from the discussion above, we will define a function that assigns a real number to every collection of n elements of $\mathbb{R}^n$. Before defining the determinant, we want to point out that the function A, for the case of elements of $\mathbb{R}^3$, represents the volume of a parallelepiped determined by three linearly independent vectors.

Definition 6.1.1. A determinant is a function $D : \underbrace{\mathbb{R}^n \times \cdots \times \mathbb{R}^n}_{n \text{ times}} \to \mathbb{R}$ that satisfies:
1. For every $i \in \{1, \ldots, n\}$ and $j \in \{1, 2, \ldots, n\} \setminus \{i\}$, $D(u_1, \ldots, u_{i-1}, u_i + u_j, u_{i+1}, \ldots, u_n) = D(u_1, \ldots, u_{i-1}, u_i, u_{i+1}, \ldots, u_n)$.
2. For every $\lambda \in \mathbb{R}$ and $i \in \{1, 2, \ldots, n\}$, $D(u_1, \ldots, u_{i-1}, \lambda u_i, u_{i+1}, \ldots, u_n) = \lambda D(u_1, \ldots, u_{i-1}, u_i, u_{i+1}, \ldots, u_n)$.
3. If $e_1, \ldots, e_n$ are the canonical vectors in $\mathbb{R}^n$, then $D(e_1, \ldots, e_n) = 1$.

Remark 6.1.1. Notice that as in the case of $\mathbb{R}^2$, given vectors $u_1, u_2, \ldots, u_n \in \mathbb{R}^n$ with coordinates $u_i = (a_{i1}, a_{i2}, \ldots, a_{in})$, we define the matrix

$$A = \begin{bmatrix} a_{11} & a_{12} & \cdots & a_{1n} \\ a_{21} & a_{22} & \cdots & a_{2n} \\ \vdots & \vdots & \ddots & \vdots \\ a_{n1} & a_{n2} & \cdots & a_{nn} \end{bmatrix},$$

and, abusing the notation and terminology, we say that the function D expresses the determinant of A.

Remark 6.1.2. If $u_i = 0$ for some $i \in \{1, 2, \ldots, n\}$, then from Definition 6.1.1(2) one has $D(u_1, \ldots, u_n) = 0$.

The following theorem establishes the main basic properties of the determinant function which will allow us to compute the determinant of a matrix.

Theorem 6.1.1. *If D is as in Definition* 6.1.1, *then:*

1. *for every $i \in \{1, 2, \ldots, n\}$ and $j \in \{1, 2, \ldots, n\} \setminus \{i\}$, $D(u_1, \ldots, u_{i-1}, u_i, u_{i+1}, \ldots, u_j, \ldots, u_n) = -D(u_1, \ldots, u_{i-1}, u_j, u_{i+1}, \ldots, u_i, \ldots, u_n)$, that is, D changes sign when interchanging any two elements,*

2. *for every $i \in \{1, 2, \ldots, n\}$, $D(u_1, \ldots, u_i, \ldots, u_n) = D(u_1, \ldots, u_i + \sum_{j \neq i} \lambda_j u_j, u_{i+1}, \ldots, u_n)$, where $\lambda_j \in \mathbb{R}$ for every $j \in \{1, 2, \ldots, n\} \setminus \{i\}$,*

3. *D is multilinear, that is:*
   (a) *for every $i \in \{1, 2, \ldots, n\}$ and $\lambda \in \mathbb{R}$, $D(u_1, \ldots, \lambda u_i, \ldots, u_n) = \lambda D(u_1, \ldots, u_i, \ldots, u_n)$,*
   (b) *for every $i \in \{1, 2, \ldots, n\}$ and any pair of vectors $u, w \in \mathbb{R}^n$, $D(u_1, \ldots, \underset{i}{u + w}, \ldots, u_n)$*
   $$= D(u_1, \ldots, \underset{i}{u}, \ldots, u_n) + D(u_1, \ldots, \underset{i}{w}, \ldots, u_n),$$

4. *$D(u_1, \ldots, u_n) = 0$ if and only if $\{u_1, \ldots, u_n\}$ is a linearly dependent set,*

5. *if for every $i \in \{1, 2, \ldots, n\}$, $u_i = (0, \ldots, 0, a_{ii}, \ldots, a_{in})$ or $u_i = (a_{i1}, \ldots, a_{ii}, 0, \ldots, 0)$, then $D(u_1, \ldots, u_i, \ldots, u_n) = a_{11} a_{22} \cdots a_{nn}$.*

Proof. (1) By applying successively properties (1) and (2) of Definition 6.1.1, we have the following equations (identify when each is applied):

$$\begin{aligned} D(u_1, \ldots, u_i, \ldots, u_j, \ldots, u_n) &= -D(u_1, \ldots, u_{i-1}, -u_i, u_{i+1}, \ldots, u_j, \ldots, u_n) \\ &= -D(u_1, \ldots, u_{i-1}, -u_i, u_{i+1}, \ldots, u_j - u_i, \ldots, u_n) \\ &= D(u_1, \ldots, u_{i-1}, -u_i, u_{i+1}, \ldots, u_i - u_j, \ldots, u_n) \\ &= D(u_1, \ldots, u_{i-1}, -u_j, u_{i+1}, \ldots, u_i - u_j, \ldots, u_n) \\ &= -D(u_1, \ldots, u_{i-1}, u_j, u_{i+1}, \ldots, u_i - u_j, \ldots, u_n) \\ &= -D(u_1, \ldots, u_{i-1}, u_j, u_{i+1}, \ldots, u_i, \ldots, u_n). \end{aligned}$$

(2) Arguing by induction, it is enough to prove

$$D(u_1, \ldots, u_{i-1}, u_i + \lambda u_j, u_{i+1}, \ldots, u_j, \ldots, u_n) = D(u_1, \ldots, u_{i-1}, u_i, u_{i+1}, \ldots, u_j, \ldots, u_n),$$

for every λ and $j \neq i$. If $\lambda = 0$, the equality is obvious. We may assume that $\lambda \neq 0$, then

$$D(u_1, \ldots, u_i, \ldots, u_j, \ldots, u_n) = \frac{1}{\lambda} D(u_1, \ldots, u_i, \ldots, \lambda u_j, \ldots, u_n)$$
$$= \frac{1}{\lambda} D(u_1, \ldots, u_i + \lambda u_j, \ldots, \lambda u_j, \ldots, u_n)$$
$$= D(u_1, \ldots, u_i + \lambda u_j, \ldots, u_j, \ldots, u_n).$$

(3) Part (a) is property 2, in Definition 6.1.1 of D. To prove (b), it is sufficient to show that

$$D(u_1 + v_1, u_2, \ldots, u_i, \ldots, u_n) = D(u_1, u_2, \ldots, u_i, \ldots, u_n) + D(v_1, u_2, \ldots, u_i, \ldots, u_n)$$

holds.

The first case. If $\{u_2, \ldots, u_n\}$ is a linearly dependent set, then we can assume that $u_2 = \sum_{i=3}^n \lambda_i u_i$, thus from property 2 already proved, we have

$$D(u_1 + v_1, u_2, \ldots, u_i, \ldots, u_n) = D\left(u_1 + v_1, u_2 - \sum_{i=3}^n \lambda_i u_i, u_3, \ldots, u_i, \ldots, u_n \right)$$
$$= D(u_1 + v_1, 0, u_3, \ldots, u_i, \ldots, u_n)$$
$$= 0.$$

In the same way, one has that each term, $D(u_1, u_2, \ldots, u_i, \ldots, u_n)$ and $D(v_1, u_2, \ldots, u_i, \ldots, u_n)$, is zero, obtaining equality in this case.

The second case. The vectors $\{u_2, \ldots, u_n\}$ are linearly independent, then there exists $w \in \mathbb{R}^n$ so that $\{w, u_2, \ldots, u_n\}$ is a basis of $\mathbb{R}^n$.

From this, we have $u_1 = aw + \sum_{i=2}^n \lambda_i u_i$ and $v_1 = bw + \sum_{i=2}^n \mu_i u_i$, for some scalars a and b. Now

$$D(u_1, u_2, \ldots, u_i, \ldots, u_n) = D\left(aw + \sum_{i=2}^n \lambda_i u_i, u_2, \ldots, u_i, \ldots, u_n \right)$$
$$= D(aw, u_2, \ldots, u_i, \ldots, u_n)$$
$$= aD(w, u_2, \ldots, u_n).$$

An analogous calculation shows that $D(v_1, u_2, \ldots, u_n) = bD(w, u_2, \ldots, u_n)$.

On the other hand, $u_1 + v_1 = (a + b)w + \sum_{i=2}^n (\lambda_i + \mu_i)u_i$. Proceeding as before, one obtains

$$D(u_1 + v_1, u_2, \ldots, u_i, \ldots, u_n) = (a + b)D(w, u_2, \ldots, u_n)$$
$$= aD(w, u_2, \ldots, u_n) + bD(w, u_2, \ldots, u_n)$$
$$= D(u_1, u_2, \ldots, u_n) + D(v_1, u_2, \ldots, u_n),$$

proving what was claimed.

(4) Assume that the vectors are linearly dependent. Without loss of generality, we may assume that $u_1 = \sum_{i=2}^{n} \lambda_i u_i$, then, using part 2, already proved, and Remark 6.1.2, we have

$$D(u_1, u_2, \ldots, u_i, \ldots, u_n) = D\left(u_1 - \sum_{i=2}^{n} \lambda_i u_i, u_2, \ldots, u_i, \ldots, u_n\right)$$

$$= D(0, u_2, \ldots, u_n)$$

$$= 0.$$

To prove the converse, we argue by contradiction, that is, assume that the vectors are linearly independent and $D(u_1, u_2, \ldots, u_i, \ldots, u_n) = 0$. Since the vectors $u_1, \ldots, u_n$ are linearly independent, they are a basis of $\mathbb{R}^n$, hence the canonical vectors can be expressed as a linear combination of $u_1, \ldots, u_n$. More precisely, for each $i \in \{1, 2, \ldots, n\}$ we have $e_i = \sum_{j=1}^{n} a_{ji} u_j$.

Using this representation of the canonical vectors and condition 3 of Definition 6.1.1, one has

$$1 = D(e_1, e_2, \ldots, e_n)$$

$$= D\left(\sum_{j=1}^{n} a_{j1} u_j, \sum_{j=1}^{n} a_{j2} u_j, \ldots, \sum_{j=1}^{n} a_{jn} u_j\right). \tag{6.2}$$

When expanding the latter expression in equation (6.2), using that D is multi-linear, property 3, already proved, one has that each term in the sum is of the form $aD(u_{i_1}, u_{i_2}, \ldots, u_{i_n})$. Now we notice that the entries in $D(u_{i_1}, u_{i_2}, \ldots, u_{i_n})$ are obtained from the set $\{u_1, \ldots, u_n\}$ and at any case the assumption $D(u_1, u_2, \ldots, u_n) = 0$ implies $D(u_{i_1}, u_{i_2}, \ldots, u_{i_n}) = 0$. From this, one concludes that $1 = 0$, a contradiction which finishes the proof of (4).

(5) We will deal with the case $u_i = (0, 0, \ldots, 0, a_{ii}, \ldots, a_{in})$, the other is approached analogously.

Notice that if $a_{ii} = 0$ for some i, then the vectors are linearly dependent (verify this). Applying property 4, we have $D(u_1, u_2, \ldots, u_i, \ldots, u_n) = 0 = a_{11} \cdots a_{ii} \cdots a_{nn}$.

If for every $i \in \{1, 2, \ldots, n\}$, $a_{ii} = 1$, then applying several times property 2, already proved, we obtain $D(u_1, u_2, \ldots, u_i, \ldots, u_n) = D(e_1, e_2, \ldots, e_n) = 1$, which is what we wanted to prove.

Assume $a_{ii} \neq 0$ for every $i \in \{1, 2, \ldots, n\}$. Define $u_i' = \frac{u_i}{a_{ii}}$ for every $i \in \{1, 2, \ldots, n\}$. Notice that the vectors u_i' are such that all their entries before the ith are zero and the ith entry is one, then from what we have observed, one has $D(u_1', u_2', \ldots, u_n') = 1$. On the other hand, $D(u_1, u_2, \ldots, u_n) = D(a_{11} u_1', a_{22} u_2', \ldots, a_{nn} u_n')$. Applying property 2 from Definition 6.1.1 and the above remark, one concludes that

$$D(u_1, u_2, \ldots, u_n) = a_{11} \cdots a_{ii} \cdots a_{nn} D(u_1', u_2', \ldots, u_n') = a_{11} \cdots a_{ii} \cdots a_{nn},$$

as it was claimed. $\qquad\square$

Before we even compute a determinant using Theorem 6.1.1, we need some terminology.

Definition 6.1.2. Let $A = (a_{ij})$ be an $n \times n$ matrix. We say that A is:
1. upper triangular if $a_{ij} = 0$ for every $i > j$;
2. lower triangular if $a_{ij} = 0$ for every $i < j$;
3. triangular if it is upper or lower triangular;
4. diagonal if $a_{ij} = 0$ for every $i \neq j$.

We recall the relationship between elementary matrices and row operations on a matrix. The elementary matrices have been denoted by E_{ij}, $E_i(r)$, and $E_{ij}(r)$. The corresponding definitions are recalled below:
1. E_{ij} is obtained by swapping the ith and jth rows in the identity matrix I_n.
2. $E_i(r)$ is obtained from I_n by multiplying the ith row by $r \neq 0$.
3. $E_{ij}(r)$ is obtained from I_n by multiplying the ith row by r and adding the result to the jth row.

We will introduce some notation to compute determinants. If A is a square matrix whose rows are obtained from the coordinates of vectors $u_1, u_2, \ldots, u_n$, then we will use the notation

$$D(u_1, u_2, \ldots, u_n) = |A| := \begin{vmatrix} a_{11} & a_{12} & \cdots & a_{1n} \\ a_{21} & a_{22} & \cdots & a_{2n} \\ \vdots & \vdots & \ddots & \vdots \\ a_{n1} & a_{n2} & \cdots & a_{nn} \end{vmatrix}.$$

With this notation, the properties listed in Theorem 6.1.1 are interpreted below and will be very useful to compute determinants.

Remark 6.1.3. Matrix properties listed in Theorem 6.1.1 can be restated as follows:
1. Interchanging two rows of a matrix changes the sign of the determinant, property 1.
2. If matrix B is obtained from A by multiplying a row of A by a scalar r and adding it to another row of A, then $|A| = |B|$, property 2.
3. If matrix B is obtained from A by multiplying a row of A by a scalar r, we have $|B| = r|A|$, property 3(a).
4. The rows of A are linearly dependent if and only if $|A| = 0$, property 4.
5. The determinant of a triangular matrix is the product of its diagonal elements, property 5.

Example 6.1.1. Compute the determinant of $A = \begin{bmatrix} 5 & 2 \\ 3 & 3 \end{bmatrix}$.

Discussion. We proceed as follows:

$$\begin{vmatrix} 5 & 2 \\ 3 & 3 \end{vmatrix} = \begin{vmatrix} 2 & -1 \\ 3 & 3 \end{vmatrix} \quad \text{(Subtracting row 2 from row 1)}$$

$$= \begin{vmatrix} 2 & -1 \\ 1 & 4 \end{vmatrix} \quad \text{(Subtracting row 1 from row 2)}$$

$$= \begin{vmatrix} 0 & -9 \\ 1 & 4 \end{vmatrix} \quad \text{(Multiplying row 2 by } -2 \text{ and adding to row 1)}$$

$$= -\begin{vmatrix} 1 & 4 \\ 0 & -9 \end{vmatrix} \quad \text{(Swapping rows)}$$

$$= 9. \quad \text{(Applying the property for triangular matrices)}$$

Example 6.1.2. Compute the determinant of $A = \begin{bmatrix} 5 & 2 & 3 \\ 3 & 3 & -1 \\ 4 & 5 & 7 \end{bmatrix}$.

Discussion. This time we compute as follows:

$$\begin{vmatrix} 5 & 2 & 3 \\ 3 & 3 & -1 \\ 4 & 5 & 7 \end{vmatrix} = \begin{vmatrix} 1 & -3 & -4 \\ 3 & 3 & -1 \\ 4 & 5 & 7 \end{vmatrix} \quad \text{(Subtracting row 3 from row 1)}$$

$$= \begin{vmatrix} 1 & -3 & -4 \\ 0 & 12 & 11 \\ 0 & 17 & 23 \end{vmatrix} \quad \text{(Which row operations were applied?)}$$

$$= \begin{vmatrix} 1 & -3 & -4 \\ 0 & 12 & 11 \\ 0 & 5 & 12 \end{vmatrix} \quad \text{(Subtracting row 2 from row 3)}$$

$$= \begin{vmatrix} 1 & -3 & -4 \\ 0 & 2 & -13 \\ 0 & 5 & 12 \end{vmatrix} \quad \text{(Which row operations were applied?)}$$

$$= \begin{vmatrix} 1 & -3 & -4 \\ 0 & 2 & -13 \\ 0 & 1 & 38 \end{vmatrix} \quad \text{(Now which operation is applied?)}$$

$$= \begin{vmatrix} 1 & -3 & -4 \\ 0 & 0 & -89 \\ 0 & 1 & 38 \end{vmatrix} \quad \text{(Why?)}$$

$$= -\begin{vmatrix} 1 & -3 & -4 \\ 0 & 1 & 38 \\ 0 & 0 & -89 \end{vmatrix} \quad \text{(Swapping rows 2 and 3)}$$

$$= 89. \quad \text{(Property of triangular matrices)}$$

Exercise 6.1.3. Compute the determinant of $A = \begin{bmatrix} 1 & a & a^2 \\ 1 & b & b^2 \\ 1 & c & c^2 \end{bmatrix}$, assuming that a, b, and c are different real numbers.

Exercise 6.1.4. Assume that $A = \begin{bmatrix} a & b \\ c & d \end{bmatrix}$. Prove that $|A| = ad - bc$.

6.1.1 Properties of the determinant

In what follows, we present some of the fundamental properties of the determinant. In the proof we use the results of Theorems 1.3.2 and 2.2.1. We strongly encourage the reader to review them.

Theorem 6.1.2. *Let A, B, and E be square matrices of the same size, with E elementary. Then the following hold:*

1. $|E| \neq 0$ and $|EA| = |E||A|$.
2. *The matrix A has an inverse if and only if $|A| \neq 0$, and in this case $|A^{-1}| = \frac{1}{|A|}$.*
3. *The determinant is multiplicative, that is, $|AB| = |A||B|$.*
4. *If A^t denotes the transpose of A, then $|A| = |A^t|$.*
5. *The rows or columns of A are linearly dependent if and only if $|A| = 0$.*

Proof. 1. The first fact is obtained directly from the definition of an elementary matrix and Definition 6.1.1; the second is obtained from Remark 6.1.3 and the definition of an elementary matrix.

2. We know, by Theorem 2.2.3, p. 51, that A has an inverse if and only if it is row equivalent to the identity, if and only if A is a product of elementary matrices.

We have that the determinant of an elementary matrix is not zero. Also if E and F are elementary, then $|EF| = |E||F|$, hence $|A| \neq 0$ if and only if A has an inverse. If A has an inverse then there are elementary matrices $F_1, F_2, \ldots, F_s$ so that $F_1 F_2 \cdots F_s A = I_n$. Using property 1, already proved, one has

$$|F_1 F_2 \cdots F_s A| = |F_1||F_2| \cdots |F_s||A| = |I_n| = 1. \tag{6.3}$$

We also know that $A^{-1} = F_1 F_2 \cdots F_s$. Applying the determinant function to this equation, using again property 1 and equation (6.3), one concludes that $|A^{-1}| = \frac{1}{|A|}$.

3. To prove this part, we notice that the product of matrices has an inverse if and only if each factor does. From this and part 2, we have that if one of A and B does not have an inverse, then neither does AB, hence $|AB| = 0$. Also, if one of A and B has no inverse, then one of $|A|$ and $|B|$ is zero, hence $|A||B| = 0 = |AB|$.

If A and B have inverses then there are elementary matrices $F_1, \ldots, F_k$ and $H_1, \ldots, H_r$ so that $A = F_1 \cdots F_k$ and $B = H_1 \cdots H_r$. From these equations and part 1, already proved,

$$|AB| = |F_1 \cdots F_k H_1 \cdots H_r| = |F_1| \cdots |F_k||H_1| \cdots |H_r| = |A||B|.$$

4. This is left to the reader as an exercise, and we suggest a route to follow:
1. Define column operation on A.
2. Examine the relationship between row operations and column operations when considering A^t.
3. Observe what occurs when applying column operations on A, starting with the first column.
4. If the row operations on A that make it row equivalent to the row reduced form R are represented by $F_1, \ldots, F_k$, that is, $F_1 \cdots F_k A = R$, then $A^t F_k^t \cdots F_1^t = R^t$.
5. A triangular matrix and its transpose have the same determinant, by Theorem 6.1.1.

5. We ask the reader to apply part 4 of Theorem 6.1.1 and part 4 above. □

6.1.2 Existence and uniqueness of the determinant

Once we have discussed the properties of a determinant function, it is important to justify its existence and uniqueness; this is done in what follows.

For the next lemma, we recommend the reader to review the content of Section A.2, p. 265. From that section we recall that the collection of all permutations of the integers $\{1, 2, \ldots, n\}$ is denoted by S_n.

Lemma 6.1.1. *Let $D : \underbrace{\mathbb{R}^n \times \cdots \times \mathbb{R}^n}_{n \text{ times}} \to \mathbb{R}$ be given by*

$$D(u_1, u_2, \ldots, u_n) = \sum_{\sigma \in S_n} \text{sign}(\sigma) a_{1\,\sigma(1)} a_{2\,\sigma(2)} \cdots a_{i\,\sigma(i)} \cdots a_{n\,\sigma(n)}, \tag{6.4}$$

where $u_i = (a_{i1}, a_{i2}, \ldots, a_{in})$, for every $i \in \{1, 2, \ldots, n\}$. Then D satisfies:
1. *For every $i < j$, $D(u_1, \ldots, u_i, \ldots, u_j, \ldots, u_n) = -D(u_1, \ldots, u_j, \ldots, u_-, \ldots, u_n)$.*
2. *For every $i \in \{1, 2, \ldots, n\}$, $D(u_1, \ldots, u_i, \ldots, u_i, \ldots, u_n) = 0$.*

Proof. We recall a couple of properties of permutations. If π and σ are elements of S_n, then from Corollary A.2.1, p. 267, we have $\text{sign}(\sigma\pi) = \text{sign}(\sigma)\text{sign}(\pi)$. The sets S_n and $\{\sigma\pi \mid \sigma \in S_n\}$ are equal.

Given $u_1, u_2, \ldots, u_n \in \mathbb{R}^n$ and $i \neq j$ elements of $\{1, 2, \ldots, n\}$, for every $k \notin \{i, j\}$, define $w_k = u_k$ and $w_i = u_j$, $w_j = u_i$. Let $\pi \in S_n$ be defined by $\pi(k) = k$ if $k \notin \{i, j\}$ and $\pi(i) = j$, that is, π is a transposition, hence $\text{sign}(\pi) = -1$.

Using what we have pointed out above,

$$D(u_1, u_2, \ldots, u_n) = \sum_{\sigma\pi \in S_n} \text{sign}(\sigma\pi) a_{1\,\sigma\pi(1)} a_{2\,\sigma\pi(2)} \cdots a_{i\,\sigma\pi(i)} \cdots a_{j\,\sigma\pi(j)} \cdots a_{n\,\sigma\pi(n)}$$

$$= -\sum_{\sigma \in S_n} \text{sign}(\sigma) a_{1\,\sigma(1)} a_{2\,\sigma(2)} \cdots a_{i\,\sigma(j)} \cdots a_{j\,\sigma(i)} \cdots a_{n\,\sigma(n)}$$

$$= -D(w_1, \ldots, w_i, \ldots, w_j \ldots, w_n)$$
$$= -D(u_1, \ldots, u_j, \ldots, u_i \ldots, u_n),$$

proving the first part. To prove the second, we notice that if $u_i = u_j$, then

$$D(u_1, \ldots, u_i, \ldots, u_i, \ldots, u_n) = -D(u_1, \ldots, u_i, \ldots, u_i, \ldots, u_n),$$

thus $D(u_1, \ldots, u_i, \ldots, u_i, \ldots, u_n) = 0$. □

Theorem 6.1.3. *There exists a unique function* $D : \underbrace{\mathbb{R}^n \times \cdots \times \mathbb{R}^n}_{n \text{ times}} \to \mathbb{R}$ *which satisfies the properties of Definition 6.1.1.*

Proof. (*Existence.*) We propose as D, the function from Lemma 6.1.1 and prove that D satisfies:

1. for every $i \in \{1, \ldots, n\}$ and $j \in \{1, 2, \ldots, n\} \setminus \{i\}$,

    $$D(u_1, \ldots, u_{i-1}, u_i + u_j, u_{i+1}, \ldots, u_n) = D(u_1, \ldots, u_{i-1}, u_i, u_{i+1}, \ldots, u_n),$$

2. for every $\lambda \in \mathbb{R}$ and $i \in \{1, 2, \ldots, n\}$,

    $$D(u_1, \ldots, u_{i-1}, \lambda u_i, u_{i+1}, \ldots, u_n) = \lambda D(u_1, \ldots, u_{i-1}, u_i, u_{i+1}, \ldots, u_n),$$

3. if $e_1, \ldots, e_n$ are the canonical vectors in $\mathbb{R}^n$, then $D(e_1, \ldots, e_n) = 1$,

which are the conditions of Definition 6.1.1.

From the definition of D, one has

$$D(u_1, u_2, \ldots, u_i + u_j, \ldots, u_j, \ldots, u_n) =$$
$$= \sum_{\sigma \in S_n} \text{sign}(\sigma) a_{1\sigma(1)} a_{2\sigma(2)} \cdots (a_{i\sigma(i)} + a_{j\sigma(j)}) \cdots a_{j\sigma(j)} \cdots a_{n\sigma(n)}$$
$$= D(u_1, \ldots, u_i, \ldots, u_j, \ldots, u_n) + D(u_1, \ldots, u_j, \ldots, u_j, \ldots, u_n)$$
$$= D(u_1, \ldots, u_i, \ldots, u_j, \ldots, u_n),$$

since $D(u_1, \ldots, u_j, \ldots, u_j, \ldots, u_n) = 0$, by Lemma 6.1.1(2), proving part 1. To prove part 2, we notice that $D(u_1, \ldots, \lambda u_i, \ldots, u_j, \ldots, u_n)$ will have λ as a factor in each term of the sum, then it can be pulled out of the sum.

Part 3 follows directly by noticing that e_i has only one nonzero term which is equal to 1.

(*Uniqueness.*) Assume that D_1 and D_2 satisfy the properties 1, 2, and 3 above. Then D_1 and D_2 satisfy properties of Theorem 6.1.1, one of which states that D_1 and D_2 are multilinear, hence their values are completely determined by their values at $(e_{\sigma(1)}, e_{\sigma(2)}, \ldots, e_{\sigma(n)})$ and D_1 and D_2 coincide at $(e_{\sigma(1)}, e_{\sigma(2)}, \ldots, e_{\sigma(n)})$, thus $D_1 = D_2$. □

Remark 6.1.4. Notice that the definition of the determinant expressed in equation (6.4) has little practical use, however, for the case of 2×2 matrices, we have

$$\begin{vmatrix} a & b \\ c & d \end{vmatrix} = ad - bc.$$

6.2 Cramer's rule, minors, and cofactors

One of the most important problems in linear algebra is to solve a system of linear equations $AX = B$. For this, there are several methods, and one of them has been discussed, namely the Gauss–Jordan reduction method. This method has several advantages, among them is the one when we try to find numerical solutions of a given system of linear equations. On the other hand, when the coefficients of the system are parameters that depend on time, for instance, in the case of an economics model, the Gauss–Jordan method could not be the best to describe the processes. In this instance, if we can express the solutions in a closed form, it could be very useful to analyze them. This can be achieved with a method called *Cramer's rule*.

We will need a little bit of terminology to formulate the result.

Given a square matrix $A = (a_{ij})$ and two integers $1 \le i, j \le n$, the (i,j)th minor M_{ij} of A is defined as the determinant of the matrix obtained from A by deleting row i and column j. We also define the cofactor C_{ij} of the element a_{ij} as $C_{ij} := (-1)^{i+j} M_{ij}$.

The *classical adjoint* matrix of A is defined by

$$\text{Adj}(A) := \begin{bmatrix} C_{11} & C_{21} & \cdots & C_{n1} \\ C_{12} & C_{22} & \cdots & C_{n2} \\ \vdots & \vdots & \ddots & \vdots \\ C_{1n} & C_{2n} & \cdots & C_{nn} \end{bmatrix}. \tag{6.5}$$

For example, if $A = \begin{bmatrix} a_{11} & a_{12} \\ a_{21} & a_{22} \end{bmatrix}$, then $M_{11} = a_{22}, M_{12} = a_{21}, M_{21} = a_{12}, M_{22} = a_{11}, C_{11} = a_{22}$, $C_{12} = -M_{12} = -a_{21}, C_{21} = -M_{21} = -a_{12}$, and $C_{22} = M_{22} = a_{11}$. From this, one has that the classical adjoint of A is

$$\text{Adj}(A) = \begin{bmatrix} a_{22} & -a_{12} \\ -a_{21} & a_{11} \end{bmatrix}.$$

From above and the definition of the classical adjoint of A, together with Remark 6.1.4, we obtain

$$A \cdot \text{Adj}(A) = \begin{bmatrix} a_{11} & a_{12} \\ a_{21} & a_{22} \end{bmatrix} \begin{bmatrix} a_{22} & -a_{12} \\ -a_{21} & a_{11} \end{bmatrix} = \begin{bmatrix} |A| & 0 \\ 0 & |A| \end{bmatrix} = |A| I_2.$$

With the notation and terminology as above, we can state and prove a couple of results that are needed in the discussion. The first is a method to evaluate determinants, called "by minors and cofactors", while the second is the background to establish Cramer's rule.

We will present two different proofs of Cramer's rule. The second proof is due to D. E. Whitford and M. S. Klamkin, [28]. In Exercise 12, p. 162, another proof of Cramer's rule is sketched.

Theorem 6.2.1. *Let A be an $n \times n$ matrix, C_{ij} as defined above, then*

$$|A| = a_{i1}C_{i1} + a_{i2}C_{i2} + \cdots + a_{in}C_{in} = a_{1j}C_{1j} + a_{2j}C_{2j} + \cdots + a_{nj}C_{nj}, \tag{6.6}$$

for any $i, j \in \{1, 2, \ldots, n\}$.

Proof. Since $|A| = |A^t|$, it is enough to show that $|A| = a_{i1}C_{i1} + a_{i2}C_{i2} + \cdots + a_{in}C_{in}$ for any $i \in \{1, 2, \ldots, n\}$.

We have

$$|A| = \sum_{\sigma \in S_n} \text{sign}(\sigma) a_{1\,\sigma(1)} a_{2\,\sigma(2)} \cdots a_{i\,\sigma(i)} \cdots a_{n\,\sigma(n)}. \tag{6.7}$$

We notice that each term in equation (6.7) is formed by taking exactly one entry from the ith row $(a_{i1}, a_{i2}, \ldots, a_{in})$, thus we can write

$$|A| = a_{i1}C'_{i1} + a_{i2}C'_{i2} + \cdots + a_{in}C'_{in}, \tag{6.8}$$

hence, the proof will be completed if we prove that $a_{ij}C'_{ij} = a_{ij}C_{ij} = a_{ij}(-1)^{i+j}M_{ij}$ for every $i, j \in \{1, 2, \ldots, n\}$. We start by considering the case $i = j = n$. Then the terms containing a_{nn} are $a_{nn}C'_{nn} = a_{nn}\sum_{\sigma \in S_n} \text{sign}(\sigma) a_{1\,\sigma(1)} a_{2\,\sigma(2)} \cdots a_{i\,\sigma(i)} \cdots a_{n-1\,\sigma(n-1)}$, where the sum runs over those σ such that $\sigma(n) = n$. This collection of σ's can be considered as the elements in S_{n-1}, hence

$$a_{nn}C'_{nn} = a_{nn}\sum_{\sigma \in S_n} \text{sign}(\sigma) a_{1\,\sigma(1)} a_{2\,\sigma(2)} \cdots a_{i\,\sigma(i)} \cdots a_{n-1\,\sigma(n-1)} = a_{nn}(-1)^{n+n}|M_{nn}|.$$

For the case of other i and j, we move the ith row to the nth row and then the jth column to the nth column. By doing this, $|A|$ and C'_{ij} have changed $n - i + n - j$ times, thus from the former case, we have $a_{ij}C'_{ij} = a_{ij}(-1)^{n-i+n-j}|M_{ij}| = a_{ij}(-1)^{-i-j}|M_{ij}| = a_{ij}C_{ij}$. $\square$

Lemma 6.2.1. *Let A be an $n \times n$ matrix, and let B be the matrix obtained from A by replacing the ith row with $(b_{i1}, \ldots, b_{in})$. Then $|B| = b_{i1}C_{i1} + \cdots + b_{in}C_{in}$. Furthermore, for $i \neq j$ we have $a_{j1}C_{i1} + \cdots + a_{jn}C_{in} = 0$ and $a_{1j}C_{1i} + \cdots + a_{nj}C_{ni} = 0$.*

Proof. From Theorem 6.2.1, developing the determinant of B along the ith row, we have $|B| = b_{i1}C_{i1} + \cdots + b_{in}C_{in}$. Now, if A' is the matrix obtained from A by replacing the ith row with the jth row, then A' has two equal rows, hence its determinant is zero, due to Lemma 6.1.1(2). Applying the same reasoning to A^t and using that $|A| = |A^t|$, we obtain $a_{1j}C_{1i} + \cdots + a_{nj}C_{ni} = 0$. $\square$

Theorem 6.2.2 (Cramer's rule). *If A is a square matrix, then* $\text{Adj}(A)\cdot A = A\cdot\text{Adj}(A) = |A|I_n$. *In particular, if A has an inverse, $A^{-1} = \frac{\text{Adj}(A)}{|A|}$.*

Proof. Let $B = A\cdot\text{Adj}(A) = (b_{ij})$. By the definitions of $\text{Adj}(A)$ and the product of matrices, we have $b_{ij} = a_{i1}C_{j1} + a_{i2}C_{j2} + \cdots + a_{in}C_{jn}$.

Now, from Theorem 6.2.1 and Lemma 6.2.1, we have $b_{ij} = |A|$ if $i = j$, and $b_{ij} = 0$ otherwise, thus $B = |A|I_n$. The other part of the theorem follows readily, since the inverse is unique. □

Recall that if A is a square matrix that has an inverse and B is a column vector with entries $b_1, b_2, \ldots, b_n$, then $AX = B$ has a unique solution. In fact, multiplying equation $AX = B$ by A^{-1}, leads to $A^{-1}AX = A^{-1}B$; from this and the definition of $\text{Adj}(A)$, we have

$$X = A^{-1}B = \frac{\text{Adj}(A)}{|A|}B = \frac{1}{|A|}\text{Adj}(A)B = \frac{1}{|A|}\begin{bmatrix} C_{11} & C_{21} & \cdots & C_{n1} \\ C_{12} & C_{22} & \cdots & C_{n2} \\ \vdots & \vdots & \ddots & \vdots \\ C_{1n} & C_{2n} & \cdots & C_{nn} \end{bmatrix}B.$$

Expanding the product and writing the components of X, one obtains

$$x_i = \frac{b_1 C_{1i} + b_2 C_{2i} + \cdots + b_n C_{ni}}{|A|}, \quad \text{for every } i \in \{1, 2, \ldots, n\}. \tag{6.9}$$

According to (6.6), equation (6.9) can be expressed as

$$x_i = \frac{|A_i|}{|A|}, \tag{6.10}$$

where matrix A_i is obtained from A by substituting the ith column of A by column B.

Remark 6.2.1. Equation (6.10) is known as *Cramer's rule* to solve a system of linear equations. We have called Cramer's rule Theorem 6.2.2, since this result directly implies (6.10).

6.2.1 Cramer's rule (Whitford and Klamkin)

The proof of Cramer's rule that we are presenting here is essentially that in [28].

Theorem 6.2.3. *Let A be an $n \times n$ matrix so that $|A| \neq 0$. Let $AX = B$ be a system of equations, then $X = (x_1, x_2, \ldots, x_n)$ is a solution if and only if $x_i = \frac{|A_i|}{|A|}$, where A_i is obtained by replacing the ith column of A by B.*

Proof. Assume that $X = (x_1, x_2, \ldots, x_n)$ is a solution of $AX = B$, where the entries of A are a_{ij}. For every $i \in \{1, 2, \ldots, n\}$, one has

$$x_i \begin{vmatrix} a_{11} & a_{12} & \cdots & a_{1i} & \cdots & a_{1n} \\ a_{21} & a_{22} & \cdots & a_{2i} & \cdots & a_{2n} \\ \vdots & \vdots & \ddots & \vdots & \ddots & \vdots \\ a_{n1} & a_{n2} & \cdots & a_{ni} & \cdots & a_{nn} \end{vmatrix} = \begin{vmatrix} a_{11} & a_{12} & \cdots & a_{1i}x_i & \cdots & a_{1n} \\ a_{21} & a_{22} & \cdots & a_{2i}x_i & \cdots & a_{2n} \\ \vdots & \vdots & \ddots & \vdots & \ddots & \vdots \\ a_{n1} & a_{n2} & \cdots & a_{ni}x_i & \cdots & a_{nn} \end{vmatrix}. \tag{6.11}$$

For $j \in \{1, 2, \ldots, n\} \setminus \{i\}$, multiplying the jth column of A by x_j and adding it to the ith column of the matrix in the right-hand side of (6.11), the entries of the ith column become $b_1, b_2, \ldots, b_n$, hence

$$x_i \begin{vmatrix} a_{11} & a_{12} & \cdots & a_{1i} & \cdots & a_{1n} \\ a_{21} & a_{22} & \cdots & a_{2i} & \cdots & a_{2n} \\ \vdots & \vdots & \ddots & \vdots & \ddots & \vdots \\ a_{n1} & a_{n2} & \cdots & a_{ni} & \cdots & a_{nn} \end{vmatrix} = \begin{vmatrix} a_{11} & a_{12} & \cdots & a_{1i}x_i & \cdots & a_{1n} \\ a_{21} & a_{22} & \cdots & a_{2i}x_i & \cdots & a_{2n} \\ \vdots & \vdots & \ddots & \vdots & \ddots & \vdots \\ a_{n1} & a_{n2} & \cdots & a_{ni}x_i & \cdots & a_{nn} \end{vmatrix}$$

$$= \begin{vmatrix} a_{11} & a_{12} & \cdots & b_1 & \cdots & a_{1n} \\ a_{21} & a_{22} & \cdots & b_2 & \cdots & a_{2n} \\ \vdots & \vdots & \ddots & \vdots & \ddots & \vdots \\ a_{n1} & a_{n2} & \cdots & b_n & \cdots & a_{nn} \end{vmatrix}.$$

Since $|A| \neq 0$, from the last equation it follows that $x_i = \frac{|A_i|}{|A|}$, as claimed.

Conversely, assuming that $x_i = \frac{|A_i|}{|A|}$, we will prove that $X = (x_1, x_2, \ldots, x_n)$ is a solution of $AX = B$. This equation is equivalent to

$$a_{j1}x_1 + a_{j2}x_2 + \cdots + a_{jn}x_n = b_j, \quad j \in \{1, 2, \ldots, n\}. \tag{6.12}$$

For a given $j \in \{1, 2, \ldots, n\}$, consider the $(n+1) \times (n+1)$ matrix

$$C_j = \begin{bmatrix} a_{j1} & a_{j2} & \cdots & a_{ji} & \cdots & a_{jn} & b_j \\ a_{11} & a_{12} & \cdots & a_{1i} & \cdots & a_{1n} & b_1 \\ \vdots & \vdots & \ddots & \vdots & \ddots & \vdots & \vdots \\ a_{n1} & a_{n2} & \cdots & a_{ni} & \cdots & a_{nn} & b_n \end{bmatrix}. \tag{6.13}$$

Notice that C_j is a singular matrix, since it has two equal rows, the first and the $(j+1)$th. Hence $|C_j| = 0$. Computing the determinant of C_j by minors and cofactors, along the first row, one has $0 = |C_j| = a_{j1}|M_1| + a_{j2}|M_2| + \cdots + a_{jn}|M_n| + b_j|A|$, where M_k is the matrix obtained from C_j by deleting the first row and column k. One can verify that $|M_k| = (-1)^k |A_k|$, for every $k \in \{1, 2, \ldots, n\}$ and that the first n terms in $|C_j|$ have the same sign which is different from the sign of $b_j|A|$, hence $a_{j1}|A_1| + a_{j2}|A_2| + \cdots + a_{jn}|A_n| = b_j|A|$. From this equation, one has

$$a_{j1}\frac{|A_1|}{|A|} + \cdots + a_{jn}\frac{|A_n|}{|A|} = b_j, \tag{6.14}$$

proving the assertion. $\qquad\square$

6.3 Determinants and differential equations

In this section we will present a proof of the test that uses the wronskian to determine if a set of n differentiable functions is linearly independent. We are assuming that the set of real-valued functions, which have derivatives up to order n on an interval I, is a vector space.

Consider the differential equation

$$y' + p(x)y = 0, \tag{6.15}$$

where the function $p(x)$ is continuous on I. Let $P(x)$ be a primitive of $p(x)$ in I, that is, $P'(x) = p(x)$ for every $x \in I$. Multiplying both sides of (6.15) by $e^{P(x)}$ and applying the rule to compute the derivative of a product, one has

$$(e^{P(x)}y)' = e^{P(x)}y' + p(x)e^{P(x)}y = e^{P(x)}(y' + p(x)y) = 0, \tag{6.16}$$

concluding that $e^{P(x)}y = k$, for some constant k.

The latter equation is equivalent to

$$y = ke^{-P(x)}. \tag{6.17}$$

The function $P(x)$ could be defined by $P(x) := \int_{x_0}^{x} p(t)dt$, for some $x_0 \in I$. With this choice of $P(x)$, the constant k is obtained by evaluating y at x_0, that is, equation (6.17) implies $y(x_0) = ke^{-P(x_0)} = ke^0 = k$.

Lemma 6.3.1. *Let* $\{f_{ij}\}$ *be a set of* n^2 *differentiable functions on I. For each $x \in I$, we define*

$$G(x) := \begin{vmatrix} f_{11}(x) & \cdots & f_{1n}(x) \\ f_{21}(x) & \cdots & f_{2n}(x) \\ \vdots & \ddots & \vdots \\ f_{n1}(x) & \cdots & f_{nn}(x) \end{vmatrix}.$$

Then $G(x)$ is differentiable and

$$G'(x) := \begin{vmatrix} f_{11}'(x) & \cdots & f_{1n}'(x) \\ f_{21}(x) & \cdots & f_{2n}(x) \\ \vdots & \ddots & \vdots \\ f_{n1}(x) & \cdots & f_{nn}(x) \end{vmatrix} + \begin{vmatrix} f_{11}(x) & \cdots & f_{1n}(x) \\ f_{21}'(x) & \cdots & f_{2n}'(x) \\ \vdots & \ddots & \vdots \\ f_{n1}(x) & \cdots & f_{nn}(x) \end{vmatrix} + \cdots + \begin{vmatrix} f_{11}(x) & \cdots & f_{1n}(x) \\ f_{21}(x) & \cdots & f_{2n}(x) \\ \vdots & \ddots & \vdots \\ f_{n1}'(x) & \cdots & f_{nn}'(x) \end{vmatrix}.$$

Proof. Let us recall that the determinant of a matrix can be expressed as

$$G(x) = \sum_{\sigma \in S_n} \text{sign}(\sigma)f_{1\sigma(1)}(x)\cdots f_{n\sigma(n)}(x), \tag{6.18}$$

see (6.7), p. 154.

Computing the derivative in both sides of (6.18), and using Leibniz rule to compute the derivative of the product of n functions, if follows that

$$
G'(x) = \sum_{\sigma \in S_n} \text{sign}(\sigma)(f_{1\sigma(1)} \cdots f_{n\sigma(n)})'(x)
$$

$$
= \sum_{\sigma \in S_n} \text{sign}(\sigma)\left(\sum_{i=1}^{n} f'_{i\sigma(i)} f_{1\sigma(1)} \cdots \hat{f}_{i\sigma(i)} \cdots f_{n\sigma(n)} \right)(x)
$$

$$
= \sum_{i=1}^{n} \left(\sum_{\sigma \in S_n} \text{sign}(\sigma) f'_{i\sigma(i)} f_{1\sigma(1)} \cdots \hat{f}_{i\sigma(i)} \cdots f_{n\sigma(n)} \right)(x),
$$

where the notation $\hat{f}_{i\sigma(i)}$ means that this factor is deleted.

Notice that each sum

$$
\sum_{\sigma \in S_n} \text{sign}(\sigma) f'_{i\sigma(i)}(x) f_{1\sigma(1)}(x) \cdots \hat{f}_{i\sigma(i)}(x) \cdots f_{n\sigma(n)}(x)
$$

is the determinant of the matrix A_i which has been obtained from $G(x)$ by substituting the ith row by $(f'_{i1}, \ldots, f'_{in})$, that is, $G'(x) = \sum_{i=1}^{n} |A_i|$, as it was claimed. □

Let $g_1(x), g_2(x), \ldots, g_n(x)$ be functions n-times differentiable on I. The jth derivative of g_i will be denoted by $g_i^{(j)}$, and recall that $g_i^{(0)} = g_i$, by definition. Setting $f_{ij}(x) := g_i^{(j)}$ for $i \in \{1, 2, \ldots, n\}$ and $j \in \{0, 1, \ldots, n-1\}$, the function $G(x)$ is expressed as

$$
G(x) := \begin{vmatrix} g_1(x) & \cdots & g_n(x) \\ g_1'(x) & \cdots & g_n'(x) \\ \vdots & \ddots & \vdots \\ g_1^{(n-1)}(x) & \cdots & g_n^{(n-1)}(x) \end{vmatrix},
$$

called the wronskian of $g_1(x), g_2(x), \ldots, g_n(x)$ and denoted by $W(g_1, g_2, \ldots, g_n)(x)$.

Lemma 6.3.2. *If $g_1(x), g_2(x), \ldots, g_n(x)$ are functions, n-times differentiable on I, then the function $W(g_1, g_2, \ldots, g_n)(x)$ is differentiable, and one has*

$$
W'(g_1, g_2, \ldots, g_n)(x) = \begin{vmatrix} g_1(x) & \cdots & g_n(x) \\ g_1'(x) & \cdots & g_n'(x) \\ \vdots & \ddots & \vdots \\ g_1^{(n-2)}(x) & \cdots & g_n^{(n-2)}(x) \\ g_1^{(n)}(x) & \cdots & g_n^{(n)}(x) \end{vmatrix}.
$$

Proof. From Lemma 6.3.1, one has $W'(g_1, g_2, \ldots, g_n)(x) = G'(x) = \sum_{i=1}^{n} |A_i|$, with A_i the matrix obtained from $W(g_1, g_2, \ldots, g_n)$ by replacing the ith row by the derivatives in the same row. Then matrix A_i has a repeated row for every $i \in \{1, 2, \ldots, (n-1)\}$, hence $|A_i| = 0$. From this, it follows that $W'(g_1, g_2, \ldots, g_n)(x) = G'(x) = |A_n|$, as needed. □

Theorem 6.3.1. *Let $a_n(x), a_{n-1}(x), \ldots, a_0(x)$ be continuous functions on a closed interval I. Additionally, assume that $a_n(x) \neq 0$ for every $x \in I$. Consider the homogeneous differential equation of order n,*

$$a_n(x)y^{(n)} + a_{n-1}(x)y^{(n-1)} + \cdots + a_1(x)y' + a_0(x)y = 0. \tag{6.19}$$

If $g_1, g_2, \ldots, g_n$ are solutions of (6.19), then $\{g_1, g_2, \ldots, g_n\}$ is a linearly independent set if and only if there is $x_0 \in I$ such that $W(g_1, g_2, \ldots, g_n)(x_0) \neq 0$.

Proof. Let $c_1, c_2, \ldots, c_n$ be scalars so that $h = c_1 g_1 + c_2 g_2 + \cdots + c_n g_n$ is the zero function on I. From this we have that $h^{(j)} = c_1 g_1^{(j)} + c_2 g_2^{(j)} + \cdots + c_n g_n^{(j)}$ is also the zero function on I for every $j \in \{0, 1, 2, \ldots, n-1\}$. If A is the matrix whose determinant is $W(g_1, g_2, \ldots, g_n)$, then $(c_1, c_2, \ldots, c_n)$ is a nontrivial solution of $AX = 0$ if and only if $W(g_1, g_2, \ldots, g_n)(x) = 0$ for every $x \in I$. To finish the proof, we will prove that $W(g_1, g_2, \ldots, g_n)(x) \neq 0$ for every $x \in I$ if and only if there is $x_0 \in I$, so that $W(g_1, g_2, \ldots, g_n)(x_0) \neq 0$. This last claim will follow from $W(g_1, g_2, \ldots, g_n)(x)$ being a solution of (6.15), with $p(x) = \frac{a_{n-1}(x)}{a_n(x)}$.

From Lemma 6.3.2, one has

$$W'(g_1, g_2, \ldots, g_n)(x) = \begin{vmatrix} g_1(x) & \cdots & g_n(x) \\ g_1'(x) & \cdots & g_n'(x) \\ \vdots & \ddots & \vdots \\ g_1^{(n-2)}(x) & \cdots & g_n^{(n-2)}(x) \\ g_1^{(n)}(x) & \cdots & g_n^{(n)}(x) \end{vmatrix}. \tag{6.20}$$

By assumption, each g_i is a solution of (6.19), and then

$$g_i^{(n)}(x) = -\frac{1}{a_n(x)}(a_{n-1}g_i^{(n-1)} + \cdots + a_0 g_i)(x), \tag{6.21}$$

for every $i \in \{1, 2, \ldots, n\}$.

Substituting (6.21) into the last row of (6.20), we notice that this row is obtained by multiplying the nth row of $W(g_1, g_2, \ldots, g_n)$ by $-a_{n-1}$ and then subtracting, from the same row, the ith row of $W(g_1, g_2, \ldots, g_n)$ multiplied by $\frac{a_i}{a_n}$, for every $i \in \{0, 1, 2, \ldots, n-2\}$. From this and using the fact that the determinant does not change when adding a multiple of a row, we have

$$W'(g_1, g_2, \ldots, g_n)(x) = -\frac{a_{n-1}}{a_n} \begin{vmatrix} g_1(x) & \cdots & g_n(x) \\ g_1'(x) & \cdots & g_n'(x) \\ \vdots & \ddots & \vdots \\ g_1^{(n-1)}(x) & \cdots & g_n^{(n-1)}(x) \end{vmatrix}$$

$$= -\frac{a_{n-1}}{a_n} W(g_1, g_2, \ldots, g_n)(x), \tag{6.22}$$

proving that $W(g_1, g_2, \ldots, g_n)(x)$ is a solution of $y' + \frac{a_{n-1}}{a_n}y = 0$. More precisely, from the discussion at the beginning of this section, we have

$$W(g_1, g_2, \ldots, g_n)(x) = W(g_1, g_2, \ldots, g_n)(x_0)e^{\int_{x_0}^{x} p(t)dt}, \quad (6.23)$$

for some $x_0 \in I$ and $p(t) = \frac{a_{n-1}(t)}{a_n(t)}$. Since the function $\exp(x)$ is not zero for every x, $W(g_1, g_2, \ldots, g_n)(x) = 0$ for every x if and only if $W(g_1, g_2, \ldots, g_n)(x_0) = 0$, for some $x_0 \in I$. □

Remark 6.3.1. Notice that in the proof we only used that a_n and a_{n-1} are continuous. We did not use any other condition on the coefficients.

6.4 Exercises

1. Compute the following determinants by using minors and cofactors:

   (a) $\begin{vmatrix} 2 & -2 & 2 \\ 0 & 1 & 3 \\ -1 & 2 & 2 \end{vmatrix}$, (b) $\begin{vmatrix} a & 0 & b \\ 0 & a & 1 \\ -1 & 0 & 1 \end{vmatrix}$, (c) $\begin{vmatrix} 1 & -2 & 0 & 1 \\ 0 & -1 & 0 & 1 \\ -1 & 0 & 1 & 0 \\ -1 & 0 & 1 & 0 \end{vmatrix}$.

2. Let (a, b) and (c, d) be two different points in $\mathbb{R}^2$. Prove that the equation of the line passing through them can be expressed by

$$\begin{vmatrix} x & y & 1 \\ a & b & 1 \\ c & d & 1 \end{vmatrix} = 0.$$

3. Prove that the area of a triangle whose vertices are (a, b), (c, d), and (e, f) is given by the absolute value of

$$\frac{1}{2}\begin{vmatrix} a & b & 1 \\ c & d & 1 \\ e & f & 1 \end{vmatrix}.$$

4. Prove that the volume of a tetrahedron with vertices at (a_1, b_1, c_1), (a_2, b_2, c_2), (a_3, b_3, c_3), and (a_4, b_4, c_4) is given by

$$\frac{1}{6}\begin{vmatrix} a_1 & b_1 & c_1 & 1 \\ a_2 & b_2 & c_2 & 1 \\ a_3 & b_3 & c_3 & 1 \\ a_4 & b_4 & c_4 & 1 \end{vmatrix}.$$

5. Compute the volume of the parallelepiped determined by the given vectors. Draw pictures for each case:

   (a) $(1, 2, 3)$, $(3, -4, 6)$ and $(1, 1, 1)$;

   (b) $(1, 0, 3)$, $(0, 1, 6)$ and $(1, 1, 1)$;

   (c) $(1, \pi, 3)$, $(\sqrt{2}, -4, 6)$ and $(1, -1, 1)$.

6. Compute the area of the parallelograms determined by the given vectors. Draw a picture:
   (a) $(1, 2), (3, -4)$;
   (b) $(1, 0), (0, 1)$;
   (c) $(\pi, 6), (\sqrt{2}, -4)$.

7. Let U, V, and W be vectors in $\mathbb{R}^3$. Prove that the volume of the parallelepiped determined by these three vectors, see Figure 6.2, is $|\langle U \times V, W \rangle|$, where $U \times V$ denotes the cross product of U and V. From a geometric consideration, it must be clear that $\langle U \times V, W \rangle = \langle U, V \times W \rangle = D(U, V, W)$, with D the determinant function.

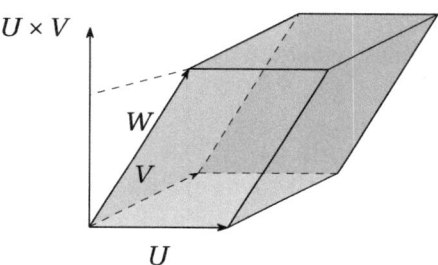

Figure 6.2: Volume of a parallelepiped.

8. Using Cramer's rule, solve the following systems of linear equations in terms of t. Do not forget to consider the case when the coefficient matrix is singular.
   (a)
   $$2(t + 1)x + 3y = 1$$
   $$3x + 4(t^2 + 1)y = 3.$$
   (b)
   $$2tx + 3y + z = 1,$$
   $$2x - ty + z = 0,$$
   $$-x + 3y + t^2 z = -1.$$

9. Let $(a_1, a_2, a_3), (b_1, b_2, b_3)$, and (c_1, c_2, c_3) be noncollinear points in $\mathbb{R}^3$. Prove that the equation of the plane that passes through them is given by

$$\begin{vmatrix} x & y & z & 1 \\ a_1 & a_2 & a_3 & 1 \\ b_1 & b_2 & b_3 & 1 \\ c_1 & c_2 & c_3 & 1 \end{vmatrix} = 0.$$

10. Let A be an $n \times n$ matrix. Assume that its columns, considered as elements of $\mathbb{R}^n$, are orthogonal. What is the value of $|A|$?

11. Let A be a square matrix, diagonal by blocks. How is the determinant of A obtained in terms of the determinant of its blocks?

12. In this exercise a proof of Cramer's rule, due to S. Robinson [22], is presented. Let A be an $n \times n$ matrix whose columns are labeled $A_1, A_2, \ldots, A_n$.

    (a) If A has an inverse, denoted by A^{-1}, then $A^{-1}A_j = e_j$, the jth canonical vector, written as a column vector.

    (b) Let X_j be the matrix obtained from the identity I_n by replacing the jth column with the transpose of $X = (x_1, x_2, \ldots, x_n)$. Prove that $|X_j| = x_j$.

    (c) Let C_j be the matrix obtained from A by replacing the jth column with the transpose of $B = (b_1, b_2, \ldots, b_n)$. Prove that $A^{-1}C_j = X_j$. From this conclude that $x_j = |A^{-1}||C_j|$, or $x_j = \frac{|C_j|}{|A|}$.

13. According to your opinion, state three of the most important theorems presented in this chapter. Explain.

14. If A and B are similar matrices, prove that $|A| = |B|$.

15. Assume that A is a 2×2 matrix. Show that $\text{Adj}(\text{Adj}(A)) = A$.

16. Let A be an $n \times n$ matrix. Prove that there are at most n values of $c \in \mathbb{R}$ so that the matrix $B = A - cI_n$ is singular. Find an example so that the matrix B is singular for exactly n values of c.

17. Let $A, B, C,$ and D be square matrices that commute. Additionally, assume that A and C have inverses. If $M = \begin{bmatrix} A & B \\ C & D \end{bmatrix}$, then show $|M| = |AD - BC|$.

18. Let $x_1, x_2, \ldots, x_n$ be real numbers and let $A = (a_{ij})$ be defined by the entries $a_{ij} = x_i^{j-1}$, where $i, j \in \{1, 2, \ldots, n\}$. Prove that

$$|A| = (-1)^{\frac{n(n-1)}{2}} \prod_{1 \le i < j \le n} (x_i - x_j).$$

This determinant is called *Vandermonde's determinant*.

19. Use the result in Exercise 18 to prove

$$\begin{vmatrix} 1 & 1 & 1 & \cdots & 1 \\ 1 & 2 & 2^2 & \cdots & 2^{n-1} \\ 1 & 3 & 3^2 & \cdots & 3^{n-1} \\ \vdots & \vdots & \vdots & \ddots & \vdots \\ 1 & n & n^2 & \cdots & n^{n-1} \end{vmatrix} = 1!2!3! \cdots (n-1)!$$

20. Recall that an antisymmetric matrix satisfies $A^t = -A$. Let A be an $n \times n$ antisymmetric matrix, with n odd. Prove that A is singular.

21. Given $x_1, x_2, \ldots, x_n$ real numbers. Prove

$$\begin{vmatrix} 1 + x_1 & x_2 & x_3 & \cdots & x_n \\ x_1 & 1 + x_2 & x_3 & \cdots & x_n \\ x_1 & x_2 & 1 + x_3 & \cdots & x_n \\ \vdots & \vdots & \vdots & \ddots & \vdots \\ x_1 & x_2 & x_3 & \cdots & 1 + x_n \end{vmatrix} = 1 + x_1 + x_2 + \cdots + x_n.$$

22. A square matrix A is called *idempotent* if $A^2 = A$. What are the possible values of $|A|$ if A is idempotent?

23. Let A be a 2×2 matrix and assume that $A^2 = 0$. Prove that for every scalar c, $|cI - A| = c^2$. Does the same result hold if A is $n \times n$ and $A^k = 0$ for some positive integer k?

24. Compute the following determinant:

$$\begin{vmatrix} a & a + h & a + 2h & \cdots & \cdots & a + (n-1)h \\ -a & a & 0 & \cdots & \cdots & 0 \\ 0 & -a & a & \cdots & \cdots & 0 \\ 0 & 0 & -a & a & \cdots & 0 \\ \vdots & \vdots & \vdots & \ddots & \vdots & \vdots \\ 0 & 0 & 0 & \cdots & a & 0 \\ 0 & 0 & 0 & \cdots & -a & a \end{vmatrix}.$$

25. Without performing the calculations, show that the following equations hold:

(a) $\begin{vmatrix} 1 & a & bc \\ 1 & b & ca \\ 1 & c & ab \end{vmatrix} = \begin{vmatrix} 1 & a & a^2 \\ 1 & b & b^2 \\ 1 & c & c^2 \end{vmatrix};$

(b) $\begin{vmatrix} a-b & b-c & c-a \\ 1 & 1 & 1 \\ a & b & c \end{vmatrix} = \begin{vmatrix} a & b & c \\ 1 & 1 & 1 \\ b & c & a \end{vmatrix}.$

26. Assume that $n > 2$ is an integer and A is the $n \times n$ matrix shown below. For which values of n is A singular?

$$A = \begin{bmatrix} 1 & 1 & 0 & 0 & \cdots & 0 & 0 \\ 0 & 1 & 1 & 0 & \cdots & 0 & 0 \\ 0 & 0 & 1 & 1 & \cdots & 0 & 0 \\ \vdots & \vdots & \vdots & \vdots & \ddots & \vdots & \vdots \\ 0 & 0 & 0 & 0 & \cdots & 1 & 1 \\ 1 & 0 & 0 & 0 & \cdots & 0 & 1 \end{bmatrix}.$$

27. Let A be an $n \times n$ matrix. Prove that $|\operatorname{Adj}(A)| = |A|^{n-1}$. If A is nonsingular, use the previous equality to show $\operatorname{Adj}(\operatorname{Adj}(A)) = |A|^{n-2}A$, if $n \geq 2$.

28. Given a matrix $A = \begin{bmatrix} a_{11} & a_{12} & \cdots & a_{1n} \\ a_{21} & a_{22} & \cdots & a_{2n} \\ \vdots & \vdots & \ddots & \vdots \\ a_{n1} & a_{n2} & \cdots & a_{nn} \end{bmatrix}$ and $d_1, d_2, \ldots, d_n$ real numbers, define $B = \begin{bmatrix} d_1 a_{11} & d_1 a_{12} & \cdots & d_1 a_{1n} \\ d_2 a_{21} & d_2 a_{22} & \cdots & d_2 a_{2n} \\ \vdots & \vdots & \ddots & \vdots \\ d_n a_{n1} & d_n a_{n2} & \cdots & d_n a_{nn} \end{bmatrix}$. Represent $|B|$ in terms of $|A|$.

29. Let A be an $n \times n$ matrix and denote its columns by $A_1, A_2, \ldots, A_n$ as elements of $\mathbb{R}^n$. Assume that $\|A_i\| = 1$ and $\langle A_i, A_j \rangle = 0$ if $i \neq j$. Prove that $|A| = \pm 1$.

30. Let A be as in Exercise 29 and consider vectors $Y_1, Y_2, \ldots, Y_n$ in $\mathbb{R}^n$. If B is the matrix whose columns are $AY_1, AY_2, \ldots, AY_n$, what is $|B|$ in terms of $|A|$ and of $Y_1, Y_2, \ldots, Y_n$? What is the geometric interpretation of this result?

31. Let $A = (a_{ij})$ be an $n \times n$ matrix whose columns $A_1, A_2, \ldots, A_n$ can be considered as elements of $\mathbb{R}^n$. Prove that the absolute value of $|A|$ is $\leq \|A_1\| \|A_2\| \cdots \|A_n\|$, where $\|A_i\|$ denotes the norm of A_i. In particular, if $|a_{ij}| \leq 1$ for every $i, j \in \{1, 2, \ldots, n\}$, then the absolute value of $|A|$ is $\leq n^{n/2}$.
    (*Hint.*) Reduce the problem to the case when the columns are linearly independent and have norm 1. The previous inequality is called *Hadamard's inequality*.

32. A square matrix H is called an *Hadamard matrix* if its entries are ± 1 and its rows are orthogonal. Is there an Hadamard matrix if n is odd? After you have given an answer, prove that if an Hadamard matrix of size n exists, then $n = 1, 2, 4k$, with k a positive integer. The converse is an open problem, that is, it is not known if for every positive integer k, there is an Hadamard matrix of size $4k$.

33. If H is an Hadamard matrix, prove:
    (a) $HH^t = nI_n$. (*Hint.*) Since the rows are orthogonal, each row, as a vector in $\mathbb{R}^n$, has norm $\sqrt{n}$, etc.;
    (b) $|H| = \pm n^{n/2}$.

34. Let V be a vector space of dimension $n > 1$ and let $\alpha_1, \alpha_2, \ldots, \alpha_n$ be a basis of V. Define a linear transformation $T : V \to V$ by $T(\alpha_1) = \alpha_2$, $T(\alpha_2) = \alpha_1 + \alpha_3$, $T(\alpha_3) = \alpha_2 + \alpha_4, \ldots, T(\alpha_{n-1}) = \alpha_{n-2} + \alpha_n$, and $T(\alpha_n) = 0$. Find the matrix associated to T with respect to the given basis. If A is this matrix, prove that $|A| = \cos(\frac{n\pi}{2})$.

35. Choose a fixed real number b and let A be an $n \times n$ matrix with entries a_{ij} that satisfy $a_{ii} = b$ for $i \in \{1, 2, \ldots, n\}$ and $a_{ij} = 1$ if $i \neq j$. Prove that $|A| = (b - 1)^{n-1}(b - 1 + n)$. Obtain $|A|$ for $b = n + 1$.

36. Let $n > 2$ be a positive integer and let $x_1, x_2, \ldots, x_n$ be real numbers. Let A be the $n \times n$ matrix whose entries are defined by $a_{ij} = \sum_{k \neq j} x_k^i$, $i \in \{1, 2, \ldots, n-2\}$; $a_{n-1 j} = x_j$, $j \in \{1, 2, \ldots, n\}$, and $a_{nj} = 1$ for all $j \in \{1, 2, \ldots, n\}$. Prove that $|A| = 0$.
    (*Hint.*) Start with $n = 3$.

37. Assume that A and B are matrices of sizes $m \times n$ and $n \times m$, respectively. Prove that if $m < n$, then $|BA| = 0$. Under which conditions on m and n, is it also true that $|AB| = 0$?

7 Eigenvalues and eigenvectors without determinants

In this chapter we will discuss the equation $T(a) = \lambda a$, with T an operator acting on a finite-dimensional vector space V, λ a scalar, and $a \in V \setminus \{0\}$. The equivalent matrix formulation of this problem is to study the equation $AX = \lambda X$, where A is a square matrix, λ a scalar, and $X \neq 0$. The condition on X is equivalent to saying that the matrix $A - \lambda I_n$ is singular. This last condition can be formulated by saying that the determinant of $A - \lambda I_n$ is zero, hence the eigenvalue problem can be formulated as: Find the values of λ so that $\det(A - \lambda I_n) = p(\lambda) = 0$.

The usual route to approach the eigenvalue problem is to study the equation $\det(A - \lambda I_n) = p(\lambda) = 0$. We will be taking a different approach to be discussed in what follows. Please, prepare yourself to see different scenarios from those that appear when using the characteristic polynomial!

We encourage the reader to consult Appendix A in order to review, or to find the meaning of terms and results that are needed to discussed the content of this chapter.

7.1 Minimal polynomial

In this section we discuss the existence and basic properties of the minimal polynomial of a matrix A. The same ideas can be applied to an operator T, acting on a finite-dimensional vector space. Actually, we will interchange the roles of a matrix and an operator, since their actions can be identified when needed. We also propose a method to compute the minimal polynomial, see Section 7.1.3, p. 175.

7.1.1 Existence of minimal polynomial and eigenvalues

In the literature, several terms are used to refer to scalars that satisfy the equation $AX = \lambda X$, where A is an $n \times n$ matrix, λ is a scalar, and X is an nonzero element of $\mathbb{R}^n$. The conditions $X \neq 0$ and $AX = \lambda X$ imply that the matrix $A - \lambda I_n$ is singular, where I_n is the $n \times n$ identity matrix. In this case the scalar λ is called an eigenvalue, a characteristic value, or a proper value. In what follows, we adopt the term *eigenvalue*. Since the term eigenvalue is of such importance, we introduce it formally in the following definition.

Definition 7.1.1. Let A be an $n \times n$ matrix, and let I_n be the $n \times n$ identity matrix. If λ is a scalar so that $A - \lambda I_n$ is singular, then λ is called an eigenvalue of A. When λ is an eigenvalue of A, and $X \in \mathbb{R}^n \setminus \{0\}$ satisfies $(A - \lambda I_n)X = 0$, X is called an eigenvector associated to λ.

Although we will not use the characteristic polynomial of a matrix to discuss eigenvalues, as a matter of reference we give its definition.

https://doi.org/10.1515/9783111135915-007

Definition 7.1.2. Let A be an $n \times n$ matrix, and let I_n be the $n \times n$ identity matrix. The polynomial $f_A(x) = \det(A - xI_n)$ is called the characteristic polynomial of A.

The following example will illustrate the ideas to be presented.

Example 7.1.1. Consider the matrix $A = \begin{bmatrix} 7 & 15 \\ -4 & -9 \end{bmatrix}$. How do we find an eigenvalue of A?

Discussion. We are looking for a nonzero element $X \in \mathbb{R}^2$ and a scalar λ so that $AX = \lambda X$. Notice that this equation holds if and only if X and AX are linearly dependent. Hence, to start our search for eigenvalues and eigenvectors let us choose any vector, for instance, $X = \begin{bmatrix} 1 \\ 0 \end{bmatrix}$. From this choice of X, we have $AX = \begin{bmatrix} 7 & 15 \\ -4 & -9 \end{bmatrix} \begin{bmatrix} 1 \\ 0 \end{bmatrix} = \begin{bmatrix} 7 \\ -4 \end{bmatrix}$. It is straightforward to verify that X and AX are linearly independent, hence we have not found an eigenvalue or an eigenvector, yet. On the other hand, since $\mathbb{R}^2$ has dimension 2, we have that X, AX, and A^2X are linearly dependent. From a direct calculation, we have that $A^2X = A(AX) = \begin{bmatrix} 7 & 15 \\ -4 & -9 \end{bmatrix} \begin{bmatrix} 7 \\ -4 \end{bmatrix} = \begin{bmatrix} -11 \\ 8 \end{bmatrix}$, and we can verify that $A^2X = 3X - 2AX$. This last equation can be written as

$$(A^2 + 2A - 3I_2)X = 0, \tag{7.1}$$

where I_2 is the 2×2 identity matrix. Also, from equation (7.1) we see that $A(A^2 + 2A - 3I_2)X = (A^2 + 2A - 3I_2)AX = 0$. The latter equation is true, since A commutes with any of its powers. Applying the result of Exercise 26, p. 115, we conclude that $A^2 + 2A - 3I_2$ is the zero matrix, given that $\{X, AX\}$ is a basis of $\mathbb{R}^2$. How do we use the fact that $A^2 + 2A - 3I_2 = 0$? Well, we can notice that

$$A^2 + 2A - 3I_2 = (A - I_2)(A + 3I_2) = 0. \tag{7.2}$$

We claim that $A - I_2$ and $A + 3I_2$ are singular. To justify the claim, we argue by contradiction. If $A + 3I_2$ is not singular, then it has an inverse and, multiplying in (7.2) by $(A + 3I_2)^{-1}$, one has $(A - I_2) = (A - I_2)(A + 3I_2)(A + 3I_2)^{-1} = (A + 3I_2)^{-1}0 = 0$, or $A - I_2 = 0$, which is not true.

Likewise, if $A - I_2$ is not singular then, multiplying in (7.2) by $(A - I_2)^{-1}$, we obtain that $A + 3I = 0$, which is false. From this we have that $A - I_2$ and $A + 3I_3$ are singular. Hence, we have found two eigenvalues of A, namely $\lambda = 1$ and $\lambda = -3$.

Remark 7.1.1. We want to highlight a couple of ideas from Example 7.1.1. Given a matrix A, to search for its eigenvalues, proceed as follows: pick any nonzero element X and check if X and AX are linearly dependent. If that is the case, we have found and eigenvalue and an eigenvector. Otherwise, consider the elements X, AX, A^2X and ask if they are linearly dependent. If that were the case, we could construct a polynomial of degree 2 and its roots would be eigenvalues. In case that X, AX, A^2X are linearly independent, this process will stop in at most n steps, since $X, AX, \ldots, A^nX$ are linearly dependent. From the linear dependence of $X, AX, \ldots, A^nX$, we obtain a polynomial in A, say $P(A)$, which is going to help us find the eigenvalues of A. In what follows, this ideas will be developed to introduce the main object of the discussion – the minimal polynomial of a matrix.

Example 7.1.1 can be generalized as follows. If $X \in \mathbb{R}^n$ is not zero and A is any $n \times n$ matrix, then the elements $X, AX, A^2X, \ldots, A^nX$ are linearly dependent, since the dimension of $\mathbb{R}^n$ is n. From this we have that there are scalars $a_0, a_1, \ldots, a_n$, not all zero, such that

$$a_0 X + a_1 AX + a_2 A^2 X + \cdots + a_n A^n X = 0. \tag{7.3}$$

Defining the polynomial $h(x) = a_0 + a_1 x + a_2 x^2 + \cdots + a_n x^n$, equation (7.3) can be written as $h(A)X = 0$, where we have declared $h(A) = a_0 I + a_1 A + a_2 A^2 + \cdots + a_n A^n$. Notice that $h(x)$ depends on A and X.

If for some $X \neq 0$, $h(x)$ factorizes as

$$h(x) = (x - \lambda_1)(x - \lambda_2) \cdots (x - \lambda_n), \tag{7.4}$$

then $h(A) = (A - \lambda_1 I_n)(A - \lambda_2 I_n) \cdots (A - \lambda_n I_n)$ is singular, since $h(A)X = 0$ and $X \neq 0$. We recall that the product of matrices is singular if and only if at least one of the factors is singular. Applying this to $h(A) = (A - \lambda_1 I_n)(A - \lambda_2 I_n) \cdots (A - \lambda_n I_n)$, we obtain that necessarily one of the factors $A - \lambda_j I_n$ is singular, that is, some λ_j is an eigenvalue of A.

Remark 7.1.2. Notice that when the scalars are allowed to be complex numbers, the previous argument would prove that every matrix has eigenvalues, since in the complex numbers every nonconstant polynomial with real coefficients factorizes as in (7.4), see Appendix A, Theorem A.5.3, p. 279.

We have pointed out that $h(x)$ depends on X. In what follows, we will prove that there is a monic and minimal-degree polynomial, $m(x)$, which depends only on A and $m(A) = 0$.

The following discussion shows two ways to approach the existence of $m(x)$ with the announced properties.

Let $\{X_1, X_2, \ldots, X_n\}$ be a basis of $\mathbb{R}^n$, and let $h_j(x)$ denote the polynomial associated to X_j as in the previous discussion. Recall that from the construction of $h_j(x)$, it has degree at most n. Defining

$$g(x) = h_1(x) h_2(x) \cdots h_n(x) \tag{7.5}$$

and using the fact that each $h_j(x)$ has degree at most n, applying Remark A.4.1, p. 272, we obtain $\deg(g(x)) \leq n^2$. We also have that $g(A)X_j = 0$ for every $j \in \{1, 2, \ldots, n\}$. Now applying the result of Exercise 26, p. 115, we conclude that $g(A) = 0$, the zero matrix.

So far, we have proved that there is a polynomial of positive degree which has A as a root. Now, we choose a monic polynomial of minimum degree which has A as a root. This polynomial will play a crucial role in the discussion of eigenvalues, as we will see shortly. Actually we will prove, in Theorem 7.1.2, that the roots of this polynomial are the eigenvalues of the associated matrix.

By the identification that can be done between a matrix and an operator, in what follows, we will be dealing with an operator T, acting on a finite-dimensional vector space. The reader is invited, as a good exercise, to translate the statements for operators to matrices and provide the corresponding proofs.

Theorem 7.1.1. *Let V be a vector space of dimension n, and let $T : V \to V$ be a linear operator, then there exists a monic polynomial of minimum degree, $m(x)$, such that:*
1. *$m(T) = 0$ and $\deg(m(x)) \le n^2$.*
2. *If $f(x)$ is another polynomial such that $f(T) = 0$, then $m(x)$ divides $f(x)$.*

Furthermore, a polynomial satisfying properties 1 and 2 is unique.

Proof. Existence of $m(x)$. From Theorem 5.4.3, p. 138, we have that the space of linear operators in V has dimension n^2, hence the set $\{I, T, T^2, \ldots, T^{n^2}\}$ is linearly dependent. Here I denotes the identity operator. From this, there are scalars $a_0, a_1 \ldots, a_{n^2}$, not all zero, so that $a_0 I + a_1 T + \cdots + a_{n^2} T^{n^2} = 0$. Defining the polynomial $f(x) = a_0 + a_1 x + \cdots + a_{n^2} x^{n^2}$, we have that $f(T) = 0$, the zero operator; also, notice that $\deg(f(x)) \le n^2$. It is clear that $f(x)$ has positive degree, and then the set $N = \{k \in \mathbb{N} \mid k = \deg(f(x))$ and $f(T) = 0\}$ is not empty. Applying the well-ordering principle, Principle A.1.3, p. 263, we have that there is $k_0 \in N$ so that $k_0 \le k$ for every $k \in N$. Let $m_1(x)$ be a polynomial of degree k_0 that satisfies $m_1(T) = 0$, and let a be its leading coefficient. Setting $m(x) = \frac{1}{a} m_1(x)$, we have $m(T) = 0$, and $m(x)$ has minimum degree and is monic. Also $\deg(m(x)) \le \deg(f(x)) \le n^2$, proving property 1.

To prove property 2, let $f(x)$ be a polynomial such that $f(T) = 0$. By the division algorithm, Theorem A.4.1, p. 272, there are polynomials $q(x)$ and $r(x)$ such that

$$f(x) = m(x)q(x) + r(x), \quad \text{with } r(x) = 0 \text{ or } \deg(r(x)) < \deg(m(x)). \tag{7.6}$$

Evaluating equation (7.6) in T and using that $f(T) = 0$ and $m(T) = 0$ yields $r(T) = 0$. Since $m(x)$ is of minimum degree that has T as a root, $r(x) = 0$, proving that $m(x)$ divides $f(x)$.

The uniqueness of $m(x)$ follows from part 2 as follows.

If $m(x)$ and $g(x)$ are monic polynomials of minimum degree that have T as a root, then by part 2, these polynomials divide each other. That is, $m(x) = q(x)g(x)$ and $g(x) = m(x)q_1(x)$ for some polynomials $q(x)$ and $q_1(x)$. From this we have $m(x) = q(x)q_1(x)m(x)$. Now, canceling out $m(x)$ on both sides, we obtain $1 = q(x)q_1(x)$. This implies that $q(x)$ and $q_1(x)$ are constant, and as $m(x)$ and $g(x)$ are monic, we conclude that $q(x) = q_1(x) = 1$, finishing the proof of the theorem. $\square$

Definition 7.1.3. The polynomial of Theorem 7.1.1 is called the minimal polynomial of T and will be denoted by $m_T(x)$.

Exercise 7.1.1. Prove that if A and B are similar matrices, that is, they represent an operator T, then $m_A(x) = m_B(x)$.

Before we dive deeper into the discussion of the minimal polynomial, it is important to prove that the roots of the minimal polynomial play the same role as the roots of the characteristic polynomial when searching for eigenvalues. More precisely, we have

Theorem 7.1.2. *Let V be a vector space of dimension n and let $T : V \to V$ be an operator. Then $T - cI$ is singular if and only if c is a root of the minimal polynomial of T. In particular, T is singular if and only if 0 is a root of the minimal polynomial of T.*

Proof. Assume that $T - cI$ is singular. If c is not a root of $m_T(x)$, then $x - c$ and $m_T(x)$ are relatively prime. From Corollary A.4.1, p. 274, there are polynomials $g(x)$ and $h(x)$ such that

$$1 = (x - c)g(x) + m_T(x)h(x). \tag{7.7}$$

Substituting $x = T$ in equation (7.7), one has $I = (T - cI)g(T) + m_T(T)h(T)$, which implies that $I = (T - cI)g(T)$, contradicting that $T - cI$ is singular.

Conversely, assume that c is a root of $m_T(x)$, then $m_T(x) = (x - c)^r q(x)$ for some integer $r \geq 1$ and some polynomial $q(x)$ of degree less than $\deg(m_T(x))$. Evaluating at T, one has $0 = m_T(T) = (T - cI)^r q(T)$.

If $T - cI$ is not singular, then multiplying by $(T - cI)^{-1}$ this equation, we obtain

$$0 = (T - cI)^{-1}0 = (T - cI)^{-1}(T - cI)^r q(T) = (T - cI)^{r-1}q(T). \tag{7.8}$$

Then T is a zero of the polynomial $(x - c)^{r-1}q(x)$ which has degree less than $\deg(m_T(x))$, which is the polynomial of minimum degree that has T as a root, a contradiction. The last part of the theorem follows by taking $c = 0$. $\qquad\square$

The previous result is valid for matrices, as well as for operators. The proof runs along the same lines.

An important corollary of Theorem 7.1.2 is the following.

Corollary 7.1.1. *The minimal and characteristic polynomial of an $n \times n$ matrix A have the same irreducible factors.*

Proof. From Corollary A.5.3, p. 280, we have that the only irreducible polynomials over the real numbers are linear or quadratic. On the other hand, from Definition 7.1.2, we have that a scalar c is a root of the characteristic polynomial $f_A(x)$ if and only if $A - cI$ is singular. Hence, Theorem 7.1.2 guarantees that $m_A(x)$ and $f_A(x)$ have the same linear factors. Assume that $p(x)$ is a quadratic irreducible factor of $m_A(x)$, then $p(x)$ has no real roots. From Theorem A.5.4, p. 280, its roots are complex numbers, say z and its conjugate $\bar{z}$. Applying Theorem 7.1.2, we have that $A - zI_n$ and $A - \bar{z}I_n$ are singular, thus z and $\bar{z}$ are roots of $f_A(x)$. Thus from the factor theorem, Theorem A.5.1 p. 278, $(x - z)(x - \bar{z}) = p(x)$ divides $f_A(x)$.

Let $q(x)$ be a quadratic irreducible factor of $f_A(x)$. A similar reasoning as above shows that $q(x)$ divides $m_A(x)$, finishing the proof. □

To have an idea of how big the degree of $m_T(x)$ is, we discuss the case when V has dimension 2. The generalization of this special, but interesting, case is Theorem 7.1.4.

Example 7.1.2. Let V be a vector space of dimension 2 and let $T : V \to V$ be an operator. Then $m_T(x)$ has degree at most two.

Case I. Assume that for every $\alpha \in V$, the vectors α and $T(\alpha)$ are linearly dependent. This condition is equivalent to saying that for every α there is a scalar c_α so that $T(\alpha) = c_\alpha \alpha$. We will show that c_α does not depend on α, that is, $T(\alpha) = c\alpha$ for every α, or equivalently, $(T - cI)(\alpha) = 0$ for every α, thus $T - cI$ is the zero operator, proving that the minimum polynomial of T has degree one.

To show this, it is sufficient to prove that for any pair of linearly independent vectors α and β, one has $c_\alpha = c_\beta$.

We have $T(\alpha + \beta) = c(\alpha + \beta)$. On the other hand, $T(\alpha + \beta) = T(\alpha) + T(\beta) = c_\alpha \alpha + c_\beta \beta$. From this equation, it follows that $c\alpha + c\beta = c_\alpha \alpha + c_\beta \beta$, or $(c - c_\alpha)\alpha + (c - c_\beta)\beta = 0$. Hence $c = c_\alpha = c_\beta$, since α and β are linearly independent. If $\gamma \in V$, then $\gamma = x\alpha + y\beta$, therefore $T(\gamma) = xT(\alpha) + yT(\beta) = xc\alpha + yc\beta = c(x\alpha + y\beta) = c\gamma$.

Summarizing, we have proved that there is a constant c such that $T(\gamma) = c\gamma$ for every $\gamma \in V$. This is equivalent to $(T - cI)(\gamma) = 0$ for every $\gamma \in V$, that is, $T - cI$ is the zero operator, thus, the minimal polynomial of T is $x - c$.

Case II. There is $\alpha \in V$ such that α and $T(\alpha)$ are linearly independent. Since V has dimension 2, α, $T(\alpha)$, $T^2(\alpha)$ are linearly dependent, hence there are scalars a and b such that $T^2(\alpha) = aT(\alpha) + b\alpha$.

We claim that the minimal polynomial of T is $m_T(x) = x^2 - ax - b$. This is equivalent to showing that $S = T^2 - aT - bI$ is the zero operator, or equivalently, that S vanishes on the basis $\{\alpha, T(\alpha)\}$.

From $T^2(\alpha) = aT(\alpha) + b\alpha$ and the definition of S, we obtain $S(\alpha) = T^2(\alpha) - aT(\alpha) - b\alpha = 0$. Also, $S(T(\alpha)) = (T^2 - aT - bI)(T(\alpha)) = T(T^2(\alpha) - aT(\alpha) - b\alpha) = T(0) = 0$, proving what was claimed, that is, S vanishes on the basis $\{\alpha, T(\alpha)\}$. From Cases I and II, one has that $m_T(x)$ has degree at most two.

Example 7.1.3. Find the minimal polynomial of $T : \mathbb{R}^2 \to \mathbb{R}^2$ given by

$$T(x,y) = (2x - y, 3x + y).$$

One has $T(1,0) = (2,3)$, then $(1,0)$ and $T(1,0) = (2,3)$ are linearly independent. Also

$$T^2(1,0) = T(T(1,0)) = T(2,3) = (1,9).$$

Since $(1, 0)$ and $T(1, 0) = (2, 3)$ are linearly independent, we need to find scalars a and b so that $T^2(1, 0) = aT(1, 0) + b(1, 0)$. Solving this equation, we find that $a = 3$ and $b = -5$, thus the minimal polynomial of T is $m_T(x) = x^2 - 3x + 5$.

Example 7.1.4. Find the minimal polynomial of $T : \mathbb{R}^3 \to \mathbb{R}^3$ given by

$$T(x, y, z) = (2x + y, 2y + z, x + y + 2z).$$

One has that $T(1, 0, 0) = (2, 0, 1)$, $T^2(1, 0, 0) = T(2, 0, 1) = (4, 1, 4)$ and the vectors $(1, 0, 0), (2, 0, 1), (4, 1, 4)$ are linearly independent, as can be verified by considering the row reduced form of the matrix whose columns are those vectors. Verify it as an exercise. We also have $T^3(1, 0, 0) = (9, 6, 13)$. This vector is a linear combination of $(1, 0, 0)$, $T(1, 0, 0) = (2, 0, 1)$, $T^2(1, 0, 0) = (4, 1, 4)$, more precisely, one finds that

$$T^3(1, 0, 0) = 6(4, 1, 4) - 11(2, 0, 1) + 7(1, 0, 0)$$
$$= 6T^2(1, 0, 0) - 11T(1, 0, 0) + 7(1, 0, 0). \tag{7.9}$$

From equation (7.9), we have that the minimal polynomial of T is $m_T(x) = x^3 - 6x^2 + 11x - 7$, since one can verify that $m_T(T)$ vanishes on the basis $\{(1, 0, 0), T(1, 0, 0), T^2(1, 0, 0)\}$.

An important concept that generalizes the eigenvector, Definition 7.1.1, p. 165, is given in the next definition.

Definition 7.1.4. Let V be a vector space and let $T : V \to V$ be an operator. A subspace W of V is called *T-invariant*, if $T(W) = \{T(w) \mid w \in W\} \subseteq W$. When there is no confusion, we say only that W is *invariant*, instead of T-invariant.

If W is a proper (neither 0 nor V) invariant subspace of V, then T restricted to W is an operator acting on a vector space of dimension less than $\dim(V)$. Additionally, if $V = U \oplus W$, with W and U being T-invariant subspaces, then the action of T on V can be thought as the action on U plus the action on W. This point of view has several advantages as we will see in the following discussion.

We close this section with a useful remark that connects eigenvectors and T-invariant subspaces.

Remark 7.1.3. Let V be a finite-dimensional vector space and let $T : V \to V$ be an operator. Then V contains a T-invariant subspace of dimension 1 if and only if the minimal polynomial of T has a linear factor.

The proof of this remark follows readily from Theorem 7.1.2.

7.1.2 Annihilators and minimal polynomial

If V is a vector space of dimension n, and $T : V \to V$ is an operator, we have proved that the degree of $m_T(x)$ is at most n^2. In what follows we will prove, among other properties, that $m_T(x)$ has degree at most n. To advance into the discussion, we introduce the annihilator polynomial of an element $a \in V$, which will play an important role to compute the minimal polynomial. One of the basic properties of this polynomial is that its degree is $\leq n$. A more elaborated result that we will prove is that there is an $a \in V$ so that its annihilator polynomial is equal to $m_T(x)$. From this, we immediately obtain that the degree of $m_T(x)$ is at most n, a significant improvement to the bound that we have so far.

We start the discussion with a lemma.

Lemma 7.1.1. *Let V be a finite-dimensional vector space and let T be an operator on V. Suppose that W is a T-invariant subspace. If $a \in V \setminus W$, then there exists a monic polynomial $g(x)$, of minimal degree, that satisfies the following properties:*
1. *$g(T)(a) \in W$,*
2. *if $f(x)$ is another polynomial such that $f(T)(a) \in W$, then $g(x)$ divides $f(x)$,*
3. *the polynomial $g(x)$ is unique,*
4. *if $g(x)$ has degree l, then $a, T(a), \ldots, T^{l-1}(a)$ are linearly independent.*

Proof. Let $m_T(x)$ be the minimal polynomial of T. By definition, $m_T(T)$ is the zero operator, hence $m_T(T)(a) = 0$, which implies that $m_T(T)(a) \in W$. Hence there are polynomials of positive degree, $h(x)$, that satisfy $h(T)(a) \in W$. From the set of these polynomials, we can choose one with minimal positive degree and monic; denote it by $g(x)$. From this, we have proved the existence of $g(x)$ which satisfies property 1, that is, $g(T)(a) \in W$.

Next, we will prove that $g(x)$ also satisfies properties 2, 3, and 4.

Let $f(x)$ be a polynomial that satisfies $f(T)(a) \in W$. By the division algorithm, Theorem A.4.1, p. 272, there exist polynomials $q(x)$ and $r(x)$ such that

$$f(x) = q(x)g(x) + r(x), \quad \text{with } r(x) = 0 \text{ or } \deg(r(x)) < \deg(g(x)). \tag{7.10}$$

From equation (7.10), one has $f(T)(a) = q(T)g(T)(a) + r(T)(a)$. Since $g(T)(a) \in W$ and W is T-invariant, $q(T)g(T)(a) \in W$, hence $r(T)(a) \in W$. The choice of $g(x)$ and the condition $\deg(r(x)) < \deg(g(x))$ imply that $r(x) = 0$, proving that $g(x)$ divides $f(x)$, and with this property 2 is proven.

To prove property 3, assume that there are polynomials $g(x)$ and $g_1(x)$ which are monic of minimal degree and satisfying $g(T)(a), g_1(T)(a) \in W$. From property 2, these polynomials divide each other. Now, the condition of both being monic implies that $g(x) = g_1(x)$, proving the uniqueness of $g(x)$.

To prove property 4, consider the equation $b_0 a + b_1 T(a) + \cdots + b_{l-1} T^{l-1}(a) = 0$. If one of the scalars $b_0, b_1, \ldots, b_{l-1}$ is not zero, then the polynomial $h(x) = b_0 + b_1 x + \cdots + b_{l-1} x^{l-1}$ has degree less than l, the degree of $g(x)$, and satisfies $h(T)(a) = 0 \in W$, contradicting the choice of $g(x)$. This contradiction shows that $a, T(a), \ldots, T^{l-1}(a)$ are linearly independent, finishing the proof of the lemma. $\square$

Definition 7.1.5. Let V, T, W, α, and $g(x)$ be as in Lemma 7.1.1. The polynomial $g(x)$ is called the T-annihilator of α with respect to W. When $W = \{0\}$, it is simply called the T-annihilator of α and will be denoted by $A_\alpha(x)$.

Remark 7.1.4. If V is a vector space of dimension n, T is an operator on V, and $\alpha \in V$, then the degree of $A_\alpha(x)$ is less than or equal to n.

Proof. We have that the elements $\alpha, T(\alpha), T^2(\alpha), \ldots, T^n(\alpha)$ are linearly dependent, hence there are scalars $a_0, a_1, a_2, \ldots, a_n$, not all zero such that

$$a_0\alpha + a_1 T(\alpha) + a_2 T^2(\alpha) + \cdots + a_n T^n(\alpha) = 0. \tag{7.11}$$

From equation (7.11), we have that the polynomial $f(x) = a_0 + a_1 x + a_2 x^2 + \cdots + a_n x^n$ satisfies $f(T)(\alpha) = 0$ and has degree at most n. Also, by Lemma 7.1.1, $A_\alpha(x)$ divides $f(x)$, thus $A_\alpha(x)$ has degree less than or equal to n. □

Exercise 7.1.2. Assume that $T : \mathbb{R}^3 \to \mathbb{R}^3$ is given by $T(x, y, z) = (x + y - z, 2x - y, y - 2z)$ and let $\alpha = (1, 0, 1)$. Find $A_\alpha(x)$.

Lemma 7.1.2. *Let V be a finite-dimensional vector space and let $T : V \to V$ be an operator. If $\beta \in V \setminus \{0\}$, then the dimension of the subspace spanned by $\{\beta, T(\beta), T^2(\beta), \ldots\}$ is equal to the degree of $A_\beta(x)$.*

Proof. Set $A_\beta(x) = x^k + a_{k-1}x^{k-1} + \cdots + a_1 x + a_0$. We will prove that a basis for the subspace spanned by $\{\beta, T(\beta), T^2(\beta), \ldots\}$ is $\{\beta, T(\beta), T^2(\beta), \ldots, T^{k-1}(\beta)\}$. The condition $A_\beta(T)(\beta) = 0$ implies that $T^k(\beta)$ belongs to the span of $\{\beta, T(\beta), T^2(\beta), \ldots, T^{k-1}(\beta)\}$. By induction one proves that for every $i \in \{1, 2, 3, \ldots\}$, the element $T^{k+i}(\beta)$ belongs to the span of

$$\{\beta, T(\beta), T^2(\beta), \ldots, T^{k-1}(\beta)\},$$

showing that $\{\beta, T(\beta), T^2(\beta), \ldots, T^{k-1}(\beta)\}$ spans the same subspace as $\{\beta, T(\beta), T^2(\beta), \ldots\}$.

From Lemma 7.1.1, we have that $\{\beta, T(\beta), T^2(\beta), \ldots, T^{k-1}(\beta)\}$ is linearly independent, thus the proof is completed. □

Definition 7.1.6. Let $T : V \to V$ be an operator, and let $\beta \in V$. The subspace spanned by $\{\beta, T(\beta), T^2(\beta), \ldots\}$ is called the T-cyclic subspace spanned by β and it will be denoted by $C(\beta, T)$.

Lemma 7.1.3. *If V is a vector space of finite dimension, $T : V \to V$ is an operator, α and β are elements of V such that $A_\alpha(x)$ and $A_\beta(x)$ are relatively prime, then*

$$A_{\alpha+\beta}(x) = A_\alpha(x)A_\beta(x).$$

Proof. With the assumptions on α and β, we will show that the cyclic subspaces spanned by α and β, respectively, form a direct sum, that is, $C(\alpha, T) \cap C(\beta, T) = \{0\}$. If $\gamma \in C(\alpha, T) \cap$

$C(\beta, T) \setminus \{0\}$, then $A_\gamma(x)$ has positive degree and $\gamma = f_1(T)(\alpha) = f_2(T)(\beta)$, for some polynomials $f_1(x)$ and $f_2(x)$. From this, one has $A_\alpha(T)(\gamma) = A_\alpha(T)(f_1(T)(\alpha)) = f_1(T)(A_\alpha(T)(\alpha)) = f_1(T)(0) = 0$. Likewise, one shows that $A_\beta(T)(\gamma) = 0$, therefore $A_\gamma(x)$ divides both $A_\alpha(x)$ and $A_\beta(x)$, which contradicts that $A_\alpha(x)$ and $A_\beta(x)$ are relatively prime. To avoid the contradiction, we must have $\gamma = 0$.

By evaluating, one verifies that $A_\alpha(T)A_\beta(T)(\alpha + \beta) = 0$. This equation implies that $A_{\alpha+\beta}(x)$ divides $A_\alpha(x)A_\beta(x)$. We also have that $A_{\alpha+\beta}(T)(\alpha+\beta) = A_{\alpha+\beta}(T)(\alpha)+A_{\alpha+\beta}(T)(\beta) = 0$, that is, $A_{\alpha+\beta}(T)(\alpha) = -A_{\alpha+\beta}(T)(\beta) \in C(\alpha, T) \cap C(\beta, T) = \{0\}$. From this one concludes that $A_\alpha(x)$ and $A_\beta(x)$ both divide $A_{\alpha+\beta}(x)$. Now, since $A_\alpha(x)$ and $A_\beta(x)$ are relatively prime, applying Corollary A.4.3, p. 275, their product also divides $A_{\alpha+\beta}(x)$, implying that $A_{\alpha+\beta}(x) = A_\alpha(x)A_\beta(x)$. $\qquad\square$

Let V be a finite-dimensional vector space, and let $T : V \to V$ be an operator. If $\alpha \in V$, then, applying Lemma 7.1.1, we have that $A_\alpha(x)$ divides $m_T(x)$, thus $\deg(A_\alpha(x)) \le \deg(m_T(x)) \le n^2$. From this, one concludes that there is $\alpha_0 \in V$ so that $\deg(A_\beta(x)) \le \deg(A_{\alpha_0}(x))$ for every $\beta \in V$.

The next lemma establishes a more precise relationship between $A_{\alpha_0}(x)$ and $A_\beta(x)$, for any $\beta \in V$.

Lemma 7.1.4. *Let $\alpha_0 \in V$ be such that $\deg(A_\beta(x)) \le \deg(A_{\alpha_0}(x))$ for every $\beta \in V$. Then $A_\beta(x)$ divides $A_{\alpha_0}(x)$.*

Proof. From Theorem A.4.7, p. 277, we can represent $A_{\alpha_0}(x)$ and $A_\beta(x)$ as a product of irreducible polynomials. We may choose $p_1(x), \dots, p_r(x)$ to represent both. More precisely, write $A_{\alpha_0}(x) = p_1^{e_1}(x)p_2^{e_2}(x) \cdots p_r^{e_r}(x)$ and $A_\beta(x) = p_1^{a_1}(x)p_2^{a_2}(x) \cdots p_r^{a_r}(x)$. If the conclusion of the lemma were false, then $a_i > e_i$ for some i. Set $h(x) = \frac{A_\beta(x)}{p_i^{a_i}(x)}$, $\gamma_1 = p_i^{e_i}(T)(\alpha_0)$, and $\gamma_2 = h(T)(\beta)$. It is clear that the T-annihilator of γ_1 is $\frac{A_{\alpha_0}(x)}{p_i^{e_i}(x)}$ and the T-annihilator of γ_2 is $p_i^{a_i}(x)$. These polynomials are relatively prime. Hence, applying Lemma 7.1.3, it follows that the T-annihilator of $\gamma_1+\gamma_2$ is $\frac{A_{\alpha_0}(x)}{p_i^{e_i}(x)}p_i^{a_i}(x)$ which has degree greater than $\deg(A_{\alpha_0}(x))$, contrary to the choice of α_0. The proof of the lemma is completed. $\qquad\square$

Theorem 7.1.3. *Let V be a finite-dimensional vector space. If $T : V \to V$ is an operator, then there is $\alpha \in V$ such that $A_\alpha(x) = m_T(x)$.*

Proof. Let $\{\alpha_1, \alpha_2, \dots, \alpha_n\}$ be a basis of V, and let $\alpha \in V$ be such that the degree of its annihilator is maximal among the degrees of all annihilators of elements of V. By Lemma 7.1.4, we have that $A_{\alpha_i}(x)$ divides $A_\alpha(x)$. Therefore $A_\alpha(T)(\alpha_i) = 0$ for every $i \in \{1, 2, \dots, n\}$. From this, one has that $A_\alpha(T) = 0$, thus necessarily $m_T(x) = A_\alpha(x)$, finishing the proof. $\qquad\square$

Remark 7.1.5. Let V be a vector space of dimension n and let $T : V \to V$ be an operator. If $\alpha \in V$, then the set $\{\alpha, T(\alpha), T^2(\alpha), \dots, T^n(\alpha)\}$ is linearly dependent, consequently the annihilator of α has degree $\le n$.

Now, we are ready to improve a bound for the degree of the minimal polynomial of an operator.

Theorem 7.1.4. *If $T : V \to V$ is an operator, with $\dim(V) = n$, then the minimal polynomial of T has degree at most n.*

Proof. Direct from Theorem 7.1.3 and Remark 7.1.5. □

7.1.3 A minimal polynomial algorithm

In this section we present a procedure to compute the minimal polynomial of a matrix. The main idea is to construct an element whose annihilator is the minimal polynomial of T, Theorem 7.1.3. In Chapter 8, Section 8.2, a second method to compute the minimal polynomial will be presented. The latest method uses the Smith normal form of the matrix $A - xI_n$. This method has a little disadvantage, since the calculations are performed with matrices whose entries are polynomials. However, it has a point in its favor – this method allows us to obtain the invariant factors of the matrix A.

For polynomials $m_1(x), m_2(x), \ldots, m_k(x)$, the least common multiple will be denoted by $\mathrm{LCM}\{m_1(x), m_2(x), \ldots, m_k(x)\}$, see Definition A.4.3, p. 275, and subsequent results to consult properties of the LCM of polynomials.

Theorem 7.1.5. *Let V be vector space of dimension n and let $T : V \to V$ be an operator. Assume that $\{a_1, a_2, \ldots, a_n\}$ is a basis of V. Let $m_i(x)$ denote the annihilator of a_i, for $i \in \{1, 2, \ldots, n\}$, and let $m_T(x)$ denote the minimal polynomial of T. Then $m_T(x)$ is the least common multiple of $m_1(x), m_2(x), \ldots, m_n(x)$.*

Proof. Let $p_1(x), \ldots, p_r(x)$ be irreducible polynomials so that $m_T(x)$ and each $m_i(x)$ can be written as a product of those irreducible polynomials. More precisely, assume that $m_i(x) = p_1^{e_{1i}} \cdots p_r^{e_{ri}}$, $i \in \{1, 2, \ldots, n\}$ and $m_T(x) = p_1^{l_1} \cdots p_r^{l_r}$. From the representations of each $m_i(x)$, one has that $\mathrm{LCM}\{m_1(x), m_2(x), \ldots, m_n(x)\} = p_1^{d_1} \cdots p_r^{d_r} = g(x)$, where $d_j = \max\{e_{ji}\}$ and the "max" is taken over the index $i \in \{1, 2, \ldots, n\}$. We know that $m_i(x)$ divides $m_T(x)$, hence $e_{ji} \leq l_j$, for all $j \in \{1, 2, \ldots, r\}$ and $i \in \{1, 2 \ldots, n\}$. Thus $d_j \leq l_j$ for all $j \in \{1, 2, \ldots, r\}$, proving that $g(x)$ divides $m_T(x)$. We also have that $p_1^{d_1}(T) \cdots p_r^{d_r}(T)(a_i) = 0$ for all $i \in \{1, 2, \ldots, n\}$, hence $p_1^{d_1}(T) \cdots p_r^{d_r}(T) = 0$, therefore $m_T(x)$ divides $p_1^{d_1}(T) \cdots p_r^{d_r}(T)$, finishing the proof. □

In Theorem 7.1.3, we proved that if $T : V \to V$ is an operator, then there is $\alpha \in V$ so that $A_\alpha(x) = m_T(x)$. This result will be improved in the sense that we will propose an algorithm to construct such an element. This is done in Theorem 7.1.6. Before proving such a theorem, we need a lemma.

Lemma 7.1.5. *With the assumptions as in Theorem 7.1.5, let α and β be elements of V with $m_1(x) = A_\alpha(x)$ and $m_2(x) = A_\beta(x)$, their corresponding annihilators. Then there is $\gamma \in V$ so that $A_\gamma(x) = \mathrm{LCM}\{m_1(x), m_2(x)\}$.*

Proof. If $\gcd(m_1(x), m_2(x)) = 1$, then Lemma 7.1.3 guarantees that $\gamma = \alpha + \beta$ has annihilator $m_1(x)m_2(x)$, which in this case is the least common multiple of $m_1(x)$ and $m_2(x)$. We may assume that $\gcd(m_1(x), m_2(x)) = d(x)$ has positive degree. Let $m_1(x) = n_1(x)d(x)$ and $m_2(x) = n_2(x)d(x)$, thus $\gcd(n_1(x), m_2(x)) = 1$. Also, the annihilator of $d(T)(a)$ is $n_1(x)$. Set $\gamma = d(T)(a) + \beta$. From the cited result, the annihilator of γ is $n_1(x)m_2(x) = \frac{m_1(x)}{d(x)}m_2(x) = \text{LCM}\{m_1(x), m_2(x)\}$, finishing the proof. □

The following theorem provides a method to find an element in V whose annihilator is the minimal polynomial of the operator T.

Theorem 7.1.6. *Let $T : V \to V$ be an operator, and let $\{a_1, a_2, \ldots, a_n\}$ be a basis of V. If $m_i(x) = A_{a_i}(x)$ for every $i \in \{1, 2 \ldots, n\}$, then there is $\gamma \in V$ so that*

$$A_\gamma(x) = \text{LCM}\{m_1(x), m_2(x), \ldots, m_n(x)\} = m_T(x).$$

Proof. By Lemma 7.1.5, there is $\gamma_1 \in V$ so that $A_{\gamma_1}(x) = \text{LCM}\{m_1(x), m_2(x)\}$. Call $M_1(x) = A_{\gamma_1}(x)$. By a recursive process, we can construct $\gamma_2 \in V$ so that $A_{\gamma_2}(x) = \text{LCM}\{M_1(x), m_3(x)\}$. Set $M_2(x) = A_{\gamma_2}(x)$. Continuing the process, we may assume that we have constructed $\gamma_{n-2} \in V$ so that $M_{n-2}(x) = A_{\gamma_{n-2}}(x) = \text{LCM}\{M_{n-3}(x), m_{n-1}(x)\}$. A final application of Lemma 7.1.5 shows that there is $\gamma_{n-1} \in V$ such that

$$
\begin{aligned}
M_{n-1}(x) &= A_{\gamma_{n-1}}(x) \\
&= \text{LCM}\{M_{n-2}(x), m_n(x)\} \\
&= \text{LCM}\{m_1(x), m_2(x), \ldots, m_n(x)\} \quad \text{(Theorem A.4.5, p. 276)} \\
&= m_T(x) \quad \text{(Theorem 7.1.5)}.
\end{aligned}
$$

To finish the proof, set $\gamma = \gamma_{n-1}$. □

The discussion above is the foundation of the following algorithm, which allows us to compute the minimal polynomial of a matrix. We present the algorithm, as well as its code in SageMath.

Algorithm 7.1.1.

Require: Square matrix A of size $n \times n$, a basis of $\mathbb{R}^n$ $\{v_1, v_2, \ldots, v_n\}$

1: Let $m(x) = 1$
2: **for** $i \in \{1, 2, \ldots, n\}$ **do**
3:     **if** $m = 0$ **then**
4:         *break*
5:     **end if**
6:     $m(x) = \text{LCM}(\text{ann}(v_i, x), m(x))$
7: **end for**
8: *return* $m(x)$

In order to show how Algorithm 7.1.1 works, we include its SageMath code.

```
def annihilator( v, T ):
    if not T.is_square():
        # If T is not a square matrix, just raise
        # an exception.
        raise TypeException("Not a square matrix.")
    n = T.nrows()
    # The highest power of T so far
    highest_power = identity_matrix(T.base_ring(),n)
    # The following list stores applications of powers
    # of T to v so far.
    applications = [v]
    V=VectorSpace(T.base_ring(),n)
    # This list will store the annihilator's coefficients.
    polynomial_coefficients = []
    # In this loop we iteratively add powers of T applied to
    # v to `applications`, and we stop until we have detected
    # the elements of `applications` are linearly dependent.
    for i in range(n):
        highest_power = T * highest_power
        applications.append(highest_power*v)
        coeffs_list = V.linear_dependence( applications )
        # Note that the list `coeffs_list` is non-empty (and thus not None)
        # iff the elements of `applications` are linearly dependent.
        if coeffs_list:
            # Given this iterative process, `coeffs_list` has a unique
            # element, and the coefficients in that element will define
            # the coefficients of the annihilator.
            polynomial_coefficients = coeffs_list[0].list()
            break
    return PolynomialRing( T.base_ring(), 'x')(polynomial_coefficients).monic()

def minpoly( A, basis ):
    R = PolynomialRing( A.base_ring(), 'x')
    # We begin the iterative process with polynomial 1.
    min_poly = R(1)
    # We now iteratively compute the least common multiple of the
    # annihilators of the elements in the basis, this will result
    # in the minimal polynomial of A.
    for v in basis:
        if min_poly(A).is_zero():
            break
        min_poly = min_poly.lcm(annihilator(v,A))
    return min_poly
```

7.1.4 Computing minimal polynomials with SageMath

In this subsection we show how to apply Algorithm 7.1.1, using SageMath.

Recall that Algorithm 7.1.1 needs as input an $n \times n$ matrix A and a basis of $\mathbb{R}^n$. Then, to use SageMath to compute the minimal polynomial of the matrix A, we need to declare where the entries of A will be taken from. In the following examples, the entries of a matrix, as well as the entries of elements of $\mathbb{R}^n$, will be taken from the rational numbers.

Example 7.1.5. Find the minimal polynomial of the matrix $A = \begin{bmatrix} 1 & 2 & 3 \\ 3 & 2 & 0 \\ -1 & 1 & 1 \end{bmatrix}$.

Discussion. In order to compute the minimal polynomial of A, using SageMath, we need to declare matrix A, as well as a basis for $\mathbb{R}^3$. Since we already have A, we only need to propose a basis for $\mathbb{R}^3$. Consider the subset $S = \{(1,0,2), (1,0,-1), (0,1,1)\} \subseteq \mathbb{R}^3$. Since $\mathbb{R}^3$ has dimension 3, S will be a basis of $\mathbb{R}^3$ if and only if S is linearly independent. We invite the reader to verify that S is linearly independent.

Declaration of A and S in SageMath:

```
A=Matrix(QQ,[[1,2,3],[3,2,0],[-1,1,1]])

S=[vector(QQ,[[1],[0],[2]]),vector(QQ,[[1],[0],[-1]]),
    vector(QQ,[[0],[1],[1]])]
```

Once you have declared the matrix and basis, use the following command:

```
minpoly(A,S)
```

whose output is $x^3 - 4x^2 + 2x - 11$.

This result can be phrased as follows: the minimal polynomial of $A = \begin{bmatrix} 1 & 2 & 3 \\ 3 & 2 & 0 \\ -1 & 1 & 1 \end{bmatrix}$ is $m_A(x) = x^3 - 4x^2 + 2x - 11$.

Example 7.1.6. Find the minimal polynomial of matrix $B = \begin{bmatrix} 3 & 0 & 1 \\ 1 & -2 & 4 \\ 1 & 4 & 9 \end{bmatrix}$.

Discussion. Since the matrix B is 3×3, we can use the same basis as before and declare only the matrix B:

```
B=Matrix(QQ,[[3,0,1],[1,-2,4],[1,4,9]])
```

From what has been declared, we have that the output of the command

```
minpoly(B,S)
```

is the minimal polynomial of $B = \begin{pmatrix} 3 & 0 & 1 \\ 1 & -2 & 4 \\ 1 & 4 & 9 \end{pmatrix}$, that is, $m_B(x) = x^3 - 10x^2 - 14x + 96$.

Example 7.1.7. Find the minimal polynomial of the matrix $C = \begin{bmatrix} 0 & 0 & 0 & 0 \\ 1 & 0 & 0 & 0 \\ 0 & 1 & 0 & 0 \\ 0 & 0 & 1 & 0 \end{bmatrix}$.

Discussion. Since the matrix C is 4×4, we need to declare a basis for $\mathbb{R}^4$.

Let $S = \{(1,0,2,0), (1,0,-1,0), (0,0,1,-1), (0,1,1,1)\} \subseteq \mathbb{R}^4$. We argue as in Example 7.1.5: since S has 4 elements and $\mathbb{R}^4$ has dimension 4, we only need to verify that S is linearly independent. To do so, we consider the equation

$$x(1,0,2,0) + y(1,0,-1,0) + z(0,0,1,-1) + w(0,1,1,1) = (0,0,0,0) \tag{7.12}$$

and need to show that the only solution of this equation is $x = y = z = w = 0$. Equation (7.12) is equivalent to the system of linear equations

$$\begin{aligned} x + y &= 0, \\ w &= 0, \\ 2x - y + z + w &= 0, \\ -z + w &= 0. \end{aligned} \tag{7.13}$$

It is straightforward to verify that the only solution to the system of linear equations (7.13) is $x = y = z = w = 0$. Hence S is a basis of $\mathbb{R}^4$.

To compute the minimal polynomial of C, we need to declare C and S in SageMath.

```
S = [vector(QQ,[[1],[0],[2],[0]]), vector(QQ,[[1],[0],[-1],[0]]),
     vector(QQ,[[0],[0],[1],[-1]]),vector(QQ,[[0],[1],[1],[1]])]

C= Matrix(QQ,[[0,0,0,0],[1,0,0,0],[0,1,0,0],[0,0,1,0]])
```

Again, to obtain the minimal polynomial of C, use the command

```
minpoly(C,S)
```

whose output is x^4, hence the minimal polynomial of C is $m_C(x) = x^4$.

After having presented some examples, we consider appropriate to invite the reader to solve a few exercises using SageMath.

Exercise 7.1.3. Compute the minimal polynomial of the following matrices:

1. $A = \begin{bmatrix} 2 & -3 & 1 \\ 0 & -2 & 4 \\ 1 & 4 & 7 \end{bmatrix}$;

2. $B = \begin{bmatrix} 2 & -3 & 1 & 2 & 0 & -1 \\ 0 & -2 & 4 & 0 & -2 & 1 \\ 1 & 4 & 7 & 0 & 10 & 9 \\ -3 & 8 & 4 & 5 & 2 & 4 \\ 0 & 2 & 4 & 10 & 2 & 1 \\ -3 & 2 & 8 & 7 & 5 & 4 \end{bmatrix}$;

3. $C = \begin{bmatrix} 2 & -3 & 1 & 0 \\ 0 & -2 & 4 & -1 \\ 1 & 4 & 7 & 3 \\ 2 & 3 & 0 & -2 \end{bmatrix}$.

7.2 Primary decomposition of a vector space

In what follows, the primary decomposition theorem, Theorem 7.2.1, will play a crucial role. Before we state and give a proof of this theorem, we will recall the definition of the direct sum of more than two subspaces.

Given a vector space V and a collection $W_1, W_2, \ldots, W_k$ of subspaces of V, their sum is defined by

$$W_1 + W_2 + \cdots + W_k := \{a_1 + a_2 + \cdots + a_k \mid a_i \in W_i, \text{ for } i = 1, 2, \ldots, k\}. \qquad (7.14)$$

When the condition that $a_1 + a_2 + \cdots + a_k = 0$ implies $a_i = 0$, for $i \in \{1, 2, \ldots, k\}$, holds, we say that the subspaces form a direct sum, and in this case the notation

$$W_1 \oplus W_2 \oplus \cdots \oplus W_k = \{a_1 + a_2 + \cdots + a_k \mid a_i \in W_i, \text{ for } i = 1, 2, \ldots, k\}$$

is used.

For instance, in $\mathbb{R}^3$ the subspaces $W_1 = \{(t, t, -t) : t \in \mathbb{R}\}$, $W_2 = \{(u, u/2, u) : u \in \mathbb{R}\}$, and $W_3 = \{(s, -s, s) : s \in \mathbb{R}\}$ form a direct sum. This is true since, given $(t, t, t) \in W_1$, $(u, u/2, u) \in W_2$, and $(s, -s, s) \in W_3$ such that $(t, t, t) + (u, u/2, u) + (s, -s, s) = (0, 0, 0)$, one has the system of equations

$$t + u + s = 0,$$
$$t + u/2 - s = 0,$$
$$-t + u + s = 0.$$

It can be verified that the only solution is $s = t = u = 0$.

Notice that each of the subspaces in the example has dimension 1.

Exercise 7.2.1. Let $W_1, W_2, \ldots, W_k$ be subspaces of V. Prove that $W_1, W_2, \ldots, W_k$ form a direct sum if and only if $W_i \cap \left(\sum_{j \neq i} W_j\right) = \{0\}$, for every $i \in \{1, 2, \ldots, k\}$.

The following lemma is needed in the proof of Theorem 7.2.1.

Lemma 7.2.1. *Let V be a finite-dimensional vector space, and let T be an operator on V. Assume that $m_T(x) = p_1^{r_1}(x) p_2^{r_2}(x) \cdots p_k^{r_k}(x)$ is the factorization of the minimal polynomial of T as a product of irreducible factors. For each $i \in \{1, 2, \ldots k\}$, set $f_i(x) = \prod_{j \neq i} p_j^{r_j}(x)$. Then:*

1. *There are polynomials $g_1(x), g_2(x), \ldots, g_k(x)$ such that*

$$f_1(x)g_1(x) + f_2(x)g_2(x) + \cdots + f_k(x)g_k(x) = 1.$$

2. *If for each $i \in \{1, 2, \ldots k\}$, we set $T_i = f_i(T)g_i(T)$, and then*
   (a) *$T_1 + T_2 + \cdots + T_k = I$, the identity operator,*
   (b) *if $i \neq j$, then $T_i T_j = 0$,*
   (c) *for every $i \in \{1, 2, \ldots, k\}$, $T_i^2 = T_i$.*

Proof. To prove claim 1, we note that in the representation $m_T(x) = p_1^{r_1}(x)p_2^{r_2}(x)\cdots p_k^{r_k}(x)$, the polynomials $p_i(x)$ are irreducible and pairwise different. From the definition of $f_i(x)$, its irreducible divisors are $p_j(x)$ for every $j \neq i$. From this, we have that there is no irreducible polynomial that divides all of $f_1(x), f_2(x), \ldots, f_k(x)$, hence the polynomials are relatively prime. Applying Corollary A.4.2, p. 275, there are polynomials $g_1(x), g_2(x), \ldots, g_k(x)$ such that

$$f_1(x)g_1(x) + f_2(x)g_2(x) + \cdots + f_k(x)g_k(x) = 1. \tag{7.15}$$

The proof of claim 2a follows directly by evaluating (7.15) at T.

To prove claim 2b, we notice that $m_T(x)$ divides $f_i(x)f_j(x)$ for $j \neq i$, hence $m_T(x)$ divides $f_i(x)g_i(x)f_j(x)g_j(x)$, and from this $T_iT_j = f_i(T)g_i(T)f_j(T)g_j(T) = 0$ for every $i \neq j$.

From claims 2a and 2b above, we have

$$T_iI = T_i(T_1 + T_2 + \cdots + T_i + \cdots + T_k)$$
$$T_i = T_iT_1 + T_1T_2 + \cdots + T_iT_i + \cdots + T_iT_k$$
$$= T_i^2,$$

finishing the proof of the lemma. □

Theorem 7.2.1 (Primary decomposition theorem). *Let V be a finite-dimensional vector space, and let T be an operator on V. Assume that $m_T(x) = p_1^{r_1}(x)p_2^{r_2}(x)\cdots p_k^{r_k}(x)$ is the factorization of the minimal polynomial of T as a product of irreducible factors. If W_i is the kernel of $p_i^{r_i}(T)$, then:*
1. *each W_i is T-invariant and $V = W_1 \oplus W_2 \oplus \cdots \oplus W_k$,*
2. *the minimal polynomial of T restricted to W_i is $p_i^{r_i}(x)$.*

Proof. In the proof, we will be using the notation of Lemma 7.2.1. Notice that if $\alpha \in W_i$, then $p_i^{r_i}(T)(\alpha) = 0$ and $p_i^{r_i}(T)(T(\alpha)) = T(p_i^{r_i}(T)(\alpha)) = 0$. That is, $T(\alpha) \in W_i$, or W_i is T-invariant.

From Lemma 7.2.1(2a), we have that $T_1 + T_2 + \cdots + T_k = I$, hence $T_1(V) + T_2(V) + \cdots + T_k(V) = I(V) = V$.

To finish the proof of claim 1, we will prove that $T_i(V) = W_i$, and $T_1(V), T_2(V), \ldots, T_k(V)$ form a direct sum.

If $\beta \in T_i(V)$, then $\beta = T_i(\alpha) = f_i(T)g_i(T)(\alpha)$ for some $\alpha \in V$. From this representation for β, we have $p_i^{r_i}(T)(\beta) = p_i^{r_i}(T)f_i(T)g_i(T)(\alpha) = m(T)g_i(T)(\alpha) = 0$, proving that β is in the kernel of $p_i^{r_i}(T)$, which is W_i, that is, we have shown that $T_i(V) \subseteq W_i$.

Conversely, if $\alpha \in W_i$, then the equation $T_1 + T_2 + \cdots + T_k = I$ implies $\alpha = I(\alpha) = T_1(\alpha) + T_2(\alpha) + \cdots + T_k(\alpha)$. From the definition of $f_j(x)$ (Lemma 7.2.1), we have that $p_i^{r_i}(x)$ divides $f_j(x)$ for every $j \neq i$. That is, $f_j(x) = h_i(x)p_i^{r_i}(x)$ for some $h_i(x)$, hence $f_j(T)(\alpha) = h_i(T)p_i^{r_i}(T)(\alpha) = h_i(T)(0) = 0$. From this, $T_j(\alpha) = g_j(T)f_j(T)(\alpha) = 0$ for every $j \neq i$. It now follows that $\alpha = T_i(\alpha) \in T_i(V)$. So far we have proved that $W_i = T_i(V)$.

We now show that $T_i(V) \cap (\sum_{j \neq i} T_j(V)) = \{0\}$, for every $i \in \{1, 2, \ldots, k\}$. If $\alpha \in T_i(V) \cap (\sum_{j \neq i} T_j(V))$, then α can be represented as $\alpha = T_i(\gamma) = \sum_{j \neq i} T_j(\alpha_j)$. Applying T_i to the latter equation and using that $T_i T_j = 0$ for $j \neq i$ (Lemma 7.2.1(2b)), one has $T_i(\alpha) = T_i^2(\gamma) = \sum_{j \neq i} T_i T_j(\alpha_j) = 0$. Now using that $T_i^2 = T_i$, it follows that $0 = T_i(\alpha) = T_i^2(\gamma) = T_i(\gamma) = \alpha$, as was claimed, finishing the proof of claim 1.

The proof of claim 2 is as follows: since W_i is T invariant, let $n(x)$ be the minimal polynomial of T restricted to W_i. Also, we have that $p_i^{r_i}(T)$ is the zero operator on W_i. Thus $n(x)$ divides $p_i^{r_i}(x)$. To finish the proof, notice that if $\alpha = \omega_1 + \omega_2 + \cdots + \omega_k$, with $\omega_i \in W_i$, is any element in V, then $f_i(T)(\alpha) = f_i(T)(\omega_i)$. In other words, the image of $f_i(T)$ is contained in W_i. From this, if $h(x)$ is any polynomial so that $h(E_i) = 0$, with E_i the restriction of T to W_i, then $h(T)f_i(T)(\alpha) = f_i(T)h(T)(\omega_i) = 0$. In other words, $h(T)f_i(T)$ is the zero operator on V, thus $m_T(x) = p_i^{r_i}(x)f_i(x)$ divides $h(x)f_i(x)$, hence $p_i^{r_i}(x)$ divides $h(x)$, in particular $p_i^{r_i}(x)$ divides $n(x)$. Summarizing, $p_i^{r_i}(x)$ is the minimal polynomial of T restricted to W_i. □

The direct sum decomposition of V in Theorem 7.2.1 will play an important role in what follows, it is why we state it in a definition.

Definition 7.2.1. With the notation and assumptions as in Theorem 7.2.1, the representation $V = W_1 \oplus W_2 \oplus \cdots \oplus W_k$ will be called the T-primary decomposition of V.

From Theorem 7.2.1, now it is interesting to know what more we can be said when the minimal polynomial of an operator is a power of an irreducible polynomial. The next result provides a relationship between the degree of the minimal polynomial of an operator T and the dimension of the vector space, where it acts.

Theorem 7.2.2. *Let V be a vector space of dimension n, and let T be an operator on V. If $m_T(x) = p^l(x)$ is the minimal polynomial of T, with $p(x)$ irreducible of degree r, then r divides n.*

Proof. We will apply induction on $\dim(V)$. If $\dim(V) = 1$, then a basis of V has one element, say α, hence $T(\alpha) = c\alpha$ for some c. From this we have that $T - cI$ is the zero operator, then $m_T(x) = x - c$ is irreducible of degree 1, the conclusion follows.

Now, let us assume that $\dim(V) > 1$ and that the result holds for all vector spaces of dimension less than $\dim(V)$ and for all operators whose minimal polynomial is a power of an irreducible polynomial.

From Theorem 7.1.3, p. 174, there exists $\alpha \in V$ so that $A_\alpha(x) = p^l(x)$. For this α, setting $W = C(\alpha, T)$ and applying Lemma 7.1.2, p. 173, we have that the dimension of W is rl, which is the degree of $p^l(x)$.

If $W = V$, then $n = rl$ and the proof is complete. Thus, we may assume that $W \neq V$ and choose U, a maximal T-invariant subspace which contains W and $U \neq V$. We have that the minimal polynomial of T restricted to U is a power of $p(x)$. Applying the induction assumption to U, one has that r divides its dimension.

Let $\gamma \in V \setminus U$ and $g(x)$ be as in Lemma 7.1.1 applied to U, that is, $g(T)(\gamma) \in U$, hence the same lemma implies that $g(x)$ divides $p^l(x)$, therefore $g(x) = p^k(x)$, for some k. The condition on the degree of $g(x)$ implies that $W_1 := \mathcal{L}\{\gamma, T(\gamma), \ldots, T^{kr-1}(\gamma)\}$ has dimension kr. The same condition implies that U and W_1 form a direct sum.

Claim. $U + W_1$ is T-invariant. In fact, if $\beta = u + c_0\gamma + c_1 T(\gamma) + \cdots + c_{kr-1} T^{kr-1}(\gamma) \in U + W_1$, with $u \in U$, then $T(\beta) = T(u) + c_0 T(\gamma) + \cdots + c_{kr-2} T^{kr-1}(\gamma) + c_{kr-1} T^{kr}(\gamma)$. Since $\deg(g(x)) = kr$, $c_{kr-1} T^{kr}(\gamma) \in U + W_1$. By assumption, U is T-invariant, hence $T(\beta) \in U + W_1$. Now, the maximality condition on U implies that $V = U \oplus W_1$. From this equation, we have $\dim(V) = \dim(U) + \dim(W_1)$. The conclusion follows, since r divides the dimension of W_1 and the dimension of U. $\qquad \square$

Corollary 7.2.1. *Let V be a finite-dimensional vector space, and let T be an operator on V. If $V = W_1 \oplus W_2 \oplus \cdots \oplus W_k$ is the T-primary decomposition of V, then the degree of $p_i(x)$ divides $\dim(W_i)$ for every $i \in \{1, 2, \ldots, k\}$.*

Proof. Each W_i is T-invariant and the minimal polynomial of T restricted to W_i is $p_i^{r_i}(x)$. The conclusion follows from Theorem 7.2.2. $\qquad \square$

7.2.1 Invariant subspaces of dimension one or two

The existence of eigenvalues and eigenvectors of a matrix or operator is closely related to the existence of invariant subspaces. The case of eigenvalues is an algebraic property of the real numbers, since it has to do with the existence of a linear factor of the minimal polynomial of an operator. When the scalars are allowed to be complex numbers, according to the fundamental theorem of algebra, Theorem A.5.3, p. 279, every operator has characteristic values. However, for the case of real vector spaces, the story is a little bit different. The best we can say is that there are invariant subspaces of dimension one or two.

Theorem 7.2.3. *Let V be vector space of dimension $n \geq 1$. If T is an operator on V, then V contains at least one T-invariant subspace of dimension one or two.*

Proof. Since V has dimension greater or equal to one, we can choose $\alpha \in V \setminus \{0\}$. Note that the subset $\{\alpha, T(\alpha), \ldots, T^n(\alpha)\}$ is linearly dependent. Hence, there are scalars $a_0, a_1, \ldots, a_n$, not all zero, so that $a_0\alpha + a_1 T(\alpha) + \cdots + a_n T^n(\alpha) = 0$. Set $p(x) = a_0 + a_1 x + \cdots + a_n x^n$. Notice that the assumption on α guarantees that $p(x)$ has positive degree. From Corollary A.5.3, p. 280, $p(x)$ factorizes as a product of linear or quadratic factors with real coefficients, that is, $p(x) = p_1(x)p_2(x) \cdots p_t(x)$, with each $p_i(x)$ linear or quadratic.

We have that $p(T)$ is singular, hence at least one of the operators, $p_i(T)$, is also singular. If $p_i(x) = ax + b$, with $a \neq 0$, then there is $\beta \neq 0$ so that $aT(\beta) + b\beta = 0$. From the latter equation, one concludes that the subspace spanned by β is T-invariant of dimension one. If $p_i(x) = ax^2 + bx + c$, with $a \neq 0$, then there is $\beta \neq 0$ so that $aT^2(\beta) + bT(\beta) + c\beta = 0$.

From this, one has that the subspace spanned by $\{\beta, T(\beta)\}$ is T-invariant and has dimension one or two. The irreducibility assumption on $p_i(x) = ax^2 + bx + c$ implies that the subspace spanned by $\{\beta, T(\beta)\}$ has dimension two, finishing the proof. $\square$

Theorem 7.2.4. *Let V be a vector space of odd dimension. If T is an operator on V, then T has an invariant subspace of dimension one.*

Proof. Let $n = 2m + 1$ denote the dimension of (V). We will apply induction on m. If $m = 0$, then the result is true, since V has dimension one and V itself is an invariant subspace.

Assume that $m > 0$ and the result is true for all spaces of dimension $n = 2l + 1$, with $l < m$. Let $m_T(x) = p_1^{r_1}(x)p_2^{r_2}(x)\cdots p_k^{r_k}(x)$ be the factorization, as a product of irreducible factors, of the minimal polynomial of T. From Corollary A.5.3, p. 280, each $p_i(x)$ has degree one or two. If $k = 1$, then $m_T(x) = p_1^{r_1}(x)$ has to be linear, otherwise Theorem 7.2.2 implies that the degree of $p_1(x)$, which is two, divides the dimension of V, which is impossible since V has odd dimension. From this, one concludes that $m_T(x)$ is a power of a linear polynomial, say $m_T(x) = (x - c)^s$. If $s = 1$, then $m_T(x) = x - c$ which implies that T is a multiple of the identity and any subspace of dimension one is invariant. If $s > 1$, then there is $\alpha \in V$ such that $(T - cI)^{s-1}(\alpha) \neq 0$ and $(T - cI)^s(\alpha) = 0$. Let $\beta = (T - cI)^{s-1}(\alpha)$, then $T(\beta) = c\beta$, which implies that β generates an invariant subspace of dimension one.

Assume that $k > 1$ and let $V = W_1 \oplus W_2 \oplus \cdots \oplus W_k$ be the T-primary decomposition of V. We know that each W_i is T-invariant. Since V has odd dimension, at least one of $W_1, W_2, \ldots, W_k$ has odd dimension. Without loss of generality, we may assume that W_1 has odd dimension. Since $k > 1$, $\dim(W_1) < \dim(V)$. By the induction hypothesis, W_1 has an invariant subspace of dimension one, hence so does V, finishing the proof of the theorem. $\square$

7.2.2 Cayley–Hamilton theorem

In this section we discuss the relationship between the characteristic and minimal polynomials of an operator and give a proof of the well-known and famous Cayley–Hamilton theorem. The proof is based on properties derived from the minimal polynomial. We recall the definition of the characteristic polynomial of a matrix, Definition 7.1.2. Given an $n \times n$ matrix A, the characteristic polynomial of A, denoted $f_A(x)$, is $f_A(x) = |A - xI_n|$.

Remark 7.2.1. Let V be a finite-dimensional vector space, and let T be an operator on V. If A and B are matrices that represent T with respect to some bases, then $f_A(x) = f_B(x)$.

Proof. Since A and B represent T with respect to some bases, then there is a nonsingular matrix P, the matrix that changes bases, such that $B = P^{-1}AP$, hence $f_B(x) = |B - xI_n| = |P^{-1}AP - xI_n| = |P^{-1}(A - xI_n)P| = |P^{-1}|\,|A - xI_n|\,|P| = |A - xI| = f_A(x)$. $\square$

Definition 7.2.2. Given an operator T acting on a finite-dimensional vector space V, the characteristic polynomial of T, $f_T(x)$, is defined as the characteristic polynomial of any matrix representing T in a basis of V.

Theorem 7.2.5. *Let V be a vector space of dimension n, and let $T : V \to V$ be an operator. Assume that the factorization of the minimal polynomial of T, in terms of irreducible polynomials, is given by $m_T(x) = p_1^{r_1}(x)p_2^{r_2}(x)\cdots p_k^{r_k}(x)$. For every $i \in \{1, 2, \ldots, k\}$, let W_i be the kernel of $p_i^{r_i}(T)$, and declare $d_i = \frac{\dim(W_i)}{\deg(p_i(x))}$. Then, d_i is an integer and the characteristic polynomial of T is given by $f_T(x) = (-1)^n p_1^{d_1}(x)p_2^{d_2}(x)\cdots p_k^{d_k}(x)$. Moreover, the minimal polynomial of T divides $f_T(x)$.*

Proof. Let $V = W_1 \oplus W_2 \oplus \cdots \oplus W_k$ be the T-primary decomposition of V, see Theorem 7.2.1. Since each W_i is T-invariant, there is a basis of V with respect to which the matrix representing T is diagonal by blocks. More precisely, if A_i is the matrix representing T restricted to W_i, then

$$A = \mathrm{diag}\{A_1, A_2, \ldots, A_k\}, \tag{7.16}$$

where the size of A_i is equal to $\dim(W_i) = n_i$, for every $i \in \{1, 2, \ldots, k\}$.

From equation (7.16), we have that the characteristic polynomial of A is

$$f_A(x) = |A - xI_n| = |A_1 - xI_{n_1}|\cdots|A_k - xI_{n_k}|, \tag{7.17}$$

where I_{n_i} is the identity matrix of size n_i. Hence, the characteristic polynomial of A is the product of the characteristic polynomials of $A_1, A_2, \ldots, A_k$. In what follows, we compute the characteristic polynomial of these matrices.

From part 2 of Theorem 7.2.1, p. 181, we have that the minimal polynomial of T restricted to W_i is $p_i^{r_i}(x)$, hence this is also the minimal polynomial of A_i. From Theorem 7.2.2, we have that $\deg(p_i(x)) = l_i$ divides $n_i = \dim(W_i)$, thus d_i is an integer.

On the other hand, Corollary 7.1.1 implies that $p_i^{r_i}(x)$ and $f_{A_i}(x)$ have the same irreducible factors, thus $f_{A_i}(x)$ is a power of $p_i(x)$. We know that the degree of $f_{A_i}(x)$ is the dimension of W_i, which is $n_i = l_i d_i$, thus $f_{A_i}(x) = p_i^{d_i}(x)$. From this we obtain that

$$f_T(x) = (-1)^n p_1^{d_1}(x)p_2^{d_2}(x)\cdots p_k^{d_k}(x). \tag{7.18}$$

We also have that Theorem 7.1.4 implies that $\deg(p_i^{r_i}(x)) = l_i r_i \le n_i = d_i l_i$, hence $r_i \le d_i$, thus $m_T(x)$ divides $f_T(x)$, finishing the proof of the theorem. $\square$

Theorem 7.2.5 implies directly the following well-known theorem.

Theorem 7.2.6 (Cayley–Hamilton). *Let V be a vector space of dimension n, T an operator on V. Then T is a zero of $f_T(x)$.*

Proof. From Theorem 7.2.5, we have that $m_T(x)$ divides $f_T(x)$, that is, there is a polynomial $q(x)$ so that $f_T(x) = m_T(x)q(x)$. Then $f_T(T) = m_T(T)q(T) = 0q(T) = 0$. $\square$

Definition 7.2.3. An operator $T : V \to V$ is called nilpotent, if there is a positive integer k so that $T^k = 0$. If T is nilpotent, the least integer k that satisfies $T^k = 0$ is called the nilpotency index. Likewise, a matrix A is said to be nilpotent if there is a positive integer k such that $A^k = 0$.

An interesting corollary of Theorem 7.2.5 is

Corollary 7.2.2. *If V is a vector space of dimension n and T is a nilpotent operator of nilpotency index k acting on V, then the minimal and characteristic polynomial of T are $m_T(x) = x^k$ and $f_T(x) = (-1)^n x^n$, respectively.*

Proof. If k is the nilpotency index of T, then $m_T(x) = x^k$. On the other hand, $f_T(x)$ has degree n and the same irreducible factors as $m_T(x)$, then $f_T(x) = (-1)^n x^n$. $\square$

7.3 Exercises

1. Compute the minimal polynomial of the following matrices:

   (a) $\begin{bmatrix} 1 & 0 & 0 \\ 0 & 1 & 4 \\ 0 & 3 & 0 \end{bmatrix}$, $\begin{bmatrix} 1 & 0 & 3 \\ 0 & 1 & 4 \\ 0 & 3 & 0 \end{bmatrix}$, $\begin{bmatrix} 1 & 0 & 0 \\ -1 & 1 & 0 \\ 0 & 0 & 3 \end{bmatrix}$, $\begin{bmatrix} 3 & -1 & -1 & -2 \\ 1 & 1 & -1 & -1 \\ 1 & 0 & 0 & -1 \\ 0 & -1 & -1 & 1 \end{bmatrix}$.

   (b) $A = (a_{ij})$, where $a_{ij} = \begin{cases} 1 & \text{if } j = i + 1, i \in \{1, 2, \ldots, n-1\}, \\ 0 & \text{otherwise.} \end{cases}$

2. Show that the minimal polynomial of the matrix H from the example "*Temperature distribution on thin plates*", on p. 57, is $m_H(x) = x^5 - 20x^4 + 150x^3 - 520x^2 + 816x - 448 = (x^2 - 8x + 14)(x^2 - 8x + 8)(x - 4)$. Assuming that H acts as an operator on $\mathbb{R}^9$, find the decomposition of $\mathbb{R}^9$, according to Theorem 7.2.1.

3. Assume that A is a square matrix with minimal polynomial $m_A(x)$. Prove that A has an inverse if and only if x does not divide $m_A(x)$.

4. Let A be an $n \times n$ matrix, and let λ be a scalar so that $A - \lambda I$ is nonsingular. Set $B = (A - \lambda I)^{-1}$. Prove that B is diagonalizable if and only if A is such. Moreover, A and B are simultaneously diagonalizable.
   (*Hint.*) If $\{X_1, X_2, \ldots, X_n\}$ is a basis of $\mathbb{R}^n$ consisting of eigenvectors of A, then each X_i is also an eigenvector of B. To prove the converse, notice that the eigenvalues of B are not zero. If X is an eigenvector of B associated to μ, then X is an eigenvector of A associated to $\frac{1+\lambda\mu}{\mu}$.

5. Let p be a prime number of the form $4k + 3$. Prove that there is no 2×2 symmetric matrix with rational entries whose characteristic polynomial is $x^2 - p$.

6. Let A be an $n \times n$ matrix whose entries are all equal to one. Prove that the only eigenvalues of A are n and 0.

7. Let A be an invertible matrix whose minimal polynomial is $m(x) = x^k + a_{k-1}x^{k-1} + \cdots + a_1 x + a_0$. What is the minimal polynomial of A^{-1}? Write A^{-1} as a polynomial in A. If A is diagonalizable, what can you say about A^{-1} concerning diagonalization?

8. Let $p(x)$ be a polynomial and let A be a square matrix. Assuming that c is an eigenvalue of A, prove that $p(c)$ is an eigenvalue of $p(A)$.

9. Prove that similar matrices have equal minimal polynomial.

10. Let A and B be $n \times n$ matrices. Prove that AB and BA have the same eigenvalues. (*Hint.*) The number 0 is an eigenvalue of AB if and only if it is an eigenvalue of BA.

11. Let V be a vector space and let T be an operator acting on V. We say that T is algebraic if there is a nonzero polynomial $f(x)$ such that $f(T) = 0$. If T is algebraic, then there is a polynomial $m(x)$, monic and of minimum degree, such that
    (a) $m(T) = 0$,
    (b) if $g(x)$ is another polynomial such that $G(T) = 0$, then $m(x)$ divides $g(x)$.
    The polynomial $m(x)$ is called the minimal polynomial of T. Is this polynomial the same as that in Theorem 7.1.1?

12. Let A be a triangular matrix. Show that the eigenvalues of A are the elements on its diagonal.

13. Let n be an integer ≥ 2 and let V be the vector space of all $n \times n$ matrices. For a fixed matrix A in V define $T_A : V \rightarrow V$ by $T_A(B) = AB$. Prove that T_A is an operator. If $m_A(x)$ is the minimal polynomial of A, what is the minimal polynomial of T_A?

14. Let n be an integer ≥ 2 and let V be a vector space whose dimension is n. Assume that $\{a_1, a_2, \ldots, a_n\}$ is a basis of V. If $T : V \rightarrow V$ is defined by $T(a_j) = a_{j+1}$ for $j \in \{1, 2, \ldots, n-1\}$ and $T(a_n) = 0$, what is the minimal polynomial of T?

15. Is there a 3×3 matrix whose minimal polynomial is x^2? More generally, assuming that $n > 3$, is there an $n \times n$ matrix whose minimal polynomial is x^2?

16. Let n be a positive integer and let $p(x)$ be a monic polynomial of degree $\leq n$. Is there an $n \times n$ matrix A such that $m_A(x) = p(x)$?

17. Let n be a positive integer and let A be an $n \times n$ matrix that satisfies $A^2 = A$. What are $m_A(x)$ and $f_A(x)$?

18. Let V be a vector space of dimension $n > 1$, and let $T : V \rightarrow V$ be an operator whose minimal polynomial has n different roots. Prove that T can be represented by a diagonal matrix with respect to some basis.
    (*Hint.*) Prove that the eigenvectors associated to the corresponding eigenvalues are linearly independent.

19. Is there a 3×3 matrix A, with real entries, such that $m_A(x) = x^2 + 1$?

20. Let A be a square matrix with minimal polynomial $m_A(x) = a_0 + a_1 x + \cdots + a_{k-1} x^{k-1} + x^k$. Assume that $a_0 \neq 0$. Prove that A has an inverse and express it as a polynomial in A.

21. Let A be an $n \times n$ matrix with minimal polynomial $m_A(x)$, let $p(x)$ be a polynomial and set $B = p(A)$. Prove that B is invertible if and only if $\gcd(m_A(x), p(x)) = 1$.

22. Let n be a positive even integer, and let $\{e_1, e_2, \ldots, e_n\}$ be the canonical basis of $\mathbb{R}^n$. Define $T : \mathbb{R}^n \rightarrow \mathbb{R}^n$ by $T(e_i) = \begin{cases} 0, & \text{if } i \text{ is odd,} \\ e_i & \text{otherwise.} \end{cases}$ What is the minimal polynomial of T?

23. Let V be a vector space of dimension n, and let T be an operator on V such $T^{n-1} \neq 0$ but $T^n = 0$. Prove that there is $a \in V$ such that $S = \{a, T(a), \ldots, T^{n-1}(a)\}$ is a basis of V. What is the matrix associated to T with respect to S?

24. Let V be a vector space of dimension $n > 1$, and let $T : V \to V$ be an operator. Furthermore, assume that the minimal polynomial of T has all its roots in $\mathbb{R}$ and that they are different. Prove that the following statements are equivalent:
    (a) There is $a \in V$ such that $B = \{a, T(a), \dots, T^{n-1}(a)\}$ is a basis of V.
    (b) The minimal polynomial of T has degree n.

25. Let V be a vector space of dimension n, and let $T : V \to V$ be an operator. Prove that the minimal polynomial of T has degree n if and only if there is $a \in V$ such that $S = \{a, T(a), \dots, T^{n-1}(a)\}$ is a basis of V. Furthermore, prove that, under any of the former equivalent conditions, the only operators that commute with T are polynomials in T.

26. If A is an $n \times n$ matrix, then A has rank one if and only if there is $A_1 = (a_1, a_2, \dots, a_n) \in \mathbb{R}^n \setminus \{0\}$ and scalars $c_2, c_3, \dots, c_n$ such that the rows of A are $A_1, c_2 A_1, \dots, c_n A_1$.

27. Let A be as in Exercise 26, then:
    (a) $C = (1, c_2, c_3, \dots, c_n)$ is an eigenvector of A associated to the eigenvalue $\lambda = a_1 + c_2 a_2 + \dots + c_n a_n$.
    (b) The characteristic polynomial of A is $f_A(x) = (-1)^n x^{n-1}(x - \lambda)$.
    (c) $\det(I_n + A) = 1 + \operatorname{tr}(A)$.

28. If $C, D \in \mathbb{R}^n$, then $\det(I_n + C^t D) = 1 + DC^t$.

29. If B is a nonsingular matrix, then $\det(B + C^t D) = (1 + D^t B^{-1} C^t) \det(B)$.

8 Canonical forms of a matrix

One important problem when representing a linear transformation with a matrix with respect to given bases is to find them such that the matrix is as simple as possible. In this respect, one of the simplest cases is when the matrix is diagonal. In this chapter we present results that establish conditions under which the matrix representing a linear transformation is diagonal, triangular, or block diagonal. The more general results in this line of ideas are the rational canonical form or the Jordan canonical form of a matrix. Also, at the end of the chapter we present some examples to illustrate how some of the results discussed in this chapter can be use to model hypothetical situations that involve linear algebra concepts and results.

8.1 Cyclic decomposition and the rational canonical form of a matrix

In this section we present some results that describe the structure of a vector space induced by an operator. For instance, we will discuss conditions under which an operator is diagonalizable. Also, we will find the cyclic decomposition of a vector space induced by the action of an operator and, as a consequence, the rational canonical form of a matrix representing the operator is found.

Definition 8.1.1. Given T, an operator acting on the vector space V, we say that T is diagonalizable if there is a basis of V that consists of eigenvectors of T.

An equivalent form of Definition 8.1.1 is as follows: T is diagonalizable if and only if there is a basis of V such that T is represented by a diagonal matrix with respect to this basis.

For matrices, the definition of being diagonalizable is as follows.

Definition 8.1.2. A square matrix A is diagonalizable if there is a nonsingular matrix P so that $P^{-1}AP$ is diagonal.

Remark 8.1.1. Definitions 8.1.1 and 8.1.2 are related by identifying a matrix with an operator.

Recall that the minimal polynomial of a matrix is defined as it was done for an operator. The next result establishes necessary and sufficient conditions under which an operator (matrix) is diagonalizable.

Theorem 8.1.1. *Let V be a vector space of dimension n, and let T be an operator on V. Then, T is diagonalizable if and only if the minimal polynomial of T, $m_T(x)$, can be represented as the product of different linear factors.*

https://doi.org/10.1515/9783111135915-008

Proof. Assume that T is diagonalizable, then there is a basis of V which consists of eigenvectors and in this basis T is represented by a diagonal matrix A, whose elements on the diagonal are labeled as $a_1, a_2, \ldots, a_n$. We will show that the minimal polynomial of A, which is the minimal polynomial of T, is the product of different linear factors. Without loss of generality, we may assume that $a_1, a_2, \ldots, a_r$ are the different elements on the diagonal of A. We claim that $f(x) = (x - a_1)(x - a_2) \cdots (x - a_r)$ is the minimal polynomial of A. To justify the claim, notice that $f(A) = (A - a_1 I_n)(A - a_2 I_n) \cdots (A - a_r I_n)$ is a diagonal matrix and that each element on its diagonal is of the form $(a_i - a_1)(a_i - a_2) \cdots (a_i - a_r)$, for some i. Also, $a_i \in \{a_1, a_2, \ldots, a_r\}$, thus $(a_i - a_1)(a_i - a_2) \cdots (a_i - a_r) = 0$, which implies that $f(A) = 0$. We also have that the minimal polynomial divides any other polynomial that has A as a root, and then $m_T(x)$ must be a factor of $f(x) = (x - a_1) \cdots (x - a_r)$. If one drops any of the factors of $f(x)$, for instance, the first one, then the first element on the diagonal of $(A - a_2 I_n)(A - a_3 I_n) \cdots (A - a_r I_n)$ is $(a_1 - a_2)(a_1 - a_3) \cdots (a_1 - a_r)$ which is not zero. Thus the minimal polynomial of A must be $m_A(x) = (x - a_1)(x - a_2) \cdots (x - a_r)$, proving the claim.

Conversely, assume that $m_T(x) = (x - c_1)(x - c_2) \cdots (x - c_k)$ is the minimal polynomial of T, with $c_i \neq c_j$ for $i \neq j$. From Theorem 7.2.1, p. 181, we have that $V = W_1 \oplus W_2 \oplus \cdots \oplus W_k$, where W_i is the kernel of $T - c_i I$. In other words, W_i consists of eigenvectors of T associated to c_i, together with zero. Let $\mathcal{B}_i$ be a basis of W_i, then $\mathcal{B} = \bigcup_{i=1}^{k} \mathcal{B}_i$ is a basis of V consisting of eigenvectors, concluding that T is diagonalizable. $\square$

Corollary 8.1.1. *If the characteristic polynomial of a matrix A has no repeated roots, then A is diagonalizable.*

Proof. Since the characteristic and minimal polynomials have the same irreducible factors, the assumption implies directly that the minimal polynomial of A is a product of different linear factors. The result now follows from Theorem 8.1.1. $\square$

A weaker result than Theorem 8.1.1 is when the operator is represented by a triangular matrix. Even though a triangular matrix could have many nonzero entries, several properties can be derived from it being triangular. For example, the determinant of a triangular matrix is the product of the elements on its diagonal. In what follows, we establish when an operator can be represented by a triangular matrix.

Definition 8.1.3. Let V be a finite-dimensional vector space, and let $T : V \to V$ be an operator. We say that T is *triangularizable* if there is a basis of V so that the matrix, associated to T with respect to that basis, is triangular.

Lemma 8.1.1. *Let V be a finite-dimensional vector space, and let $T : V \to V$ be an operator. Furthermore, assume that $W \neq V$ is a T-invariant subspace and that all the roots of $m_T(x)$ are in $\mathbb{R}$. Then there are $a \in V \setminus W$ and c, a root of $m_T(x)$, such that $T(a) - ca \in W$.*

Proof. By assumption, $W \neq V$, thus there is $\beta \in V \setminus W$. Let $g(x)$ be the T-annihilator of β with respect to W (Definition 7.1.5, p. 172). Also, $m_T(x) = (x - c_1)^{r_1}(x - c_2)^{r_2} \cdots (x - c_k)^{r_k}$, for every $c_j \in \mathbb{R}$. From Lemma 7.1.1, p. 172, we have that $g(x)$ divides $m_T(x)$, thus $g(x) =$

$(x-c_1)^{s_1}(x-c_2)^{s_2}\cdots(x-c_k)^{s_k}$, with $s_j \le r_j$ for every j. We also have that $g(x)$ has positive degree, so we may assume that $s_1 > 0$. Thus $g(x) = (x-c_1)h(x)$. From this we have that $g(T)(\beta) = (T-c_1I)h(T)(\beta) \in W$. Since $g(x)$ is of minimum degree, $g(T)(\beta) \in W$, and then $h(T)(\beta) = \alpha \notin W$ is the needed element. □

Theorem 8.1.2. *Let V be a vector space of dimension n, and let T be an operator on V. Then T is triangularizable if and only if $m_T(x)$ is a product of linear factors.*

Proof. Assume that all the roots of $m_T(x)$ are real. Applying Lemma 8.1.1 with $W_0 = \{0\}$, we have that there are $\alpha_1 \ne 0$ and c_1, a root of $m_T(x)$, such that $T(\alpha_1) = c_1\alpha_1$. Let W_1 be the subspace spanned by α_1 and notice that W_1 is T-invariant. If $W_1 \ne V$, apply Lemma 8.1.1 again to conclude that there are α_2 and c_2, a root of $m_T(x)$, so that $T(\alpha_2) - c_2\alpha_2 \in W_1$. By the choice of α_2, α_1 and α_2 are linearly independent. Let W_2 be the subspace spanned by $\{\alpha_1, \alpha_2\}$, then, again, W_2 is T-invariant. If $W_2 \ne V$, then applying Lemma 8.1.1, there are $\alpha_3 \notin W_2$ and c_3, a root of $m_T(x)$, so that $T(\alpha_3) - c_3\alpha_3 \in W_2$. As before, α_1, α_2, and α_3 are linearly independent and, defining W_3 as the span of these three elements, W_3 has dimension 3. Applying the process above n times, we finish constructing a basis $B = \{\alpha_1, \alpha_2, \ldots, \alpha_n\}$ of V such that $T(\alpha_j)$ is a linear combination of $\alpha_1, \alpha_2, \ldots, \alpha_{j-1}$. Translating this to the associated matrix of T with respect to the basis B, we have that such a matrix is triangular.

Conversely, assume that T is triangularizable, then there is a basis of V with respect to which T is represented by a triangular matrix with entries $a_{11}, a_{22}, \ldots, a_{nn}$ on the diagonal. Hence, the characteristic polynomial of T is $f_T(x) = (-1)^n(x-a_{11})(x-a_{22})\cdots(x-a_{nn})$. The conclusion follows from the fact that $m_T(x)$ divides $f_T(x)$. Another proof of the latter part, without invoking the characteristic polynomial, is proposed in Exercise 12, p. 187. □

Theorem 8.1.3. *Let $T : V \to V$ be an operator with V, a finite-dimensional vector space, and let $\beta \in V$ be such that $A_\beta(x) = m_T(x)$. If U is a T-invariant subspace, $C(\beta, T)$ denotes the T-cyclic subspace spanned by β, and $g : U \to C(\beta, T)$ is a linear transformation which commutes with T, then g can be extended to $g_1 : V \to C(\beta, T)$ and $g_1 \circ T = T \circ g_1$.*

Proof. If $U = V$, it suffices to take $g_1 = g$. Hence we may assume that there is $\alpha_1 \in V \setminus U$ and propose $U_1 = U + C(\alpha_1, T)$. It is clear that U_1 is T-invariant and contains U properly. We will prove that g can be extended to $g_1 : U_1 \to C(\beta, T)$ and $g_1 \circ T = T \circ g_1$.

Let $s(x)$ be the T-annihilator of α_1 with respect to U, that is, $s(T)(\alpha_1) \in U$. As $m_T(T)(\alpha_1) = 0 \in U$, we get that $s(x)$ divides $m_T(x)$, hence $m_T(x) = s(x)h(x)$, for some polynomial $h(x)$. From $s(T)(\alpha_1) \in U$ and $g : U \to C(\beta, T)$, one has that $g(s(T)(\alpha_1)) \in C(\beta, T)$, thus $g(s(T))(\alpha_1) = u(T)(\beta)$ for some polynomial $u(x)$. On the other hand, since g and T commute, $h(T)g(s(T)(\alpha_1)) = g(h(T)s(T)(\alpha_1)) = g(m_T(T)(\alpha_1)) = g(0) = 0$, hence we obtain $h(T)(g(s(T)(\alpha_1))) = h(T)(u(T)(\beta)) = 0$, and from this it follows that $m_T(x) = A_\beta(x)$ divides $h(x)u(x)$, that is,

$$h(x)u(x) = m_T(x)q(x) = s(x)h(x)q(x),$$

for some $q(x)$. This implies that $u(x) = s(x)q(x)$, thus,

$$g(s(T)(\alpha_1)) = u(T)(\beta) = s(T)q(T)(\beta). \tag{8.1}$$

Let $\beta_0 = q(T)(\beta)$ and define

$$g_1 : U_1 \to C(\beta, T), \quad \text{given by } g_1(\alpha + r(T)(\alpha_1)) = g(\alpha) + r(T)(\beta_0). \tag{8.2}$$

We will prove that the definition of g_1 is consistent in the sense that if $\alpha + r(T)(\alpha_1) = \alpha' + r_1(T)(\alpha_1)$, then $g(\alpha) + r(T)(\beta_0) = g(\alpha') + r_1(T)(\beta_0)$, which means that g_1 is well defined.

In fact, the condition $\alpha + r(T)(\alpha_1) = \alpha' + r_1(T)(\alpha_1)$ implies $\alpha' - \alpha = [r(T) - r_1(T)](\alpha_1) \in U$. Therefore $s(x)$ divides $r(x) - r_1(x)$, that is, $r(x) - r_1(x) = s(x)l(x)$, for some polynomial $l(x)$. From the latter equation, we have

$$\begin{aligned}
g([r(T) - r_1(T)](\alpha_1)) &= g(s(T)l(T))(\alpha_1) \\
&= l(T)g(s(T)(\alpha_1)) \\
&= l(T)s(T)q(T)(\beta) \\
&= l(T)s(T)(\beta_0) \\
&= (r(T) - r_1(T))(\beta_0).
\end{aligned}$$

On the other hand, $g([r(T) - r_1(T)](\alpha_1)) = g(\alpha' - \alpha) = g(\alpha') - g(\alpha)$. From this and the previous equations, one has $g(\alpha) + r(T)(\beta_0) = g(\alpha') + r_1(T)(\beta_0)$, as claimed. One can verify that g_1 is linear and commutes with T.

Continuing the argument by replacing U with U_1, the proof will finish in a finite number of steps. $\qquad \square$

Theorem 8.1.4 (Cyclic decomposition). *Let V be a finite-dimensional vector space of positive dimension, and let T be an operator on V with minimal polynomial $m_T(x)$. Then there are vectors $\alpha_1, \alpha_2, \ldots, \alpha_k$ in V such that $V = C(\alpha_1, T) \oplus C(\alpha_2, T) \oplus \cdots \oplus C(\alpha_k, T)$, with $m_i(x) = A_{\alpha_i}(x)$ dividing $m_{i-1}(x)$, for every $i \in \{2, 3, \ldots, k\}$ and $m_1(x) = m_T(x)$.*

Proof. From Theorem 7.1.3, p. 174, and the assumption on V, we have that there is $\alpha_1 \in V \setminus \{0\}$ whose annihilator is $m_1(x) := m_T(x)$. If $V = C(\alpha_1, T)$, we have finished; otherwise, setting $U = C(\alpha_1, T)$ and applying Theorem 8.1.3 with $g = $ identity, one has that there is a linear transformation $g_1 : V \to C(\alpha_1, T)$, so that g_1 restricted to $C(\alpha_1, T)$ is the identity and $T \circ g_1 = g_1 \circ T$. This condition guarantees that the kernel of g_1 is a T-invariant subspace. Let N_1 be the kernel of g_1. We will show that $V = C(\alpha_1, T) \oplus N_1$. Given $\alpha \in V$, one has that $g_1(\alpha - g_1(\alpha)) = g_1(\alpha) - g_1(g_1(\alpha))$. Since $g_1(\alpha) \in C(\alpha_1, T)$ and g_1 restricted to $C(\alpha_1, T)$ is the identity, $g_1(g_1(\alpha)) = g_1(\alpha)$, hence $\alpha - g_1(\alpha) \in N_1$, thus $\alpha - g_1(\alpha) = \gamma \in N_1$, or $\alpha = g_1(\alpha) + \gamma \in C(\alpha_1, T) + N_1$. If $\alpha \in C(\alpha_1, T) \cap N_1$, then $\alpha = g_1(\alpha) = 0$, proving that $V = C(\alpha_1, T) \oplus N_1$. From the assumption that $V \neq C(\alpha_1, T)$, we have that N_1 has positive dimension.

Now replacing V with N_1 and applying the same process to N_1, we prove that there is $a_2 \in N_1$ whose annihilator is the minimal polynomial of T restricted to N_1. If this polynomial is denoted by $m_2(x)$, then it follows that $m_2(x)$ divides $m_1(x)$ and $N_1 = C(a_2, T) \oplus N_2$. From this we have $V = C(a_1, T) \oplus C(a_2, T) \oplus N_2$. This process finishes in at most as many steps as the dimension of V and produces a cyclic decomposition of V. The divisibility condition on the polynomials $m_i(x)$, $i \in \{1, 2, \ldots, k\}$ is also obtained. □

With Theorem 8.1.5, we will conclude the discussion of the unique cyclic decomposition of a finite-dimensional vector space, as the direct sum of cyclic subspaces. In the proof we will need a lemma, whose proof is left to the reader, Exercise 5, p. 220.

Lemma 8.1.2. *Let T be an operator acting on a finite-dimensional vector space V, and let $h(x)$ be any polynomial. Then the following statements hold:*
1. *If $a \in V$, then $h(T)(C(a, T)) = C(h(T)(a), T)$.*
2. *If W_1 and W_2 are T-invariant subspaces and $V = W_1 \oplus W_2$, then $h(T)(V) = h(T)(W_1) \oplus h(T)(W_2)$.*
3. *If $\alpha, \beta \in V$ satisfy $A_\alpha(x) = A_\beta(x)$, then $A_{h(T)(\alpha)}(x) = A_{h(T)(\beta)}(x)$.*

Theorem 8.1.5 (Uniqueness of the cyclic decomposition). *Let V be a finite-dimensional vector space, and T an operator on V. Assume that $V = C(\beta_1, T) \oplus C(\beta_2, T) \oplus \cdots \oplus C(\beta_k, T) = C(a_1, T) \oplus C(a_2, T) \oplus \cdots \oplus C(a_l, T)$, and that $m_i(x)$ and $n_j(x)$ are the annihilators of β_i and a_j, respectively. Also, assume that $m_i(x)$ divides $m_{i-1}(x)$ for every $i \in \{2, 3, \ldots, k\}$, and that $n_j(x)$ divides $n_{j-1}(x)$ for every $j \in \{2, 3, \ldots, l\}$. Then $k = l$ and $m_i(x) = n_i(x)$ for every $i \in \{1, 2, \ldots, k\}$.*

Proof. Let m be the minimum of $\{l, k\}$. Without loss of generality, we may assume that $m = k$. We will apply induction on m. If $m = 1$, then we will prove that $m_1(x) = n_1(x)$ and $l = 1$. Let $y \in V$ be any element, then $y = g_1(T)(a_1) + \cdots + g_k(T)(a_k)$, for some polynomials $g_1(x), g(x), \ldots, g_k(x)$. From the assumption we have that $n_i(x)$ divides $n_{i-1}(x)$ for every $i \geq 2$, hence

$$
\begin{aligned}
n_1(T)(y) &= n_1(T)(g_1(T)(a_1) + \cdots + g_k(T)(a_k)) \\
&= n_1(T)(g_1(T)(a_1)) + n_1(T)(g_2(T)(a_2)) + \cdots + n_1(T)(g_k(T)(a_k)) \\
&= g_1(T)(n_1(T)(a_1)) + g_2(T)(n_1(T)(a_2)) + \cdots + g_k(T)(n_1(T)(a_k)) \\
&= 0 + 0 + \cdots + 0 \quad \text{(since } n_j \text{ divides } n_1 \text{ for all } j \geq 2) \\
&= 0.
\end{aligned}
$$

In particular, for $y = \beta_1$ one has that $n_1(T)(\beta_1) = 0$, hence $m_1(x)$ divides $n_1(x)$. Likewise one proves that $n_1(x)$ divides $m_1(x)$, proving that $m_1(x) = n_1(x)$.

If $l \geq 2$, then $\dim(C(a_1, T)) < \dim(V)$ and, from Lemma 7.1.2, p. 173, we have that

$$
\dim(C(\beta_1, T)) = \dim(C(a_1, T)) = \deg(m_1(x)) = \deg(n_1(x)),
$$

thus $\dim(C(\beta_1, T)) < \dim(V)$, contradicting that $V = C(\beta_1, T)$, hence $l = 1$.

Assume that $m = k > 1$ and that $m_j(x) = n_j(x)$ for every $j \in \{1, 2, \ldots, k-1\}$. We will prove that $m_k(x) = n_k(x)$ and $l = k$.

By applying statements 1 and 2 of Lemma 8.1.2 to the cyclic decomposition of V, we obtain

$$m_k(T)(V) = C(m_k(T)(\beta_1), T) \oplus C(m_k(T)(\beta_2), T) \oplus \cdots \oplus C(m_k(T)(\beta_{k-1}), T),$$
$$m_k(T)(V) = C(m_k(T)(\alpha_1), T) \oplus (m_k(T)(\alpha_2), T) \oplus \cdots \oplus C(m_k(T)(\alpha_l), T).$$
$$(8.3)$$

From the induction assumption, $m_j(x) = n_j(x)$ for every $j \in \{1, 2, \ldots, k-1\}$, hence from Lemma 8.1.2, statement 3, and Lemma 7.1.2, p. 173, one has that $C(m_k(T)(\alpha_j), T)$ and $C(m_k(T)(\beta_j), T)$ have the same dimension, for every $j \in \{1, 2, \ldots, k-1\}$. Thus, from equation (8.3), $\dim(C(m_k(T)(\alpha_j), T)) = 0$ for all $j \geq k$. From this, one has that $m_k(T)(\alpha_k) = 0$, which implies that $n_k(x)$ divides $m_k(x)$. Interchanging the roles of $m_k(x)$ and $n_k(x)$, one proves that $m_k(x)$ divides $n_k(x)$, proving that $m_k(x) = n_k(x)$.

Now consider the equation

$$V = C(\beta_1, T) \oplus C(\beta_2, T) \oplus \cdots \oplus C(\beta_k, T) = C(\alpha_1, T) \oplus C(\alpha_2, T) \oplus \cdots \oplus C(\alpha_l, T). \qquad (8.4)$$

We have proved that $\dim(C(\beta_j, T)) = \dim(C(\alpha_j, T))$ for every $j \in \{1, 2, \ldots, k\}$. Hence, by counting dimensions in (8.4), l has to be equal to k, finishing the proof of the theorem. $\qquad \square$

To continue the discussion toward the rational canonical form of a matrix, it is necessary to introduce some terms that relate polynomials and matrices.

Definition 8.1.4. Let $p(x) = c_0 + c_1 x + c_2 x^2 + \cdots + c_{l-1} x^{l-1} + x^l$ be a polynomial. The matrix

$$A(p) := \begin{bmatrix} 0 & 0 & \cdots & 0 & -c_0 \\ 1 & 0 & \cdots & 0 & -c_1 \\ 0 & 1 & \cdots & 0 & -c_2 \\ \vdots & \vdots & \ddots & \vdots & \vdots \\ 0 & 0 & \cdots & 1 & -c_{l-1} \end{bmatrix} \qquad (8.5)$$

is called the companion matrix of $p(x)$.

Remark 8.1.2. The companion matrix $A(p)$ can be considered in the following sense. Let $\{\alpha_1, \alpha_2, \ldots, \alpha_l\}$ be a basis of $\mathbb{R}^l$, and let $T : \mathbb{R}^l \to \mathbb{R}^l$ be the operator defined by $T(\alpha_i) = \alpha_{i+1}$, for $i \in \{1, 2, \ldots, (l-1)\}$ and $T(\alpha_l) = -c_0 \alpha_1 - c_1 \alpha_2 - \cdots - c_{l-1} \alpha_l$, for some scalars $c_0, c_1, \ldots, c_{l-1}$. Then $A(p)$ is the matrix that represents T with respect to the given basis.

Exercise 8.1.1. Prove that the minimal and characteristic polynomials of $A(p)$ are equal to $p(x)$.

The following is an important corollary to Theorem 8.1.5.

Corollary 8.1.2 (Rational canonical form). *Let A be an n × n matrix. Then there exists a nonsingular matrix P such that*

$$P^{-1}AP = \begin{bmatrix} A_1 & 0 & \cdots & 0 & 0 \\ 0 & A_2 & \cdots & 0 & 0 \\ 0 & 0 & \cdots & 0 & 0 \\ \vdots & \vdots & \ddots & \vdots & \vdots \\ 0 & 0 & \cdots & 0 & A_k \end{bmatrix}, \tag{8.6}$$

where each matrix A_i is of the form (8.5). Moreover, this representation of A is unique. If $m_i(x)$ is the minimal polynomial of A_i, then $m_1(x)$ is the minimal polynomial of A and $m_{i+1}(x)$ divides $m_i(x)$ for every $i \in \{1, 2, \ldots, k - 1\}$.

Proof. We may assume that A represents an operator T, acting on $\mathbb{R}^n$. From Theorem 8.1.5, there are vectors $a_1, a_2, \ldots, a_k$ in $\mathbb{R}^n$ such that

$$\mathbb{R}^n = C(a_1, T) \oplus C(a_2, T) \oplus \cdots \oplus C(a_k, T). \tag{8.7}$$

The conclusion follows by noticing that a basis of $\mathbb{R}^n$ can be obtained by choosing the basis $\{a_i, T(a_i), \ldots, T^{n_i-1}(a_i)\}$ of $C(a_i, T)$, and taking the union of those bases. The rest of the conclusion follows by carefully examining the action of T on each summand $C(a_i, T)$, in equation (8.7) and taking into account the result of Exercise 8.1.1. □

Definition 8.1.5. Let A be a square matrix. The representation of A given by (8.6) is called the rational canonical form of A. If $m_1(x), \ldots, m_k(x)$ are the minimal polynomials of the matrices $A_1, A_2, \ldots, A_k$, respectively, those polynomials are called the invariant factors of A.

Corollary 8.1.3. *Two n × n matrices are similar if and only if they have the same rational canonical form.*

Proof. Let A and B be two $n \times n$ matrices and assume that they have the same rational canonical form (8.6), then there are nonsingular matrices P an Q such that $P^{-1}AP = Q^{-1}BQ$. From this it follows directly that A and B are similar. The converse follows by noticing that A and B represent an operator in corresponding bases. □

This corollary justifies the following

Definition 8.1.6. Let T be an operator on V, and let A be any matrix representing T with respect to a basis of V. The invariant factors of A are called the invariant factors of T.

Example 8.1.1. Let $T : \mathbb{R}^4 \to \mathbb{R}^4$ be given by $T(x, y, z, w) = (x + y, z - w, 2y, y + z)$.
1. Determine the minimal polynomial of T.
2. Find the rational canonical form of T.
3. Decide if T is diagonalizable and, if so, find a basis such that T is represented by a diagonal matrix.

Discussion. To determine the minimal polynomial of T, we will compute the annihilators of the canonical basis $\{e_1, e_2, e_3, e_4\}$ of $\mathbb{R}^4$ and use Theorem 7.1.6, p. 176. It is obtained directly from the definition of T that $T(e_1) = e_1$, hence $A_{e_1}(x) = x - 1$. Notice that we have already found an eigenvalue, as well as an eigenvector of T. To compute the annihilator of e_2, we need to compute several powers of T applied to e_2:

$$T(e_2) = (1, 0, 2, 1),$$
$$T^2(e_2) = (1, 1, 0, 2),$$
$$T^3(e_2) = (2, -2, 2, 1).$$

It can be verified that $\{e_2, T(e_2), T^2(e_2), T^3(e_2)\}$ is linearly independent, hence the annihilator of e_2 has degree 4. We find that

$$T^4(e_2) = T^3(e_2) + T^2(e_2) - 3T(e_2) + 2e_2. \tag{8.8}$$

From equation (8.8), we have that $m_T(x) = x^4 - x^3 - x^2 + 3x - 2 = (x - 1)(x^3 - x + 2)$, hence we do not need to compute the annihilators of the other canonical vectors.

It can be verified, for instance, using SageMath, that the polynomial $q(x) = x^3 - x + 2$ has only one real root, which is

$$x = \left(\frac{\sqrt{78}}{9} - 1 \right)^{1/3} + \frac{1}{3(\frac{\sqrt{78}}{9} - 1)^{1/3}}.$$

From this we conclude that T is not diagonalizable, since one of the irreducible factors of $m_T(x)$ is not linear.

To obtain the rational canonical form of the matrix A, associated to T with respect to the canonical basis, we have that

$$A = \begin{bmatrix} 1 & 1 & 0 & 0 \\ 0 & 0 & 1 & -1 \\ 0 & 2 & 0 & 0 \\ 0 & 1 & 1 & 0 \end{bmatrix}.$$

Also the matrix P changing the canonical basis of $\mathbb{R}^4$ to the basis $\{e_2, T(e_2), T^2(e_2), T^3(e_2)\}$ is

$$P = \begin{bmatrix} 0 & 1 & 1 & 2 \\ 1 & 0 & 1 & -2 \\ 0 & 2 & 0 & 2 \\ 0 & 1 & 2 & 1 \end{bmatrix}.$$

Therefore the rational canonical form of A is

$$P^{-1}AP = \begin{bmatrix} 0 & 0 & 0 & 2 \\ 1 & 0 & 0 & -3 \\ 0 & 1 & 0 & 1 \\ 0 & 0 & 1 & 1 \end{bmatrix}.$$

Since e_2 is T-cyclic, that is, $C(e_2, T) = \mathbb{R}^4$, it follows that A is the companion matrix of $m_T(x)$, as expected.

Exercise 8.1.2. Using calculus techniques, draw the graph of the polynomial function $f(x) = x^3 - x + 2$, by finding the intervals where the function is increasing and decreasing. From this analysis conclude that f has only one real root, justifying part of the conclusions in Example 8.1.1.

Exercise 8.1.3. Show that the matrix $A = \begin{bmatrix} 2 & -1 & 0 & -1 \\ -1 & 2 & -1 & 0 \\ 0 & -1 & 2 & -1 \\ -1 & 0 & -1 & 2 \end{bmatrix}$ is diagonalizable and find a nonsingular matrix P such that $P^{-1}AP$ is diagonal. What are the rational canonical form of A and its invariant factors?

8.2 Other method to compute the minimal polynomial

Let A be a square matrix, say $n \times n$. From Corollary 8.1.2, there is a nonsingular matrix P such that $P^{-1}AP$ is block diagonal, explicitly

$$P^{-1}AP = \begin{bmatrix} A_1 & 0 & \cdots & 0 & 0 \\ 0 & A_2 & \cdots & 0 & 0 \\ 0 & 0 & \cdots & 0 & 0 \\ \vdots & \vdots & \ddots & \vdots & \vdots \\ 0 & 0 & \cdots & 0 & A_k \end{bmatrix},$$

where each of the matrices on the diagonal is of the form (8.5); see p. 194.
From this we have

$$P^{-1}AP - xI = \begin{bmatrix} A_1 - xI & 0 & \cdots & 0 & 0 \\ 0 & A_2 - xI & \cdots & 0 & 0 \\ 0 & 0 & \cdots & 0 & 0 \\ \vdots & \vdots & \ddots & \vdots & \vdots \\ 0 & 0 & \cdots & 0 & A_k - xI \end{bmatrix}, \tag{8.9}$$

where I represents the identity matrix of appropriate size.
For each block $A_i - xI$, we will show that, by performing elementary row or column operations, this matrix can be turned into a diagonal matrix whose $(1,1)$th entry is $m_i(x)$, the minimal polynomial of A_i and the other entries on the diagonal are equal to 1.

To make notation friendly, we will assume that

$$A_i = \begin{bmatrix} 0 & 0 & \cdots & 0 & -c_0 \\ 1 & 0 & \cdots & 0 & -c_1 \\ 0 & 1 & \cdots & 0 & -c_2 \\ \vdots & \vdots & \ddots & \vdots & \vdots \\ 0 & 0 & \cdots & 1 & -c_{l-1} \end{bmatrix},$$

where $m_i(x) = x^l + c_{l-1}x^{l-1} + \cdots + c_1 x + c_0$ represents its minimal polynomial. Writing explicitly the matrix $A_i - xI$, one has

$$A_i - xI = \begin{bmatrix} -x & 0 & \cdots & 0 & -c_0 \\ 1 & -x & \cdots & 0 & -c_1 \\ 0 & 1 & \cdots & 0 & -c_2 \\ \vdots & \vdots & \ddots & \vdots & \vdots \\ 0 & 0 & \cdots & 1 & -c_{l-1} - x \end{bmatrix}.$$

Multiplying the second row by x and adding the result to the first, one obtains

$$B_1 = \begin{bmatrix} 0 & -x^2 & \cdots & 0 & -c_0 - c_1 x \\ 1 & -x & \cdots & 0 & -c_1 \\ 0 & 1 & \cdots & 0 & -c_2 \\ \vdots & \vdots & \ddots & \vdots & \vdots \\ 0 & 0 & \cdots & 1 & -c_{l-1} - x \end{bmatrix}.$$

Multiplying the third row of B_1 by x^2 and adding the result to the first, we obtain matrix B_2 whose first row has only zeros except the third and last entries, which are $-x^3$ and $c_0 - c_1 x - c_2 x^2$, respectively. The other entries coincide with those of B_1.

Multiplying row 4 of B_2 by x^3 and adding the result to its first row, we have as result B_3, whose first row has zeros except the fourth and last entries, which are $-x^4$ and $c_0 - c_1 x - c_2 x^2 - c_3 x^3$, respectively. The rest of the rows are the same as those of B_1. Continuing this way $l - 1$ times, one obtains a matrix B_{l-1} for which the $(1, l)$th entry is $-p(x) = -c_0 - c_1 x - \cdots - c_{l-1}x^{l-1} - x^l$. The other entries in the first row of B_{l-1} are zero and the other rows coincide with those of B_1, namely,

$$B_{l-1} = \begin{bmatrix} 0 & 0 & \cdots & 0 & -p(x) \\ 1 & -x & \cdots & 0 & -c_1 \\ 0 & 1 & \cdots & 0 & -c_2 \\ \vdots & \vdots & \ddots & \vdots & \vdots \\ 0 & 0 & \cdots & 1 & -c_{l-1} - x \end{bmatrix}.$$

Applying appropriate elementary operations (which?) on the columns of B_{l-1}, we obtain

$$B = \begin{bmatrix} 0 & 0 & \cdots & 0 & -p(x) \\ 1 & 0 & \cdots & 0 & 0 \\ 0 & 1 & \cdots & 0 & 0 \\ \vdots & \vdots & \ddots & \vdots & \vdots \\ 0 & 0 & \cdots & 1 & 0 \end{bmatrix}.$$

Multiplying the last column of B by -1 and moving the result to the first column, matrix B becomes

$$C = \begin{bmatrix} p(x) & 0 & \cdots & 0 & 0 \\ 0 & 1 & \cdots & 0 & 0 \\ 0 & 0 & \cdots & 0 & 0 \\ \vdots & \vdots & \ddots & \vdots & \vdots \\ 0 & 0 & \cdots & 0 & 1 \end{bmatrix},$$

as claimed.

Notice that the elementary operations applied to block $A_i - xI$ in (8.9), p. 197, have not changed any of the remaining blocks, and the process described above can be applied to each of the blocks.

Summarizing, we have proved that the matrix $P^{-1}(A - xI)P$, hence $A - xI$, is equivalent to a diagonal matrix. More precisely, there are nonsingular matrices Q and R, with polynomial entries, so that

$$Q(A - xI)R = \begin{bmatrix} m_1(x) & 0 & \cdots & 0 & 0 & 0 & \cdots & 0 \\ 0 & m_2(x) & \cdots & 0 & 0 & 0 & \cdots & 0 \\ 0 & 0 & \cdots & 0 & 0 & 0 & \cdots & 0 \\ \vdots & \vdots & \ddots & m_k(x) & 0 & 0 & \cdots & 0 \\ 0 & 0 & \cdots & 0 & 1 & 0 & \cdots & 0 \\ 0 & 0 & \cdots & 0 & 0 & 1 & \cdots & 0 \\ \vdots & \vdots & \ddots & \vdots & \vdots & \vdots & \ddots & \vdots \\ 0 & 0 & \cdots & 0 & 0 & 0 & \cdots & 1 \end{bmatrix},$$

where $m_{i+1}(x)$ divides $m_i(x)$ for each $i \in \{1, 2, \ldots, k - 1\}$ and $m_1(x)$ is the minimal polynomial of A.

To state the algorithm that computes the invariant factors of A, we need a definition and a theorem.

Definition 8.2.1. Let N be a matrix whose entries are polynomials with coefficients in the real or the complex numbers. We say that N is in Smith normal form if the following conditions hold:

1. Every element off the diagonal is zero.
2. The diagonal of N consists of polynomials $f_1, f_2, \ldots, f_l$ such that f_j divides f_{j+1}, for every $j \in \{1, 2, \ldots, l-1\}$.

Theorem 8.2.1 appears without proof since it falls outside the scope of the book. This theorem and the previous discussion provide an algorithm to compute the invariant factors of a matrix, as well as its minimal polynomial.

Theorem 8.2.1. *Let B be a matrix with polynomial entries having real or complex coefficients. Then B is equivalent to a unique matrix in Smith normal form.*

In this last part of the section, the matrices that we are dealing with have entries from the set of real or complex polynomials, that is why we will declare the type of elementary operations allowed, in order to have nonsingular matrices:

1. Interchange rows or columns.
2. Multiply a row (column) by a polynomial and add it to another row (column).
3. Multiply a row (column) by a nonzero constant.

Algorithm 8.2.1.

Require: Square matrix A

1: Define the matrix $A_1 = A - xI$.
2: Applying row and column operations to A_1, one obtains $B = \mathrm{diag}\{p(x), C\}$, where $p(x)$ is the gcd of the elements of the first row and first column of A_1, and C is a square matrix.
3: **while** $p(x)$ does not divide every entry of C **do**
4:     Find the first column, from left to right, which contains an element which is not divisible by $p(x)$ and add this column to the first column of B.
5:     Apply *Step 2* to B
6: **end while**
7: Set $A_1 = C$ and apply the same procedure.

In a finite number of iterations, one obtains a matrix of type

$$\mathrm{diag}\{m_1(x), m_2(x), \ldots, m_k(x), 1, \ldots, 1\}, \tag{8.10}$$

where $m_i(x)$ divides $m_{i+1}(x)$. The minimal polynomial of A is $m_k(x)$.

Example 8.2.1. Find the minimal polynomial of $A = \begin{bmatrix} 1 & 0 & -1 \\ -1 & 1 & 0 \\ 0 & -1 & 1 \end{bmatrix}$.

Set $A_1 = \begin{bmatrix} 1-x & 0 & -1 \\ -1 & 1-x & 0 \\ 0 & -1 & 1-x \end{bmatrix}$. Multiplying the second row of A_1 by $1-x$ and adding the result to the first row, one has $A_2 = \begin{bmatrix} 0 & (1-x)^2 & -1 \\ -1 & 1-x & 0 \\ 0 & -1 & 1-x \end{bmatrix}$. Multiplying the second row of A_2 by -1 and swapping the result with the first row, we obtain $A_3 = \begin{bmatrix} 1 & x-1 & 0 \\ 0 & (1-x)^2 & -1 \\ 0 & -1 & 1-x \end{bmatrix}$.

Multiplying the first column of A_3 by $-(x-1)$ and adding the result to the second column, we have $A_4 = \begin{bmatrix} 1 & 0 & 0 \\ 0 & (1-x)^2 & -1 \\ 0 & -1 & 1-x \end{bmatrix}$. Multiplying the third row of A_4 by $(1-x)^2$ and adding to the second row, one has $A_5 = \begin{bmatrix} 1 & 0 & 0 \\ 0 & 0 & -1+(1-x)^3 \\ 0 & -1 & 1-x \end{bmatrix}$.

Multiplying the second column of A_5 by $(1-x)$ and adding to the third column gives $A_6 = \begin{bmatrix} 1 & 0 & 0 \\ 0 & 0 & -1+(1-x)^3 \\ 0 & -1 & 0 \end{bmatrix}$. Multiplying the second and third rows of A_6 by -1 and swapping them, we obtain $A_7 = \begin{bmatrix} 1 & 0 & 0 \\ 0 & 1 & 0 \\ 0 & 0 & (x-1)^3+1 \end{bmatrix}$.

So far we have the following information:

1. The minimal polynomial of A is

$$(x-1)^3 + 1 = x(x^2 - 3x + 3) = x^3 - 3x^2 + 3x = x\left(\left(x - \frac{3}{2}\right)^2 + \frac{21}{4}\right).$$

2. The matrix A is not diagonalizable over the real numbers, since its minimal polynomial has only one real linear factor.

3. The rational canonical form of A is $\begin{bmatrix} 0 & 0 & 0 \\ 1 & 0 & -3 \\ 0 & 1 & 3 \end{bmatrix}$, since the minimal polynomial of A has degree 3, that is, $\mathbb{R}^3$ has a cyclic vector whose annihilator is the minimal polynomial.

4. The matrix A is singular, since zero is one of its eigenvalues.

Exercise 8.2.1. Assuming that $m_1(x), \ldots, m_k(x)$ are the invariant factors of A, what is the characteristic polynomial of A?

8.3 Jordan canonical form

We recall that the kernel of an operator T is denoted by N_T.

Let V be a finite-dimensional vector space, $T : V \to V$ be a linear operator, and $m_T(x) = p_1^{e_1}(x) p_2^{e_2}(x) \cdots p_k^{e_k}(x)$ the minimal polynomial of T, factored as a product of irreducible polynomials.

From Theorem 7.2.1 (Primary decomposition theorem), one has

$$V = W_1 \oplus W_2 \oplus \cdots \oplus W_k, \quad \text{where } W_i = N_{p_i^{e_i}(T)}. \tag{8.11}$$

We also have that the minimal polynomial of T restricted to W_j is $p_j^{e_j}(x)$. Applying Theorem 8.1.4, p. 192 (Cyclic decomposition theorem) to each W_j, we obtain

$$W_j = W_{1j} \oplus W_{2j} \oplus \cdots \oplus W_{i,j}, \tag{8.12}$$

where each W_{ij} is T-cyclic with annihilator $p_j^{e_{ij}}(x)$ and the exponents satisfy $e_j = e_{1j} \geq e_{2j} \geq \cdots \geq e_{i,j}$. Also, from Theorem 7.2.2, p. 182, one has that $\dim(W_{ij}) = e_{ij}\deg(p_j(x))$.

Definition 8.3.1. The polynomials $p_j^{e_{ij}}(x), j \in \{1, 2, \ldots, k\}$, are called *elementary divisors* of T.

Lemma 8.3.1. *With notation as above, if $p_j(x)$ is linear, $W_{ij} = C(v, T)$ for some $v \in V$ and $A_v(x) = (x - c_j)^{e_{ij}}$, then $S = \{v, (T - c_j I)v, \ldots, (T - c_j I)^{e_{ij}-1}(v)\}$ is a basis of W_{ij}.*

Proof. We know that W_{ij} has dimension e_{ij}, then it is enough to show that S is linearly independent. Let

$$a_0 v + a_1(T - c_j I)v + \cdots + a_l(T - c_j I)^l v = 0, \tag{8.13}$$

with $l = e_{ij} - 1$. We have that $(T - c_j I)^{e_{ij}} v = 0$, then, applying $(T - c_j I)^l$ to equation (8.13), one obtains $a_0(T - c_j I)^l v = 0$. Also $(T - c_j I)^l v \neq 0$, then $a_0 = 0$. Continuing in this way, one concludes that all coefficients are zero. $\square$

The matrix of T, restricted to W_{ij}, with respect to the basis

$$\{v, (T - c_j I)(v), \ldots, (T - c_j I)^{e_{ij}-1}(v)\}$$

is obtained by applying T to each element of the basis:

$$
\begin{aligned}
T(v) &= c_j v + (T - c_j I)(v), \\
T(T - c_j I)(v) &= (T - c_j I)(T(v)) \\
&= (T - c_j I)(c_j v + (T - c_j I)(v)) \\
&= c_j(T - c_j I)(v) + (T - c_j I)^2(v), \\
&\quad \vdots \\
T(T - c_j I)^m(v) &= (T - c_j I)^m(c_j v + (T - c_j I)(v)) \\
&= c_j(T - c_j I)^m(v) + (T - c_j I)^{m+1}(v), \\
&\quad \vdots \\
T((T - c_j I)^{e_{ij}-1}(v)) &= c_j(T - c_j)^{e_{ij}-1}(v).
\end{aligned}
\tag{8.14}
$$

From the equations in (8.14), one has that the matrix associated to T, restricted to W_{ij}, is

$$
\begin{bmatrix}
c_j & 0 & \cdots & 0 & 0 & 0 \\
1 & c_j & \cdots & 0 & 0 & 0 \\
0 & 1 & \cdots & 0 & 0 & 0 \\
\vdots & \vdots & \ddots & \vdots & \vdots & \vdots \\
0 & 0 & \cdots & 1 & c_j & 0 \\
0 & 0 & \cdots & 0 & 1 & c_j
\end{bmatrix},
\tag{8.15}
$$

thus, the matrix of T restricted to W_j is block diagonal, where each of those blocks has the form (8.15), called a *Jordan block*. If all irreducible factors of $m_T(x)$ are linear, then the annihilator of each W_{ij} is of the form $(x - c_j)^{e_{ij}}$, hence the above process implies that the restriction of T to each W_j is block diagonal, with each of those blocks of the form (8.15). Summarizing, we have proved

Theorem 8.3.1 (Jordan canonical form). *Let V be a finite-dimensional vector space, and let T be an operator on V. Assume that $m_T(x)$ factorizes as $m_T(x) = (x - c_1)^{e_1}(x - c_2)^{e_2} \cdots (x - c_k)^{e_k}$. Then there exists a basis of V with respect to which T is represented by the matrix $J = \mathrm{diag}\{J_1, \ldots, J_k\}$, with each J_m block diagonal, $J_m = \mathrm{diag}\{J_{1m}, \ldots, J_{r_m m}\}$, and each J_{sm} a Jordan block of size e_{sm}. Also, $e_m = e_{1m} \geq e_{2m} \geq \cdots \geq e_{r_m m}$, for every $m \in \{1, 2, \ldots, k\}$.*

Definition 8.3.2. The basis in Theorem 8.3.1 is called a Jordan basis.

The formulation of Theorem 8.3.1 for matrices is

Corollary 8.3.1. *Let A be a square matrix. If the minimal polynomial of A factorizes as $m(x) = (x - c_1)^{e_1}(x - c_2)^{e_2} \cdots (x - c_k)^{e_k}$, then A is similar to a matrix $J = \mathrm{diag}\{J_1, \ldots, J_k\}$, where $J_m = \mathrm{diag}\{J_{1m}, \ldots, J_{r_m m}\}$ and each J_{sm} is a Jordan block of size e_{sm}. Also, $e_m = e_{1m} \geq e_{2m} \geq \cdots \geq e_{r_m m}$, for every $m \in \{1, 2, \ldots, k\}$.*

Definition 8.3.3. The matrix J of Corollary 8.3.1 is called the *Jordan canonical form* of A.

Remark 8.3.1. The Jordan canonical form of a matrix A is completely determined by the characteristic values of A and by the sizes of the Jordan blocks. More precisely, if the eigenvalues of A are $\{c_1, c_2, \ldots, c_k\}$ and the size of the Jordan blocks are

$$
\{(e_{11}, e_{21}, \ldots, e_{i_1 1}), (e_{12}, e_{22}, \ldots, e_{i_2 2}), \ldots, (e_{1k}, e_{2k}, \ldots, e_{ikk})\},
\tag{8.16}
$$

then the Jordan canonical form of A is uniquely determined by Corollary 8.3.1.

Definition 8.3.4. Let A be as in Remark 8.3.1. The information given by (8.16) is called the Segre characteristic of A.

Example 8.3.1. Assume that a matrix A has characteristic values $\{2, 4, -1\}$ and size of the Jordan blocks $\{(3, 2, 1), (2, 2), (2)\}$, then the Jordan canonical form of A is

$$
J = \mathrm{diag}\{J_{11}, J_{21}, J_{31}, J_{12}, J_{22}, J_{13}\},
$$

where

$$J_{11} = \begin{bmatrix} 2 & 0 & 0 \\ 1 & 2 & 0 \\ 0 & 1 & 2 \end{bmatrix}, \quad J_{21} = \begin{bmatrix} 2 & 0 \\ 1 & 2 \end{bmatrix}, \quad J_{31} = [2], \quad J_{12} = J_{22} = \begin{bmatrix} 4 & 0 \\ 1 & 4 \end{bmatrix}, \quad J_{13} = \begin{bmatrix} -1 & 0 \\ 1 & -1 \end{bmatrix},$$

explicitly

$$J = \begin{bmatrix}
2 & 0 & 0 & 0 & 0 & 0 & 0 & 0 & 0 & 0 & 0 & 0 \\
1 & 2 & 0 & 0 & 0 & 0 & 0 & 0 & 0 & 0 & 0 & 0 \\
0 & 1 & 2 & 0 & 0 & 0 & 0 & 0 & 0 & 0 & 0 & 0 \\
0 & 0 & 0 & 2 & 0 & 0 & 0 & 0 & 0 & 0 & 0 & 0 \\
0 & 0 & 0 & 1 & 2 & 0 & 0 & 0 & 0 & 0 & 0 & 0 \\
0 & 0 & 0 & 0 & 0 & 2 & 0 & 0 & 0 & 0 & 0 & 0 \\
0 & 0 & 0 & 0 & 0 & 0 & 4 & 0 & 0 & 0 & 0 & 0 \\
0 & 0 & 0 & 0 & 0 & 0 & 1 & 4 & 0 & 0 & 0 & 0 \\
0 & 0 & 0 & 0 & 0 & 0 & 0 & 0 & 4 & 0 & 0 & 0 \\
0 & 0 & 0 & 0 & 0 & 0 & 0 & 0 & 1 & 4 & 0 & 0 \\
0 & 0 & 0 & 0 & 0 & 0 & 0 & 0 & 0 & 0 & -1 & 0 \\
0 & 0 & 0 & 0 & 0 & 0 & 0 & 0 & 0 & 0 & 1 & -1
\end{bmatrix}.$$

Exercise 8.3.1. Write the elementary divisors of the matrix in Example 8.3.1.

Exercise 8.3.2. Let A be a matrix whose eigenvalues are $c_1, c_2, \ldots, c_k$ and its Segre characteristic is as in (8.16). Write the minimal and characteristic polynomials of A.

Exercise 8.3.3. Let A be a 2×2 matrix. Describe all possible Jordan canonical forms of A.

Exercise 8.3.4. Let $A = \begin{bmatrix} 2 & 0 & 0 \\ a & -1 & 0 \\ c & b & -1 \end{bmatrix}$. Determine necessary and sufficient conditions in order that A be diagonalizable and determine the Jordan canonical form of A when it is not diagonalizable.

We end this section with a definition that is useful to reformulate a diagonalization criterion.

Definition 8.3.5. Let $T : V \to V$ be a linear operator. If λ is an eigenvalue of T, the dimension of $W_\lambda = \{\alpha \in V \mid T(\alpha) = \lambda\alpha\}$ is called the geometric multiplicity of λ. If the characteristic polynomial of T is $f_T(x) = (x - \lambda)^k q(x)$ and $q(\lambda) \neq 0$, k is called the algebraic multiplicity of λ.

Exercise 8.3.5. With notation as in Definition 8.3.5, find necessary and sufficient conditions in order that the algebraic and geometric multiplicities coincide.

8.4 Matrices with nonreal eigenvalues

Under a strong assumption on the minimal polynomial $m(x)$ of a matrix, for example, when the irreducible factors of $m(x)$ are linear, we have found that the matrix is similar to its Jordan canonical form, Theorem 8.3.1. Hence a question: What is the best result that can be obtained when the minimal polynomial has pairwise different irreducible factors and some of them are quadratic?

In this section we will provide an answer to this question. That is, if the minimal polynomial of an operator T has quadratic irreducible factors and pairwise different linear factors, then T can be represented by a block diagonal matrix, whose blocks have sizes one or two. The general case, that is, when the irreducible factors of the minimal polynomial are not necessarily different, leads to what some texts call *the real Jordan canonical form* (see [20, pp. 336–339]; [15, Chapter 6, Section 3]).

8.4.1 Real 2 × 2 matrices

We start by classifying the 2×2 matrices which have no real eigenvalues.

Let

$$B = \begin{bmatrix} a & b \\ c & d \end{bmatrix}$$

be a matrix whose minimal polynomial has no real roots. We will show that B is similar to a matrix of the form

$$A = \begin{bmatrix} m & -n \\ n & m \end{bmatrix}, \quad \text{with } n \neq 0. \tag{8.17}$$

From what has been observed before, we will start the discussion considering this type of matrices.

Given a matrix A as in (8.17), set $\lambda = \sqrt{m^2 + n^2}$ and

$$R = \begin{bmatrix} \frac{m}{\lambda} & \frac{-n}{\lambda} \\ \frac{n}{\lambda} & \frac{m}{\lambda} \end{bmatrix}. \tag{8.18}$$

Defining $\sin(\theta) = \frac{n}{\lambda}$ and $\cos(\theta) = \frac{m}{\lambda}$, one has that R is a rotation matrix, equation (5.15), p. 126. From equation (8.18), one concludes that A is the product of the scaling matrix λI and the rotation matrix R, i. e.,

$$A = \begin{bmatrix} m & -n \\ n & m \end{bmatrix} = \begin{bmatrix} \lambda & 0 \\ 0 & \lambda \end{bmatrix} \begin{bmatrix} \frac{m}{\lambda} & \frac{-n}{\lambda} \\ \frac{n}{\lambda} & \frac{m}{\lambda} \end{bmatrix}.$$

In geometric terms, the result of applying A to a vector X is shown in Figure 8.1. Summarizing, given a matrix

$$A = \begin{bmatrix} m & -n \\ n & m \end{bmatrix}, \quad \text{with } n \neq 0,$$

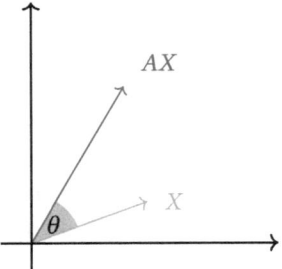

Figure 8.1: Rotation and scaling of X under the action of A.

the action of A on a vector X is a rotation by an angle $\theta = \tan^{-1}\left(\frac{n}{m}\right)$, followed by the action of the scaling matrix determined by $\lambda = \sqrt{m^2 + n^2}$.

Another important aspect to be noticed about the matrix A is the existence of its inverse, condition obtained from the assumption $n \neq 0$.

Let A be as in (8.17) and define $P = \begin{bmatrix} 1 & m \\ 0 & n \end{bmatrix}$, whose inverse is $P^{-1} = \frac{1}{n} \begin{bmatrix} n & -m \\ 0 & 1 \end{bmatrix}$. From this we obtain

$$P^{-1}AP = \frac{1}{n} \begin{bmatrix} n & -m \\ 0 & 1 \end{bmatrix} \begin{bmatrix} m & -n \\ n & m \end{bmatrix} \begin{bmatrix} 1 & m \\ 0 & n \end{bmatrix} = \begin{bmatrix} 0 & -n^2 - m^2 \\ 1 & 2m \end{bmatrix}. \tag{8.19}$$

To continue the discussion, let $B = \begin{bmatrix} a & b \\ c & d \end{bmatrix}$ be such that its minimal polynomial has no real roots. The assumption on B and the equation $B \begin{bmatrix} 1 \\ 0 \end{bmatrix} = \begin{bmatrix} a \\ c \end{bmatrix}$ show that the vectors $\begin{bmatrix} 1 \\ 0 \end{bmatrix}$ and $\begin{bmatrix} a \\ c \end{bmatrix}$ are linearly independent, hence $P_1 = \begin{bmatrix} 1 & a \\ 0 & c \end{bmatrix}$ is nonsingular with inverse $P_1^{-1} = \frac{1}{c} \begin{bmatrix} c & -a \\ 0 & 1 \end{bmatrix}$.

From these considerations, we have

$$P_1^{-1}BP_1 = \frac{1}{c} \begin{bmatrix} c & -a \\ 0 & 1 \end{bmatrix} \begin{bmatrix} a & b \\ c & d \end{bmatrix} \begin{bmatrix} 1 & a \\ 0 & c \end{bmatrix} = \begin{bmatrix} 0 & bc - ad \\ 1 & a + d \end{bmatrix}. \tag{8.20}$$

We notice that the matrix on the right-hand side of (8.20) is of the same type as that in equation (8.19).

From the assumption on B, we have that it has no real eigenvalues, equivalently, its minimal polynomial $m_B(x) = x^2 - (a+d)x + ad - bc$ has no real roots. The latter condition is equivalent to having the discriminant of $m_B(x)$ negative, that is,

$$(a - d)^2 + 4bc < 0. \tag{8.21}$$

On the other hand, we are looking for reals m and $n \neq 0$ such that the matrices A and B, as above, are similar. From the previous calculations, and using equations (8.19) and (8.20), we need to find m and n such that $m = \frac{a+d}{2}$ and $-n^2 - m^2 = bc - ad$. Substituting the value of m into $-n^2 - m^2 = bc - ad$, simplifying, and solving for n^2, one has $n^2 = \frac{-(a-d)^2 - 4bc}{4}$. Now, using (8.21), we conclude that n can be obtained from the previous equation, by taking square root. In other words, using equations (8.19) and (8.20), we have that there are real numbers m and $n \neq 0$ such that

$$PP_1^{-1}BP_1P^{-1} = \begin{bmatrix} m & -n \\ n & m \end{bmatrix} = A, \tag{8.22}$$

from where we obtain that B is similar to A. Summarizing, we have proved:

Theorem 8.4.1. *Every real* 2×2 *matrix which has nonreal eigenvalues is similar to a matrix of the form* $\left[\begin{smallmatrix} m & -n \\ n & m \end{smallmatrix}\right]$, *with* $n \neq 0$.

Remark 8.4.1. It is interesting to notice that the set of matrices of the form $\left[\begin{smallmatrix} m & -n \\ n & m \end{smallmatrix}\right]$, with $n \neq 0$, can be identified with the complex numbers. More precisely, if $z = m + in$ is a complex number, then the function $f(z) = \left[\begin{smallmatrix} m & -n \\ n & m \end{smallmatrix}\right]$ defines an isomorphism between the complex numbers $\mathbb{C}$ and the set of matrices described before. From this point of view, if z_0 is a fixed complex number and z is any other complex number, the multiplication by z_0 can be interpreted as a rotation followed by a scaling of z when the latter is represented in the complex plane.

8.4.2 Matrices whose minimal polynomial has different roots

For the reading of this section it is recommended to review Appendix A, Section A.3. In this section, i will denote the square root of -1, namely $i = \sqrt{-1}$.

Recall that a matrix A is diagonalizable if and only if the minimal polynomial can be written as the product of distinct linear factors. Thus, under this assumption, the only obstruction for a real matrix to be diagonalizable is due to the irreducible factors of degree two in $m_A(x)$. The aim of this section is to show that in this case A is similar to a matrix that is block diagonal, where each block is of size at most two. More precisely, each of the blocks of size two is determined by the irreducible factors of degree two in $m_A(x)$.

Assume that A is a matrix with entries in the complex numbers. We declare the conjugate matrix of A to be denoted by $\overline{A}$, as the matrix obtained by conjugating each of the entries of A. For instance, if $A = \left[\begin{smallmatrix} 1+i & 2-i \\ 2-2i & 2-3i \end{smallmatrix}\right]$, then $\overline{A} = \left[\begin{smallmatrix} 1-i & 2+i \\ 2+2i & 2+3i \end{smallmatrix}\right]$.

We should notice that if all entries of A are real numbers, then $A = \overline{A}$.

Let A be a real $n \times n$ matrix and assume that $\lambda \in \mathbb{C} \setminus \mathbb{R}$ is an eigenvalue of A. We will see that $\overline{\lambda}$ is also an eigenvalue of A. By the definition of an eigenvalue, there is $X \in \mathbb{C}^n \setminus \{0\}$ such that $AX = \lambda X$. From the definition of the conjugate matrix, one obtains $\overline{AX} = A\,\overline{X}$. On the other hand, $\overline{AX} = \overline{\lambda X} = \overline{\lambda}\,\overline{X}$, concluding that $A\overline{X} = \overline{\lambda}\,\overline{X}$. That is, $\overline{\lambda}$ has an associated eigenvector $\overline{X}$.

With this we have prepared the ground to formulate the following result.

Theorem 8.4.2. *Let* $T : \mathbb{R}^n \to \mathbb{R}^n$ *be an operator, and assume that the minimal polynomial of T is represented as the product of different linear factors in the form*

$$m_T(x) = (x - \lambda_1)(x - \lambda_2) \cdots (x - \lambda_l)(x - \mu_1)(x - \overline{\mu_1})(x - \mu_2)(x - \overline{\mu_2}) \cdots (x - \mu_r)(x - \overline{\mu_r}), \quad (8.23)$$

where $\lambda_j \in \mathbb{R}$ for every $j \in \{1, 2, \ldots, l\}$ and $\mu_j \in \mathbb{C} \setminus \mathbb{R}$ for every $j \in \{1, 2, \ldots, r\}$. Then, there is a basis of $\mathbb{R}^n$ with respect to which T is represented by a matrix of the form

$$D = \begin{bmatrix} \lambda_1 & 0 & \cdots & 0 & 0 & \cdots & 0 \\ 0 & \lambda_2 & \cdots & 0 & 0 & \cdots & 0 \\ \vdots & \vdots & \ddots & \vdots & \vdots & \ddots & \vdots \\ 0 & 0 & \cdots & \lambda_l & 0 & \cdots & 0 \\ 0 & 0 & \cdots & 0 & D_1 & \cdots & 0 \\ \vdots & \vdots & \ddots & \vdots & \vdots & \ddots & \vdots \\ 0 & 0 & \cdots & 0 & 0 & \cdots & D_r \end{bmatrix}, \quad (8.24)$$

where $D_j = \begin{bmatrix} a_j & -b_j \\ b_j & a_j \end{bmatrix}$, and $\mu_j = a_j + ib_j$ for every $j \in \{1, 2, \ldots, r\}$.

Proof. Let $X_1, \ldots, X_l, Y_1, \overline{Y_1}, \ldots, Y_r, \overline{Y_r}$ be eigenvectors associated to the corresponding eigenvalues. Notice that $X_k \in \mathbb{R}^n$ for every $k \in \{1, 2, \ldots, l\}$ and $Y_j \in \mathbb{C}^n$ for every $j \in \{1, 2, \ldots, r\}$.

For each $j \in \{1, 2, \ldots, r\}$ define $Z_{2j-1} = \frac{1}{2}(Y_j + \overline{Y_j})$ and $Z_{2j} = -\frac{i}{2}(Y_j - \overline{Y_j})$.

It can be verified readily that $\{X_1, \ldots, X_l, Z_1, Z_2, \ldots, Z_{2r-1}, Z_{2r}\}$ is a basis of $\mathbb{R}^n$. From this we have that $n = l + 2r$.

Also, $T(X_k) = \lambda_k X_k$ for $k \in \{1, 2, \ldots, l\}$, and

$$\begin{aligned} T(Z_{2j-1}) &= T\left(\frac{1}{2}(Y_j + \overline{Y_j})\right) \\ &= \frac{1}{2}(T(Y_j) + T(\overline{Y_j})) \\ &= \frac{1}{2}(\mu_j Y_j + \overline{\mu}\, \overline{Y_j}) \\ &= \frac{1}{2}[(a_j + ib_j)Y_j + (a_j - ib_j)\overline{Y_j}] \\ &= \frac{1}{2}[a_j(Y_j + \overline{Y_j}) + ib_j(Y_j - \overline{Y_j})] \\ &= a_j Z_{2j-1} - b_j Z_{2j}, \quad (8.25) \end{aligned}$$

for every $j \in \{1, 2, \ldots, r\}$. A calculation as in (8.25) shows that $T(Z_{2j}) = a_j Z_{2j-1} + b_j Z_{2j}$.

Defining $D_j = \begin{bmatrix} a_j & -b_j \\ b_j & a_j \end{bmatrix}$ and using the definition of the associated matrix to an operator in a given basis, the conclusion of the theorem follows. $\square$

Remark 8.4.2. Theorem 8.4.2 can be phrased for real matrices A, whose minimal polynomial has different roots (real or complex), as follows. If the minimal polynomial of A is as in Theorem 8.4.2, then there is a nonsingular matrix P such that $P^{-1}AP = D$, where D is as in (8.24).

8.5 Applications

In this section we present some applications of the theory of eigenvalues. The main objective is to illustrate how we can use some of the results developed in this chapter, when approaching hypothetical problems. We invite the reader to consult more examples in different sources.

8.5.1 Interacting species

The main ideas presented here were taken from [7]. Our contribution consists in adapting the continuous case to a discrete one to show what is called a discrete linear dynamical system and to give a more "real" meaning to the continuous case.

When analyzing the interaction of species, it is clear that the population growth of one depends on the others. The relationship among species can be of several types: cooperation, competition, and predator–prey. To understand the complex dynamics that takes place among species, we start by considering the interaction between two species and assume that data, describing how they interact, is recorded at fixed time intervals numbered as $k \in \{0, 1, 2, \ldots\}$. Let us denote by x_k and y_k the population of the two species at time k. We are interested in describing the relation that there is between the quantities x_k and y_k, depending on the type of interaction.

The model to be analyzed is called *predator–prey*. For a more complete discussion of some other interactions, see [7, pp. 332–352].

Under isolation and appropriate conditions of surviving and growing of the species, it is reasonable to assume that at time $k + 1$ the population is proportional to the population at time k.

Algebraically, this can be represented by the equations

$$\begin{aligned} x_{k+1} &= R_1 x_k, \\ y_{k+1} &= R_2 y_k. \end{aligned} \tag{8.26}$$

Assuming that the proportionality quantities R_1 and R_2 do not change with time and that the populations are known at the beginning of the experiment, the system (8.26) can be solved by back-substitution, that is, if x_0 and y_0 are the initial populations, then $x_1 = R_1 x_0$; $x_2 = R_1 x_1 = R_1 R_1 x_0 = R_1^2 x_0$. Inductively, we have $x_k = R_1^k x_0$. Likewise, $y_k = R_2^k y_0$.

In general, R_1 and R_2 depend on time and on the populations x_k and y_k. In order to formulate a model, we will assume that x_k and y_k represent the predator and prey population, respectively. We also assume that $R_1 = -a + b y_k$ and $R_2 = c - d x_k$, with a, b, c, and d positive constants. Substituting these expressions into equation (8.26), one has

$$\begin{aligned} x_{k+1} &= (-a + b y_k) x_k = -a x_k + b y_k x_k, \\ y_{k+1} &= (c - d x_k) y_k = c y_k - d x_k y_k. \end{aligned} \tag{8.27}$$

The meaning of these equations is as follows. For the first equation in (8.27), one has that without prey the predator population decreases, and with the presence of prey the predator population increases proportionally to the amount of prey. The change in predator population is due to the "natural" decrease by competition and due to the growth in the presence of prey. The increase is given by the term $by_k x_k$. The rate of change of one population interacting with another is proportional to the product of the two populations [7, p. 334].

For the second equation in (8.27), one has that without predators, the population growth is proportional to that in the previous period and the rate of change is constant, while with the predators present the net change is the difference of growth without predators and $dx_k y_k$ due to the interaction of the species.

Notice that the system represented by (8.27) is nonlinear, thus linear methods to approach the solutions are restricted. To apply linear methods, we add another assumption which is captured in the following equations:

$$\begin{aligned} x_{k+1} &= -ax_k + by_k, \\ y_{k+1} &= cy_k - dx_k, \end{aligned} \tag{8.28}$$

where $a, b, c,$ and d positive constants.

The first equation in (8.28) is interpreted as follows. The total change in the predator population is due to natural death and the term proportional to the size of prey population, with makes an increase of the population possible.

Analogously, the second equation in (8.28) can be interpreted as follows. The total change in prey is the difference between the natural growth of its population and pray eliminated by the predators. The latter term is interpreted by saying that each predator consumes a constant amount of prey during a unit of time.

With the above considerations, equations in (8.28) can be represented in matrix form

$$AX_k = X_{k+1}, \tag{8.29}$$

where $A = \begin{bmatrix} -a & b \\ -d & c \end{bmatrix}$ and $X_k = \begin{bmatrix} x_k \\ y_k \end{bmatrix}$. Assuming that the initial populations are known and denoted by x_0 and y_0, one has $X_1 = \begin{bmatrix} -a & b \\ -d & c \end{bmatrix} \begin{bmatrix} x_0 \\ y_0 \end{bmatrix}$ and

$$\begin{aligned} X_2 &= \begin{bmatrix} -a & b \\ -d & c \end{bmatrix} \begin{bmatrix} x_1 \\ y_1 \end{bmatrix} \\ &= \begin{bmatrix} -a & b \\ -d & c \end{bmatrix} \begin{bmatrix} -a & b \\ -d & c \end{bmatrix} \begin{bmatrix} x_0 \\ y_0 \end{bmatrix} \\ &= \begin{bmatrix} -a & b \\ -d & c \end{bmatrix}^2 \begin{bmatrix} x_0 \\ y_0 \end{bmatrix}. \end{aligned}$$

By an inductive argument, we have

$$X_k = \begin{bmatrix} -a & b \\ -d & c \end{bmatrix}^k \begin{bmatrix} x_0 \\ y_0 \end{bmatrix}. \tag{8.30}$$

To understand how the prey and predator populations change, we will study the powers A^k and their action on $X_0 = \begin{bmatrix} x_0 \\ y_0 \end{bmatrix}$. For this purpose, we will use the eigenvalues of A. Let E_1 denote the first canonical vector of $\mathbb{R}^2$. Since b and d are positive real numbers, we see directly that E_1 and AE_1 are linearly independent, thus the annihilator polynomial of E_1 is the minimal polynomial of A, which is obtained from the equation

$$A^2 E_1 + (a - c)AE_1 + (bd - ac)E_1 = 0, \tag{8.31}$$

obtained by performing the corresponding calculations. From equation (8.31), we have that $m_A(x) = x^2 + (a-c)x + bd - ac$ is the minimal polynomial of A, hence the eigenvalues of A are:

$$x = \frac{c - a \pm \sqrt{(a - c)^2 - 4(bd - ac)}}{2} = \frac{c - a \pm \sqrt{(a + c)^2 - 4bd}}{2}.$$

Let $\lambda_1 = \frac{c-a-\sqrt{(a+c)^2-4bd}}{2}$ and $\lambda_2 = \frac{c-a+\sqrt{(a+c)^2-4bd}}{2}$ be the eigenvalues of A. To examine the nature of the eigenvalues, we use the discriminant of $m_A(x)$ which will be denoted by

$$\Delta = (a + c)^2 - 4bd. \tag{8.32}$$

The description of X_k will be provided by examining several cases of Δ:
1. Assume that $\Delta > 0$, then $\lambda_1 \neq \lambda_2$. In this case, A is diagonalizable, and then the eigenvectors of A form a basis of $\mathbb{R}^2$. Denote the eigenvectors by V_1 and V_2, hence there are scalars μ_1 and μ_2 such that $X_0 = \mu_1 V_1 + \mu_2 V_2$. From this equation, one has

$$X_k = A^k X_0 = \mu_1 \lambda_1^k V_1 + \mu_2 \lambda_2^k V_2. \tag{8.33}$$

If $|\lambda_1|, |\lambda_2| < 1$, both species will go extinct, since λ_1^k and λ_2^k approach zero as k approaches infinity. The other cases are left to the reader as an exercise; see Exercise 30, p. 222.
2. If $\Delta < 0$, then the eigenvalues are not real. Applying Theorem A.5.4, p. 280, we have that λ_2 is the conjugate of λ_1, thus their moduli are the same.
   Also, A is diagonalizable over the complex numbers, hence there is a nonsingular complex matrix P whose columns are the eigenvectors of A. From this we have an equation analogous to (8.33). More precisely, we have $X_0 = \mu_1 V_1 + \mu_2 V_2$, with μ_1, μ_2 being complex numbers. From this, one has

$$X_k = A^k X_0 = \mu_1 \lambda_1^k V_1 + \mu_2 \lambda_2^k V_2, \tag{8.34}$$

as in the real case.

- If $|\lambda_1| = 1$, the vectors X_k rotate on a elliptical orbit, hence the species do not go extinct.
- If $|\lambda_1| < 1$, then $X_k \to 0$, and hence the species go extinct.
- If $|\lambda_1| > 1$, then X_k "escapes to infinity", which, biologically speaking, is impossible.

3. Assume that $\Delta = 0$, then $\lambda_1 = \lambda_2 = \frac{c-a}{2}$. If A were diagonalizable, then $m_A(x) = x - \lambda_1$, which is not possible, since we are assuming that b and d are positive real numbers. To describe X_k, we will use the Jordan canonical form of A. In order to find the Jordan canonical form of A, J, and the matrix P that changes A to J, we will find a Jordan basis of $\mathbb{R}^2$. We have already pointed out that E_1 is not an eigenvector, thus E_1 is a cyclic vector. Applying Lemma 8.3.1, we have that a Jordan basis is $\{E_1, (A - \lambda I_2)E_1\}$. By a direct calculation, we find that $E_2 = (A - \lambda I_2)E_1 = \left[\begin{smallmatrix} -\frac{a+c}{2} \\ -d \end{smallmatrix} \right]$, thus the matrix P that satisfies $P^{-1}AP = J$ is

$$P = \begin{bmatrix} 1 & -\frac{a+c}{2} \\ 0 & -d \end{bmatrix}, \tag{8.35}$$

whose inverse is

$$P^{-1} = \frac{1}{d} \begin{bmatrix} d & -\frac{a+c}{2} \\ 0 & -1 \end{bmatrix}. \tag{8.36}$$

From the equation $J = P^{-1}AP$, we obtain $A^k = PJ^kP^{-1}$. By induction, it can be shown that $J^k = \begin{bmatrix} \lambda_1^k & 0 \\ k\lambda_1^{k-1} & \lambda_1^k \end{bmatrix}$.

Performing the corresponding calculations, one has

$$A^k = \frac{\lambda_1^{k-1}}{d} \begin{bmatrix} d(\lambda_1 - \frac{(a+c)(k-1)}{2}) & \frac{(a+c)^2(k-1)}{4} \\ -d^2(k-1) & d\lambda_1 + \frac{d(k-1)(a+c)}{2} \end{bmatrix}. \tag{8.37}$$

From equation (8.32), we obtain $\frac{(a+b)^2}{4} = bd$; substituting this into (8.37), factorizing d, simplifying, and rearranging terms, one has

$$A^k = (k-1)\lambda_1^{k-1} \begin{bmatrix} -\frac{(a+c)}{2} & b \\ -d & \frac{(a+c)}{2} \end{bmatrix} + \lambda_1^k I_2. \tag{8.38}$$

From equation (8.38), using that $X_k = A^k X_0$ and writing the coordinates of X_k, we have

$$x_k = (k-1)\lambda_1^{k-1}\left(-\frac{a+c}{2}x_0 + by_0\right) + \lambda_1^k x_0,$$

$$y_k = (k-1)\lambda_1^{k-1}\left(-dx_0 + \frac{a+c}{2}y_0\right) + \lambda_1^k y_0. \tag{8.39}$$

In Exercise 8.5.1, we invite the reader to complete the analysis of X_k.

Exercise 8.5.1. Recall that $\lambda_1 = \frac{c-a}{2}$. Now substitute this value of λ_1 into equations of (8.39) and consider the cases $|\lambda_1| \le 1$ and $|\lambda_1| > 1$ to estimate the growth of x_k and y_k.

From the equations in (8.39) prove that for k sufficiently large, x_k is approximately equal to

$$\frac{2by_0 - (a+c)x_0}{(a+c)y_0 - 2dx_0}y_k.$$

8.5.2 A linear discrete dynamical system

There are various processes where the behavior of the process depends only on the previous stage. This happens in nature and society. The following hypothetical case describes the movement from one category to another among members of a sport club.

The sport club association "Venados de Pisaflores" (Pisaflores' Deer's) has three categories: beginners, professionals, and masters. Each member of the club must stay on a category exactly 2 years before moving to the next or leave the category by resigning. There is no penalty for resigning from the corresponding category before the expiration time. Masters must resign from the club when their time expires. The members must observe the following rules. Beginners have no duties, while each professional is obliged to bring a quantity a of beginners during his time as a professional; each master is obliged to bring b beginners before his time expires. It is estimated that a fraction c of beginners will become professionals, while a fraction d of professionals will become masters. At the starting time there are x_0, y_0, and z_0 beginners, professionals, and masters, respectively. In the long run, what fraction of beginners will become masters and what fraction of professionals will become masters?

Discussion. In period j, let us denote by x_j, y_j, and z_j the number of beginners, professionals, and masters, respectively. The number of beginners at the end of the first period is $x_1 = ay_0 + bz_0$, while at the starting point there were x_0 beginners. Analogously, the number of professionals at the end of the first period is $y_1 = cx_0$ and the number of masters at the end of the first period is $z_1 = dy_0$. These equations can be written in matrix form as

$$X_1 = \begin{bmatrix} x_1 \\ y_1 \\ z_1 \end{bmatrix} = \begin{bmatrix} 0 & a & b \\ c & 0 & 0 \\ 0 & d & 0 \end{bmatrix} \begin{bmatrix} x_0 \\ y_0 \\ z_0 \end{bmatrix}. \tag{8.40}$$

Analogously, for the next period we have

$$X_2 = \begin{bmatrix} x_2 \\ y_2 \\ z_2 \end{bmatrix} = \begin{bmatrix} 0 & a & b \\ c & 0 & 0 \\ 0 & d & 0 \end{bmatrix} \begin{bmatrix} x_1 \\ y_1 \\ z_1 \end{bmatrix} = \begin{bmatrix} 0 & a & b \\ c & 0 & 0 \\ 0 & d & 0 \end{bmatrix}^2 \begin{bmatrix} x_0 \\ y_0 \\ z_0 \end{bmatrix}. \tag{8.41}$$

In general, after $2k$ years, using an inductive reasoning, one has

$$X_k = \begin{bmatrix} x_k \\ y_k \\ z_k \end{bmatrix} = \begin{bmatrix} 0 & a & b \\ c & 0 & 0 \\ 0 & d & 0 \end{bmatrix} \begin{bmatrix} x_{k-1} \\ y_{k-1} \\ z_{k-1} \end{bmatrix} = \begin{bmatrix} 0 & a & b \\ c & 0 & 0 \\ 0 & d & 0 \end{bmatrix}^k \begin{bmatrix} x_0 \\ y_0 \\ z_0 \end{bmatrix}. \tag{8.42}$$

To answer the posed question, we need to estimate the quotients

$$\frac{x_k}{z_k} \quad \text{and} \quad \frac{y_k}{z_k} \tag{8.43}$$

when $k \to \infty$.

We need to know the behavior of X_k. In order to achieve this, we will use the same idea as in Example 8.5.2, that is, we will compute the eigenvalues of $A = \begin{bmatrix} 0 & a & b \\ c & 0 & 0 \\ 0 & d & 0 \end{bmatrix}$. Using Algorithm 8.2.1, p. 200, one finds that the minimal polynomial of A is $m_A(x) = x^3 - acx - bdc$.

An example of values for $a, b, c,$ and d are $a = 60, b = 80, c = \frac{1}{10},$ and $d = \frac{1}{2}$. For these values, we have $A = \begin{bmatrix} 0 & 60 & 80 \\ 1/10 & 0 & 0 \\ 0 & 1/2 & 0 \end{bmatrix}$ and its minimal polynomial is $m_A(x) = x^3 - 6x - 4 = (x + 2)(x^2 - 2x - 2)$. From this representation of $m_A(x)$, we obtain that the eigenvalues of A are $\lambda_1 = -2, \lambda_2 = 1 - \sqrt{3},$ and $\lambda_3 = 1 + \sqrt{3}$. Applying Theorem 8.1.1, p. 189, we have that A is diagonalizable, that is, there are eigenvectors $Y_1, Y_2,$ and Y_3, linearly independent, so that $AY_i = \lambda_i Y_i$. These eigenvectors are computed by solving the system of linear equations

$$(A - \lambda I_3)X = 0. \tag{8.44}$$

Since the system (8.44) is homogeneous, to find its solutions, it is enough to obtain the row reduced form of the matrix $A - \lambda I_3$, which is done by the process below

$$A - \lambda I_3 = \begin{bmatrix} 0 & 3 & 1 \\ 2 & 0 & 0 \\ 0 & 2 & 0 \end{bmatrix} - \begin{bmatrix} \lambda & 0 & 0 \\ 0 & \lambda & 0 \\ 0 & 0 & \lambda \end{bmatrix}$$

$$= \begin{bmatrix} -\lambda & 3 & 1 \\ 2 & -\lambda & 0 \\ 0 & 2 & -\lambda \end{bmatrix} \sim \begin{bmatrix} -\lambda & 3 & 1 \\ 1 & -\lambda/2 & 0 \\ 0 & 1 & -\lambda/2 \end{bmatrix} \sim \begin{bmatrix} -\lambda & 0 & 1 + (3/2)\lambda \\ 1 & 0 & -\lambda^2/4 \\ 0 & 1 & -\lambda/2 \end{bmatrix}$$

$$\sim \begin{bmatrix} 0 & 0 & 1 + (3/2)\lambda - \lambda^3/4 \\ 1 & 0 & -\lambda^2/4 \\ 0 & 1 & -\lambda/2 \end{bmatrix}.$$

Since λ is a root of $m_A(x)$, $\lambda^3 - 6\lambda - 4\lambda = 0$, which is equivalent to $1 + (3/2)\lambda - \lambda^3/4 = 0$. Therefore, the row reduced form of $A - \lambda I_3$ is

$$R = \begin{bmatrix} 1 & 0 & -\lambda^2/4 \\ 0 & 1 & -\lambda/2 \\ 0 & 0 & 0 \end{bmatrix}. \tag{8.45}$$

From this we have that the solutions of (8.44) are given by $x = \frac{\lambda^2}{4}z$, $y = \frac{\lambda}{2}z$, and z is a free variable. Choosing the value of $z = 1$, and setting $\lambda_1 = -2$, $\lambda_2 = 1 - \sqrt{3}$, and $\lambda_3 = 1 + \sqrt{3}$, we obtain that three linearly independent eigenvectors of A are

$$Y_1 = \begin{bmatrix} 1 \\ -1 \\ 1 \end{bmatrix}, \quad Y_2 = \begin{bmatrix} \lambda_2^2/4 \\ \lambda_2/2 \\ 1 \end{bmatrix}, \quad \text{and} \quad Y_3 = \begin{bmatrix} \lambda_3^2/4 \\ \lambda_3/2 \\ 1 \end{bmatrix}.$$

Let $X_0 = \begin{bmatrix} x_0 \\ y_0 \\ z_0 \end{bmatrix}$ be a vector that represents the number of members in each category at the initial point. Since Y_1, Y_2, and Y_3 are linearly independent, there are constants r, s, and t such that

$$X_0 = rY_1 + sY_2 + tY_3. \tag{8.46}$$

Since $X_k = A^k X_0$, from equation (8.46) and the fact that λ_j^k is an eigenvalue of A^k, we obtain

$$X_k = A^k X_0 = A^k(rY_1 + sY_2 + tY_3) = r\lambda_1^2 Y_1 + s\lambda_2^k Y_2 + t\lambda_3^k Y_3. \tag{8.47}$$

Equation (8.47) can be written using the coordinates of $X_k = (x_k, y_k, z_k)$ as

$$\begin{aligned}
x_k &= \lambda_1^k r + \frac{\lambda_2^{k+2}}{4}s + \frac{\lambda_3^{k+2}}{4}t, \\
y_k &= \lambda_1^k r + \frac{\lambda_2^{k+2}}{2}s + \frac{\lambda_3^{k+2}}{2}t, \\
z_k &= \lambda_1^k r + \lambda_2^k s + \lambda_3^k t.
\end{aligned} \tag{8.48}$$

To answer the posed question, we need to examine the behavior of the quotient

$$\begin{aligned}
\frac{x_k}{z_k} &= \frac{\lambda_1^k r + \frac{\lambda_2^{k+2}}{4}s + \frac{\lambda_2^{k+2}}{4}t}{\lambda_1^k r + \lambda_2^k s + \lambda_3^k t} \\
&= \frac{r(\frac{\lambda_1}{\lambda_3})^k + \frac{s}{4}(\frac{\lambda_2}{\lambda_3})^k \lambda_2^2 + \frac{t}{4}\lambda_3^2}{r(\frac{\lambda_1}{\lambda_3})^k + s(\frac{\lambda_2}{\lambda_3})^k + t}
\end{aligned} \tag{8.49}$$

as $k \to \infty$.

On the other hand, we have that $|\lambda_1| = |-2| < |1 + \sqrt{3}| = |\lambda_3|$ and $|\lambda_2| = |1 - \sqrt{3}| < |1 + \sqrt{3}| = |\lambda_3|$. Thus $|\frac{\lambda_j}{\lambda_3}| < 1$ for $j = 1, 2$. We also see that if $|v| < 1$, then $v^k \to 0$ as $k \to \infty$. From these facts and equation (8.49), we obtain

$$\lim_{k \to \infty} \frac{x_k}{z_k} = \frac{t \frac{\lambda_3^2}{4}}{t} = \frac{\lambda_3^2}{4} = \frac{(1 + \sqrt{3})^2}{4} = \frac{4 + 2\sqrt{3}}{4},$$

which is approximately 1.86. That is, in the long run, the number of novices is approximately twice the number of masters.

In Exercise 27, we invite the reader to answer the question concerning professionals and masters.

Exercise 8.5.2. Generalize the previous example to an association with n different categories, defining rules that regulate the moving of the members to a new category.

8.5.3 Discrete age-structured population growth

Some of the ideas in this example were borrowed from the excellent reference for mathematical models in biology [13]. Assume that a population is divided into $k+1$ age sectors $S_0, S_1, S_2, \ldots, S_k$, indexed according age, that is, individuals in sector S_j are younger than individuals in sector S_{j+1}, and S_0 are the newborns. Let $p_{0n}, p_{1n}, p_{2n}, \ldots, p_{kn}$ be the number of individuals in each sector at time n. Assume that in sector S_j there are a_j births and the fraction from sector S_j that goes to sector S_{j+1} is b_j. What is the number of individuals in each sector for time $n + 1$?

Discussion. First of all, the number of newborns is $p_{0\,n+1} = 0p_{0n} + a_1p_{1n} + a_2p_{2n} + \cdots + a_kp_{kn}$, and the number of individuals in sector S_{j+1} is $p_{j+1\,n+1} = b_jp_{j\,n}, j \in \{0, 1, 2, \ldots, k-1\}$. This system of equations can be written in matrix form as

$$\begin{bmatrix} p_{0\,n+1} \\ p_{1\,n+1} \\ p_{2\,n+1} \\ p_{3\,n+1} \\ \vdots \\ p_{k-1\,n+1} \\ p_{k\,n+1} \end{bmatrix} = \begin{bmatrix} 0 & a_1 & a_2 & a_3 & \cdots & a_{k-1} & a_k \\ b_0 & 0 & 0 & 0 & \cdots & 0 & 0 \\ 0 & b_1 & 0 & 0 & \cdots & 0 & 0 \\ 0 & 0 & b_2 & 0 & \cdots & 0 & 0 \\ \vdots & \vdots & \vdots & \vdots & \ddots & \vdots & \vdots \\ 0 & 0 & 0 & 0 & \cdots & 0 & 0 \\ 0 & 0 & 0 & 0 & \cdots & b_{k-1} & 0 \end{bmatrix} \begin{bmatrix} p_{0\,n} \\ p_{1\,n} \\ p_{2\,n} \\ p_{3\,n} \\ \vdots \\ p_{k-1\,n} \\ p_{k\,n} \end{bmatrix}. \tag{8.50}$$

Let A be the coefficient matrix in equation (8.50), then its characteristic polynomial is given by

$$f_A(x) = \begin{vmatrix} -x & a_1 & a_2 & a_3 & \cdots & a_{k-1} & a_k \\ b_0 & -x & 0 & 0 & \cdots & 0 & 0 \\ 0 & b_1 & -x & 0 & \cdots & 0 & 0 \\ 0 & 0 & b_2 & -x & \cdots & 0 & 0 \\ \vdots & \vdots & \vdots & \vdots & \ddots & \vdots & \vdots \\ 0 & 0 & 0 & 0 & \cdots & -x & 0 \\ 0 & 0 & 0 & 0 & \cdots & b_{k-1} & -x \end{vmatrix}. \tag{8.51}$$

Matrix A is called a *Leslie matrix*.

From Exercise 31, p. 222, we have

$$f_A(x) = (-1)^{k+1}(x^{k+1} - b_0 a_1 x^k - b_0 b_1 a_2 x^{k-1} - \cdots - b_0 b_1 \cdots b_{k-2} a_{k-1} x - b_0 b_1 \cdots b_{k-2} a_k). \tag{8.52}$$

Applying Descartes' rule of signs, Theorem A.5.5, p. 281, we obtain directly that $f_A(x)$ has only one positive root. Let λ be the real positive root of $f_A(x)$ and assume that X_0 is an eigenvector associated to λ. Furthermore, assume that $|\mu| \le |\lambda|$ for every root of $f_A(x)$. With these assumptions, how would the age distribution be related with the entries of X_0, the vector of initial population?

8.5.4 Oscillation of masses coupled with springs

We will consider the oscillation of n coupled masses with springs as shown in Figure 8.2.

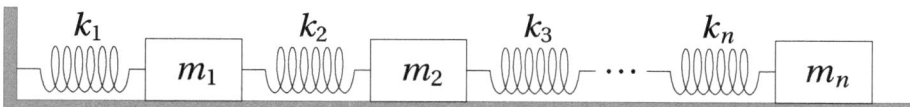

Figure 8.2: Coupled masses with springs.

We will assume that there is no friction on the table and that the masses of the springs are negligible compared with each mass m_i. We will also assume that the forces pointing to the left are negative, while the forces pointing to the right are positive.

To illustrate the main ideas, we will start by considering $n = 2$.

Before we start the discussion, let us set the notation. If y is a function of the time variable t, its second derivative will be denoted by $\ddot{y}$.

If y_1 and y_2 are the displacements of masses m_1 and m_2 from the equilibrium point, then two forces act on m_1, one due to the spring with elasticity constant k_1 and the other due to the spring with elasticity constant k_2. From Hooke's law, those forces have magnitudes $-k_1 y_1$ and $k_2(y_2 - y_1)$, respectively. There is only one force acting on m_2, namely that due to the deformation of spring with elasticity constant k_2. This force is the opposite of that acting on m_1.

From the above considerations and using Newton's second law, we have

$$m_1\ddot{y}_1 = -k_1 y_1 + k_2(y_2 - y_1),$$
$$m_2\ddot{y}_2 = -k_2(y_2 - y_1).$$

(8.53)

Exercise 8.5.3. For the system of equations in (8.53), prove that:

1. $m_1 m_2 y_1^{(4)} + ((k_1 + k_2)m_2 + k_2 m_1)\ddot{y}_1 + k_1 k_2 y_1 = 0,$
2. $m_1 m_2 y_2^{(4)} + ((k_1 + k_2)m_2 + k_1 m_2)\ddot{y}_2 + k_1 k_2 y_2 = 0.$

Solve the equations above for the case $m_1 = m_2 = 1$. (Hint) Consider the characteristic equations, which in this case turns out to be $x^4 + (k_1 + 2k_2)x^2 + k_1 k_2 = 0$. What are the characteristic roots if k_1, k_2 are positive integers and $k_1^2 + 4k_2^2$ is a square?

In the discussion we are assuming that m_1 and m_2 are not zero, hence system (8.53) becomes

$$\ddot{y}_1 = -\frac{k_1 + k_2}{m_1}y_1 + \frac{k_2}{m_1}y_2,$$
$$\ddot{y}_2 = -\frac{k_2}{m_2}y_2 + \frac{k_2}{m_2}y_1.$$

(8.54)

Defining $Y = \begin{bmatrix} y_1 \\ y_2 \end{bmatrix}$, the system of equations (8.54) can be represented in matrix form as

$$\ddot{Y} = \begin{bmatrix} -\frac{k_1+k_2}{m_1} & \frac{k_2}{m_1} \\ -\frac{k_2}{m_2} & \frac{k_2}{m_2} \end{bmatrix} \begin{bmatrix} y_1 \\ y_2 \end{bmatrix}.$$

(8.55)

Instead of continuing discussing equation (8.55), we will generalize to the case $n > 2$. We will describe the forces acting on mass m_i for every $i \in \{1, 2, \ldots, n\}$.

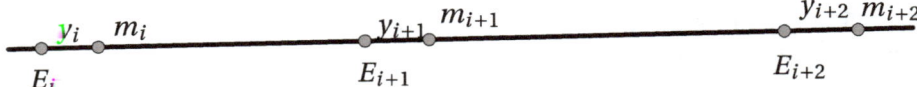

Figure 8.3: Equilibrium points E_i and positions of each m_i.

In Figure 8.3, reference points are as follows: points E_i, E_{i+1}, and E_{i+2} indicate the equilibrium positions of masses m_i, m_{i+1}, and m_{i+2}, respectively. The displacement of each mass is indicated by y_i, y_{i+1}, and y_{i+2}; l_i is the distance from E_i to E_{i+1}, and l_i' is the distance between masses m_i and m_{i+1} when moving at any given moment. The deformation of spring i is $l_i' - l_i$. From the considerations above, we have the following equations:

$$l_1 = y_1 + x_1, \qquad l_1' = x_1 + y_2,$$
$$l_2 = y_2 + x_2, \qquad l_2' = x_2 + y_3,$$
$$\vdots \qquad\qquad \vdots \tag{8.56}$$
$$l_{n-1} = y_{n-1} + x_{n-1} \qquad l_{n-1}' = x_{n-1} + y_n,$$

from where we obtain

$$l_i' - l_i = y_{i+1} - y_i. \tag{8.57}$$

Mass m_1 is under the action of springs 1 and 2, hence $m_1 \ddot{y}_1 = -k_1 y_1 + k_2(y_2 - y_1)$; mass m_2 is under the action of springs 2 and 3, and the force due to spring 2 is the opposite acting on m_1, then $m_2 \ddot{y}_2 = -k_2(y_2 - y_1) + k_3(y_3 - y_2)$. In general, for $2 \le i \le n-1$, we have

$$m_i \ddot{y}_i = -k_i(y_i - y_{i-1}) + k_{i+1}(y_{i+1} - y_i)$$
$$= k_i y_{i-1} - (k_i + k_{i+1})y_i + k_{i+1} y_{i+1}. \tag{8.58}$$

For $i = n$, one has $m_n \ddot{y}_n = -k_n(y_n - y_{n-1})$.

The system (8.58) can be written in matrix form as $\ddot{Y} = AY$, where A is the matrix

$$A = \begin{bmatrix} -\frac{k_1+k_2}{m_1} & \frac{k_1}{m_1} & 0 & 0 & \cdots & 0 & 0 \\ \frac{k_2}{m_2} & -\frac{k_2+k_3}{m_2} & \frac{k_3}{m_2} & 0 & \cdots & 0 & 0 \\ 0 & \frac{k_3}{m_3} & -\frac{k_3+k_2}{m_3} & \frac{k_4}{m_3} & \cdots & 0 & 0 \\ \vdots & \vdots & \vdots & \vdots & \ddots & \vdots & \vdots \\ 0 & 0 & 0 & 0 & \cdots & \frac{k_n}{m_n} & -\frac{k_n}{m_n} \end{bmatrix}.$$

From Corollary 8.3.1 we know that there is a nonsingular complex matrix P so that $P^{-1}AP = \mathrm{diag}\{J_1, \dots, J_k\}$ is the Jordan canonical form of A. We are assuming that the vector

$$Y = \begin{bmatrix} y_1(t) \\ \vdots \\ y_n(t) \end{bmatrix}$$

belongs to $\mathbb{R}^n$. Given such a P, for each t there is $Z(t)$ so that $Y(t) = PZ(t)$, $Z(t) \in \mathbb{C}^n$. Since $Y(t)$ has second derivative, so does $Z(t)$ and $\ddot{Y}(t) = P\ddot{Z}(t)$. From this we have $AY = \ddot{Y}(t) = P\ddot{Z}(t)$, hence $APZ = P\ddot{Z}(t)$, therefore $P^{-1}APZ = JZ = \ddot{Z}$.

We have that each matrix J_i is of the form $J_i = \mathrm{diag}\{J_{i_1}, \dots, J_{i_{s(i)}}\}$ and then the problem has been reduced to solving

$$\begin{bmatrix} c & 0 & 0 & \cdots & 0 & 0 \\ 1 & c & 0 & \cdots & 0 & 0 \\ 0 & 1 & c & \cdots & 0 & 0 \\ \vdots & \vdots & \vdots & \ddots & \vdots & \vdots \\ 0 & 0 & 0 & \cdots & c & 0 \\ 0 & 0 & 0 & \cdots & 1 & c \end{bmatrix} \begin{bmatrix} z_1 \\ z_2 \\ z_3 \\ \vdots \\ z_{n-1} \\ z_n \end{bmatrix} = \begin{bmatrix} \ddot{z}_1 \\ \ddot{z}_2 \\ \ddot{z}_3 \\ \vdots \\ \ddot{z}_{n-1} \\ \ddot{z}_n \end{bmatrix}, \tag{8.59}$$

which is equivalent to

$$\ddot{z}_1 = cz_1, \tag{8.60}$$

$$\ddot{z}_2 = z_1 + cz_2, \tag{8.61}$$

$$\ddot{z}_3 = z_2 + cz_3, \tag{8.62}$$

$$\vdots$$

$$\ddot{z}_n = z_{n-1} + cz_n,$$

and this system can be solved easily, by first solving (8.60) then solving (8.61), and so on.

Remark 8.5.1. If A is diagonalizable, the problem reduces to solving a system of the form $\ddot{z} = cz$. The new system is called noncoupled.

Remark 8.5.2. The matrix of the original system is called tridiagonal and there are various methods to deal with such matrices (Tridiagonal matrices).

8.6 Exercises

1. Assuming that A is diagonalizable and letting k be a positive integer, is A^k diagonalizable? If A^k is diagonalizable for some integer k, is A diagonalizable as well?
2. Let A be a diagonalizable matrix with eigenvalues $\lambda_1, \lambda_2, \ldots, \lambda_n$ and $q(x)$ a nonconstant polynomial. Define $B := q(A)$. Is B diagonalizable? If your answer is yes, which matrix is similar to B?
3. Let A be a 6×6 matrix whose minimal polynomial is $x^2(x-3)^3$. Determine the possible Jordan canonical forms of A.
4. Let $T : V \to V$ be an operator, where V is a finite-dimensional vector space. Assume that $V = W \oplus W_1$, where W and W_1 are T-invariant subspaces. Prove that T is diagonalizable if and only if the restriction of T to W and W_1 is diagonalizable.
5. Prove Lemma 8.1.2, p. 193.
6. Let T and T_1 be operators on V which commute and are diagonalizable. Prove that there is a basis of V with respect to which both are represented by diagonal matrices.
   (*Hint.*) Let r be an eigenvalue of T and $W_r = \{a \in V : T(a) = ra\}$ the eigenspace determined by r. Use $TT_1 = T_1T$ to show that W_r is T_1-invariant. Since T is diagonal-

izable, V is the direct sum of the eigenspaces of T. Now use Exercise 4 to find a basis of V with respect to which T and T_1 are represented by diagonal matrices.

7. Let T and T_1 be operators on V which commute and are triangularizable. Prove that they are triangularizable simultaneously.

8. Prove that the matrix in Exercise 1b, p. 186, is similar to its transpose.
   (*Hint.*) Consider A as the matrix of the operator T with respect to the basis $\{a_1, a_2, \dots, a_n\}$, where $T(a_i) = a_{i+1}$ for $i \in \{1, 2, \dots, n-1\}$ and $T(a_n) = 0$.

9. Use the previous exercise to prove that every matrix is similar to its transpose.

10. Let A be a diagonalizable matrix. Prove that all the eigenvalues of A are equal if and only if $A = cI$, for some c.

11. Let $p(x)$ be a monic polynomial. Are there nonsimilar matrices such that $p(x)$ is their minimal polynomial?

12. Prove that the function f defined in Remark 8.4.1, p. 207, is an isomorphism, that is, f is bijective and satisfies $f(z + w) = f(z) + f(w)$ and $f(zw) = f(z)f(w)$ for every z and w in $\mathbb{C}$.

13. Are there real symmetric matrices whose minimal polynomial is $x^4 - 1$?

14. Let A and B be 3×3 nilpotent matrices. Prove that A is similar to B if and only if they have the same minimal polynomial.

15. Assume that the minimal and characteristic polynomials of A are $(x-3)^2$ and $(x-3)^3$, respectively. Determine the possible Jordan canonical forms of A.

16. Let $g(x)$ be the polynomial in equation (7.5), p. 167. Prove that $g(x)$ has degree $\leq n^2$ and that $g(A) = 0$, where A is the matrix used to construct $g(x)$.

17. Let $p(x)$ be a monic irreducible polynomial with integer coefficients. If A is an $n \times n$ matrix whose minimal polynomial is $p(x)$, answer the following questions:
    (a) Is it possible that A does not have integer entries?
    (b) Is A diagonalizable?
    (c) What is the degree of $p(x)$ in order that A be symmetric?
    (d) Is it possible that $p(x)$ be of degree 13?

18. If A is an $n \times n$ diagonalizable matrix with linearly independent eigenvectors X_1, $X_2, \dots, X_n$, then $P^{-1}AP = \text{diag}\{\lambda_1, \lambda_2, \dots, \lambda_n\}$, where $AX_i = \lambda_i X_i$ and P is the matrix whose columns are $X_1, X_2, \dots, X_n$.

19. Let A be an $n \times n$ matrix. Assume that $c_1, c_2, \dots, c_k$ are different eigenvalues of A associated to eigenvectors $X_1, X_2, \dots, X_k$, that is, $AX_i = c_i X_i$, for every $i \in \{1, 2, \dots, k\}$. Prove that $X_1, X_2, \dots, X_k$ are linearly independent.

20. Given λ, a nonzero scalar, let A be an $n \times n$ matrix, and define $B = \lambda A$. Assume that the minimal polynomial of A is $m(x) = x^r + a_{r-1}x^{r-1} + \cdots + a_1 x + a_0$. Prove that the minimal polynomial of B is $m_1(x) = x^r + \lambda a_{r-1}x^{r-1} + \cdots + \lambda^{r-1}a_1 x + \lambda^r a_0$. What is the relationship between the roots of $m(x)$ and the roots of $m_1(x)$?

21. Let D be the matrix in Exercise 28, p. 66, and $A = \begin{bmatrix} 9 & 7 & 4 & 1 \\ 7 & 19/2 & 7 & 4 \\ 4 & 7 & 19/2 & 7 \\ 1 & 4 & 7 & 9 \end{bmatrix}$. Notice that $D = \frac{1}{10^3}A$.
    Use the previous exercise and SageMath to compute the eigenvectors of D, knowing those of A.

22. In relation to Exercise 8.5.2, p. 216, perform a complete discussion of the case $n = 2$.

23. Let $A = \left[\begin{smallmatrix} 0.25 & 0.95 \\ 0.75 & 0.05 \end{smallmatrix}\right]$ be the matrix that describes the migration of population in Example 1.1.3, p. 6. Show that the minimal polynomial of A is $m(x) = (x - 1)(x + \frac{7}{10})$. Find a matrix P so that $P^{-1}AP = \left[\begin{smallmatrix} 1 & 0 \\ 0 & -\frac{7}{10} \end{smallmatrix}\right]$. From this equation, one has $A^k = P\left[\begin{smallmatrix} 1 & 0 \\ 0 & -(\frac{7}{10})^k \end{smallmatrix}\right]P^{-1}$. Use this representation of A^k to perform an analysis of the migration.

24. In relation with Exercise 23, notice that the sum of the entries of each column of A is 1, also one of its eigenvalues is 1. Assume that a matrix A is $n \times n$ with entries ≥ 0 so that the sum of the elements in each column is 1. Prove that one of the eigenvalues of A is 1. Matrices of this type are called *stochastic matrices*. They represent discrete stochastic processes with a finite number of states. Some authors call a stochastic matrix the transpose of A.

25. Prove that the product of stochastic matrices is stochastic, particularly, if A is stochastic then so is A^k for any $k \geq 1$.

26. Suppose that A is a 2×2 stochastic matrix. What is the sum of the entries of each column of A^{-1}? Is the matrix A^{-1} stochastic as well?

27. Review Example in Section 8.5.2 and compute $\lim_{k \to \infty} \frac{x_k}{y_k}$. Give an interpretation of the number of professionals vs the number of masters in the club.

28. Let p be a real number and let $A_p = \left[\begin{smallmatrix} 0 & 3 & 2 \\ p & 0 & 0 \\ 0 & p & 0 \end{smallmatrix}\right]$. Show that the minimal polynomial of A_p is $x^3 - 3px - 2p$. For which values of p is A_p diagonalizable?

29. Given the matrix $A = \left[\begin{smallmatrix} 0 & 5 & 7 \\ 3 & 0 & 0 \\ 0 & 6 & 0 \end{smallmatrix}\right]$, is it diagonalizable? Justify your answer.

30. Complete the analysis of case 1, p. 211, Section 8.5.1.

31. Show that the determinant in equation (8.51) is given by $f_A(x) = (-1)^{k+1}(x^{k+1} - b_0a_1x^k - b_0b_1a_2x^{k-1} - \cdots - b_0b_1 \cdots b_{k-2}a_{k-1}x - b_0b_1 \cdots b_{k-2}a_k)$.

32. Assume that the possible work characteristics of a person are: professional, qualified worker, or nonqualified worker. Assume that the children of a professional are distributed as follows: 80 % are professionals, 10 % are a qualified workers, and 10 % are nonqualified workers. For a qualified worker, his children are distributed as follows: 60 % are professionals, 20 % are qualified workers, and 20 % are nonqualified workers. For a nonqualified worker, 50 % of his children are nonqualified workers 25 % are professionals, and 25 % are qualified workers. Assume that each person has at least one child. Use a matrix to describe the distribution of the kind of work characteristics of children after several generations. Find the probability that, choosing randomly a great-grandchild of a nonqualified worker, he/she would be a professional.

33. Can you use the ideas in the previous exercise to model the propagation of a contagious disease? For example, COVID-19. Explain.

9 Euclidean vector spaces

The Euclidean vector spaces $\mathbb{R}^2$ and $\mathbb{R}^3$ are those that we usually use to represent problems that arise from the "real world." One of the characteristics of these spaces is that we can "do geometry" on them. Our point of view is that two of the main concepts in Euclidean geometry are distance and angle. Using the algebraic structure of a vector space, together with an inner product, we will be able to define geometric concepts and establish results such as a generalization of the Pythagorean theorem.

9.1 Geometric aspects of a vector space

In this section we define the concept of an inner product and prove some important results that allow us to discuss geometric problems. Among the results that we will discuss are the Pythagorean theorem, the Cauchy–Schwarz inequality, and the triangle inequality, which allow us to study concepts related to functions such as limit, continuity and differentiability, among many others.

Definition 9.1.1. We say that the real vector space V is euclidean if there is a function $\langle \cdot, \cdot \rangle : V \times V \to \mathbb{R}$, called inner product, scalar product, or dot product, that satisfies:
1. (*symmetry*) for every $\alpha, \beta \in V$, $\langle \alpha, \beta \rangle = \langle \beta, \alpha \rangle$,
2. (*linearity in the first entry*) for every $\alpha, \beta, \gamma \in V$ and $a, b \in \mathbb{R}$, $\langle a\alpha + b\beta, \gamma \rangle = a\langle \alpha, \gamma \rangle + b\langle \beta, \gamma \rangle$,
3. (*positivity*) for every $\alpha \in V$, $\langle \alpha, \alpha \rangle \geq 0$, and $\langle \alpha, \alpha \rangle = 0$ only if $\alpha = 0$.

Remark 9.1.1. Properties 1 and 2 of Definition 9.1.1 imply that an inner product is linear in the second entry as well.

Example 9.1.1. Let V be the space of $n \times n$ real matrices. Define in $V \times V$ the function $\langle \cdot, \cdot \rangle$ by $\langle A, B \rangle = \operatorname{tr}(AB^t)$. We will show that with this function V is euclidean.

Using the definition of the transpose of a matrix, it is straightforward to verify the first two properties. We will justify only the third one. If A has entries a_{ij} and $C = AA^t$, then the entries c_{ii} of C are obtained by multiplying the ith row of A by the ith column of A^t, that is, $c_{ii} = a_{i1}^2 + a_{i2}^2 + \cdots + a_{in}^2$, hence

$$\langle A, A \rangle = \operatorname{tr}(AA^t) = \sum_{i,j=1}^{n} a_{ij}^2 \geq 0,$$

therefore $\langle A, A \rangle = 0$ if and only if $a_{ij} = 0$, for every $i, j \in \{1, 2, \ldots, n\}$, that is, if and only if $A = 0$.

In order to discuss one of the most important examples of euclidean vector spaces, we need to recall several results from calculus. We state them without any further explanation. However, a good reference to review this or other topics from calculus is [24].

https://doi.org/10.1515/9783111135915-009

Remark 9.1.2. In this remark we state the results from calculus needed to understand the next example.

1. If f and g are continuous functions, then so is fg.
2. If f is continuous at x_0 and $f(x_0) > 0$, then there is $\delta > 0$ such that $f(x) > 0$ for all $x \in (x_0 - \delta, x_0 + \delta)$.
3. If f is integrable on $[a, b]$ and $f(x) \geq 0$, then $\int_a^b f \geq 0$.
4. If f is continuous on $[a, b]$, then f is integrable on $[a, b]$.
5. If f is integrable on $[a, b]$ and $c \in (a, b)$, then $\int_a^b f = \int_a^c f + \int_c^b f$.

Example 9.1.2. Let V be the space of continuous functions on the interval $[0, 1]$. Defining in $V \times V$ the function $\langle \cdot, \cdot \rangle$ by $\langle f, g \rangle = \int_0^1 fg$, we will show that V is euclidean.

As in the previous example, we will verify only the third property, since the others are verified directly using properties of integrals. We have that $f^2(x) \geq 0$ for every $x \in [0, 1]$, then property 3 in Remark 9.1.2 implies $\langle f, f \rangle = \int_0^1 f^2 \geq 0$. We need to prove that $\langle f, f \rangle = \int_0^1 f^2 = 0$ implies $f = 0$. To prove what is needed, it is enough to prove that if g is continuous, $g(x) \geq 0$ for every $x \in [0, 1]$, and $\int_0^1 g = 0$, then $g(x) = 0$ for every $x \in [0, 1]$. We argue by contradiction, assume that there is $x_0 \in (0, 1)$ such that $g(x_0) > 0$. Let $h(x) = g(x) - \frac{g(x_0)}{2}$, then $h(x_0) > 0$. From Remark 9.1.2(2), there is $\delta > 0$ such that $h(x) = g(x) - \frac{g(x_0)}{2} > 0$ for every $x \in (x_0 - \delta, x_0 + \delta)$, thus $g(x) > \frac{g(x_0)}{2}$ for every $x \in (x_0 - \delta, x_0 + \delta)$. From this inequality and properties from Remark 9.1.2, we have

$$0 = \int_0^1 g = \int_0^{x_0-\delta} g + \int_{x_0-\delta}^{x_0+\delta} g + \int_{x_0+\delta}^1 g \geq \int_{x_0-\delta}^{x_0+\delta} g \geq \int_{x_0-\delta}^{x_0+\delta} \frac{g(x_0)}{2} = g(x_0)\delta > 0,$$

which is impossible, hence $g(x) = 0$ for every $x \in [0, 1]$.

Definition 9.1.2. Let V be a euclidean vector space with inner product $\langle \cdot, \cdot \rangle$.

1. We say that $\alpha, \beta \in V$ are orthogonal if $\langle \alpha, \beta \rangle = 0$.
2. Given $\alpha \in V$, we define its norm by $\|\alpha\| = \sqrt{\langle \alpha, \alpha \rangle}$.

Remark 9.1.3. If a is a scalar, then $\|a\alpha\| = |a|\|\alpha\|$.

Proof. By definition, $\|a\alpha\| := \sqrt{\langle a\alpha, a\alpha \rangle} = \sqrt{a^2 \langle \alpha, \alpha \rangle} = |a|\sqrt{\langle \alpha, \alpha \rangle}$. □

Theorem 9.1.1 (Pythagorean theorem). *Let V be a euclidian vector space and let $\alpha, \beta \in V$. Then α is orthogonal to β if and only if $\|\alpha + \beta\|^2 = \|\alpha\|^2 + \|\beta\|^2$.*

The geometric meaning of the result is illustrated in Figure 9.1.

Proof. By the definition of the norm and the bilinearity property of the inner product, we have $\|\alpha + \beta\|^2 = \langle \alpha + \beta, \alpha + \beta \rangle = \|\alpha\|^2 + 2\langle \alpha, \beta \rangle + \|\beta\|^2$. Now applying the definition of orthogonality, α is orthogonal to β if and only if $\langle \alpha, \beta \rangle = 0$. The conclusion follows. □

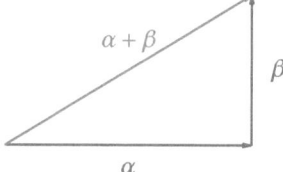

Figure 9.1: Pythagorean theorem.

Theorem 9.1.2 (Cauchy–Schwarz inequality). *If V is a euclidean vector space and $\alpha, \beta \in V$, then $|\langle \alpha, \beta \rangle| \leq \|\alpha\| \|\beta\|$, with equality if and only if α and β are linearly dependent.*

Proof. Fixing $\alpha, \beta \in V$ and taking any $x \in \mathbb{R}$, one has that

$$f(x) = \langle \alpha + x\beta, \alpha + x\beta \rangle = \|\alpha\|^2 + 2\langle \alpha, \beta \rangle x + \|\beta\|^2 x^2 \tag{9.1}$$

defines a nonnegative quadratic function of x, hence its discriminant d satisfies

$$d = 4\langle \alpha, \beta \rangle^2 - 4\|\alpha\|^2 \|\beta\|^2 \leq 0,$$

thus $f(x)$ has a double real zero if and only if $d = 0$. Equivalently, this latter occurs if and only if $4\langle \alpha, \beta \rangle^2 - 4\|\alpha\|^2 \|\beta\|^2 = 0$, and this in turn occurs if and only if $|\langle \alpha, \beta \rangle| = \|\alpha\| \|\beta\|$. Then $f(x) = \langle \alpha + x\beta, \alpha + x\beta \rangle = 0$ if and only if $\alpha + x\beta = 0$; if and only if $|\langle \alpha, \beta \rangle| = \|\alpha\| \|\beta\|$. Arguing as before, one has $f(x) = \langle \alpha + x\beta, \alpha + x\beta \rangle > 0$ if and only if $4\langle \alpha, \beta \rangle^2 - 4\|\alpha\|^2 \|\beta\|^2 < 0$, that is, α and β are linearly independent if and only if $|\langle \alpha, \beta \rangle| < \|\alpha\| \|\beta\|$.

Other Proof. The idea of this proof is shown in Figure 9.2. From Definition 3.3.4, p. 83, we know that $\alpha = u + \frac{\langle \alpha, \beta \rangle}{\|\beta\|^2} \beta$, with u and $\frac{\langle \alpha, \beta \rangle}{\|\beta\|^2} \beta$ orthogonal. Applying Theorem 9.1.1 and Remark 9.1.3, we have

$$\|\alpha\|^2 = \|u\|^2 + \left\| \frac{\langle \alpha, \beta \rangle}{\|\beta\|^2} \beta \right\|^2 \geq \left\| \frac{\langle \alpha, \beta \rangle}{\|\beta\|^2} \beta \right\|^2 = \frac{|\langle \alpha, \beta \rangle|^2}{\|\beta\|^4} \|\beta\|^2 = \frac{|\langle \alpha, \beta \rangle|^2}{\|\beta\|^2}.$$

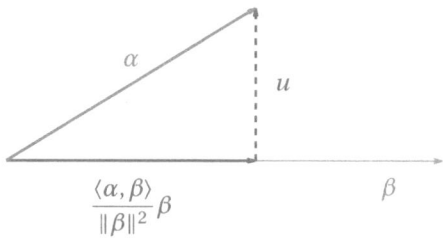

Figure 9.2: Orthogonal projection of a vector along another one.

Hence $\|a\|^2 \geq \frac{|\langle a, \beta \rangle|^2}{\|\beta\|^2}$. From this the conclusion follows by taking square root and multiplying by $\|\beta\|$.

Notice that there is equality if and only if $u = 0$, and this occurs if and only if α coincides with its projection along β, implying that α and β are linearly dependent.

The advantage of the second proof is that it works for complex vector spaces, the argument does not use the order in the real numbers. □

Corollary 9.1.1 (Distance from a point to a hyperplane). *For a hyperplane $H \subset \mathbb{R}^n$, whose equation is $a_1 x_1 + \cdots + a_n x_n + b = 0$ and $P = (b_1, \ldots, b_n) \notin H$, the distance from P to H, $d(P, H)$ is given by*

$$d(P, H) = \frac{|a_1 b_1 + \cdots + a_n b_n + b|}{\sqrt{a_1^2 + \cdots + a_n^2}}. \tag{9.2}$$

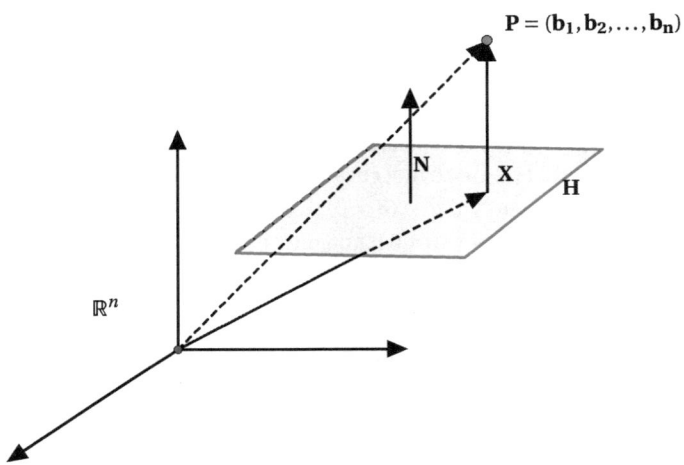

Figure 9.3: The distance from point P to hyperplane H is $\|B\|$.

Proof. Let $X \in H$ be such that $B = P - X$ is orthogonal to H, hence the norm of B is the distance from P to H, as is illustrated in Figure 9.3. We have that $N = (a_1, a_2, \ldots, a_n)$ is also orthogonal to H, since N is orthogonal to the hyperplane $H_1 = \{X \in \mathbb{R}^n \mid \langle N, X \rangle = 0\}$ which is parallel to H. Also, H_1 has dimension $n-1$, hence B and N are linearly dependent. Applying Theorem 9.1.2, we have

$$
\begin{aligned}
\|B\|\|N\| &= |\langle N, B \rangle| \\
&= |\langle N, P - X \rangle| \\
&= |\langle N, P \rangle - \langle N, X \rangle| \\
&= |a_1 b_1 + \cdots + a_n b_n + b|, \quad \text{since } \langle N, X \rangle = -b.
\end{aligned}
$$

From this equation, we have

$$d(P,H) = \|B\| = \frac{|a_1 b_1 + \cdots + a_n b_n + b|}{\sqrt{a_1^2 + \cdots + a_n^2}}, \tag{9.3}$$

finishing the proof of the corollary. □

Theorem 9.1.3 (Triangle inequality). *Let V be a euclidean vector space, and let α and β be elements in V. Then $\|\alpha + \beta\| \leq \|\alpha\| + \|\beta\|$.*

Proof. We have $\|\alpha + \beta\|^2 = \langle \alpha + \beta, \alpha + \beta \rangle = \|\alpha\|^2 + 2\langle \alpha, \beta \rangle + \|\beta\|^2 \leq \|\alpha\|^2 + 2|\langle \alpha, \beta \rangle| + \|\beta\|^2$. Using the Cauchy–Schwarz inequality on the right-hand side of the previous inequality, it follows that

$$\|\alpha + \beta\|^2 \leq \|\alpha\|^2 + 2\|\alpha\|\,\|\beta\| + \|\beta\|^2 = (\|\alpha\| + \|\beta\|)^2.$$

The conclusion follows by taking the square root in this inequality. □

Definition 9.1.3. Let V be a euclidian vector space and $S \subseteq V$. It is said that S is an orthonormal set if:
1. for every $\alpha, \beta \in S$ with $\alpha \neq \beta$, $\langle \alpha, \beta \rangle = 0$,
2. for every $\alpha \in S$, $\langle \alpha, \alpha \rangle = 1$.

When S is a basis of V, it is called an orthonormal basis. If S satisfies only property 1, S is called an orthogonal set.

The Gram–Schmidt process is an important and useful result for constructing orthogonal sets which are used in several applications.

Lemma 9.1.1. *Let V be a euclidean vector space. If $S = \{a_1, a_2, \ldots, a_k\}$ is an orthogonal subset of $V \setminus \{0\}$, then S is linearly independent.*

Proof. Assume that $a_1 \alpha_1 + a_2 \alpha_2 + \cdots + a_k \alpha_k = 0$, then

$$\begin{aligned}
0 &= \langle a_1 \alpha_1 + a_2 \alpha_2 + \cdots + a_k \alpha_k, \alpha_i \rangle \\
&= a_1 \langle \alpha_1, \alpha_i \rangle + a_2 \langle \alpha_2, \alpha_i \rangle + \cdots + a_i \langle \alpha_i, \alpha_i \rangle + \cdots + a_k \langle \alpha_k, \alpha_i \rangle \\
&= a_i \langle \alpha_i, \alpha_i \rangle,
\end{aligned}$$

for every $i \in \{1, 2, \ldots, k\}$. Since $\alpha_i \neq 0$, thus $a_i = 0$ for every $i \in \{1, 2, \ldots, k\}$, proving the assertion. □

Theorem 9.1.4 (Gram–Schmidt process). *Every finite-dimensional euclidian vector space has an orthonormal basis.*

Proof. Let V be a finite-dimensional euclidean vector space. We will use induction on the dimension of V to prove the theorem. Denote the dimension of V by n. If $n = 1$ and a is a generator of V, then $\frac{a}{\|a\|}$ has norm 1 and generates V.

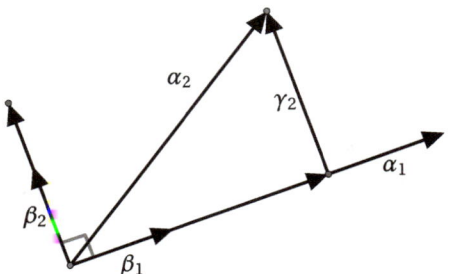

Figure 9.4: Orthonormalization of a_1 and a_2.

To illustrate the construction in the case $n = 2$, we refer to Figure 9.4, and invite the reader to consult it while reading the following argument. Let $\{a_1, a_2\}$ be a basis of V and define $\beta_1 = \frac{a_1}{\|a_1\|}$. It is straightforward to verify that $\{\beta_1, a_2\}$ is a basis of V. We need to construct $\beta_2 \neq 0$ so that $\langle \beta_1, \beta_2 \rangle = 0$. To this end, we start by finding $\gamma_2 \neq 0$ such that $\langle \gamma_2, \beta_1 \rangle = 0$, then normalizing this vector will produce the basis that satisfies the required conditions. Let $\gamma_2 = b_2 a_2 + b_1 \beta_1$ and notice that if γ_2 is orthogonal to β_1, then $a\gamma_2$ is also orthogonal to β_1 for any a, hence we may assume that $b_2 = 1$. From the condition $\langle \gamma_2, \beta_1 \rangle = 0$, one has $0 = \langle a_2 + b_1\beta_1, \beta_1 \rangle = \langle a_2, \beta_1 \rangle + b_1 \langle \beta_1, \beta_1 \rangle$. Since β_1 has norm 1, $b_1 = -\langle a_2, \beta_1 \rangle$. By the construction procedure, $\gamma_2 = a_2 - \langle a_2, \beta_1 \rangle \beta_1$ is orthogonal to β_1. Defining $\beta_2 = \frac{\gamma_2}{\|\gamma_2\|}$, we have that $S = \{\beta_1, \beta_2\}$ is an orthonormal set. Now, Lemma 9.1.1 implies that S is a basis.

Assume that $n > 2$ and the result to be true for all euclidean vector spaces of dimension less than n. Set $r = n - 1$ and suppose that the vectors $\beta_1, \beta_2, \ldots, \beta_r$ are orthonormal and span the same subspace as $a_1, a_2, \ldots, a_r$. Then $\{\beta_1, \beta_2, \ldots, \beta_r, a_n\}$ is a basis of V. Let us construct an element β_n such that $\langle \beta_n, \beta_i \rangle = 0$, for every $i \in \{1, 2, \ldots, n - 1\}$ and $\{\beta_1, \beta_2, \ldots, \beta_{n-1}, \beta_n\}$ spans V.

If γ_n is an element spanned by $\{\beta_1, \beta_2, \ldots, \beta_r, a_n\}$, then $\gamma_n = b_n a_n + b_r \beta_r + \cdots + b_1 \beta_1$. We may assume that $b_n = 1$, additionally, if $\langle \gamma_n, \beta_i \rangle = 0$ for every $i \in \{1, 2, \ldots, n - 1\}$, then $0 = \langle a_n, \beta_i \rangle + b_i \langle \beta_i, \beta_i \rangle$. Since each β_i has norm 1, $b_i = -\langle a_n, \beta_i \rangle$, that is, one has

$$\gamma_n = a_n - \langle a_n, \beta_r \rangle \beta_r - \cdots - \langle a_n, \beta_1 \rangle \beta_1. \tag{9.4}$$

Let $\beta_n = \frac{\gamma_n}{\|\gamma_n\|}$, then the set $\{\beta_1, \beta_2, \ldots, \beta_r, \beta_n\}$ is orthonormal and generates V. $\square$

Example 9.1.3. Let $S = \{(1, 1, -1), (0, 1, 2), (1, -2, 1)\}$ be a subset of $\mathbb{R}^3$. Show that S is linearly independent and orthonormalize it.

Discussion. Let $A = \begin{bmatrix} 1 & 1 & -1 \\ 0 & 1 & 2 \\ 1 & -2 & 1 \end{bmatrix}$ be the matrix whose rows are the given vectors, then applying row operations on A, one has that it is row equivalent to $R = \begin{bmatrix} 1 & 1 & -1 \\ 0 & 1 & 2 \\ 0 & 0 & 8 \end{bmatrix}$, which is nonsingular. Thus, the given vectors are linearly independent. Let $\alpha_1 = (1, 1, -1)$, $\alpha_2 = (0, 1, 2)$, and $\alpha_3 = (1, -2, 1)$. To start the Gram–Schmidt process, define $\beta_1 = \frac{1}{\|\alpha_1\|}\alpha_1 = \frac{1}{\sqrt{3}}(1, 1, -1)$. Having defined β_1, let

$$\gamma_2 = (0, 1, 2) - \langle (0, 1, 2), \frac{1}{\sqrt{3}}(1, 1, -1)\rangle \frac{1}{\sqrt{3}}(1, 1, -1)$$

$$= \frac{1}{3}(1, 4, 5),$$

and define $\beta_2 = \frac{\gamma_2}{\|\gamma_2\|} = \frac{1}{\sqrt{42}}(1, 4, 5)$. Continuing the process, let

$$\gamma_3 = (1, -2, 1) - \left\langle (1, -2, 1), \frac{1}{\sqrt{42}}(1, 4, 5)\right\rangle \frac{1}{\sqrt{42}}(1, 4, 5) - \left\langle (1, -2, 1), \frac{1}{\sqrt{3}}(1, 1, -1)\right\rangle \frac{1}{\sqrt{3}}(1, 1, -1)$$

$$= \frac{4}{7}(3, -2, 1).$$

After defining γ_3, let $\beta_3 = \frac{\gamma_3}{\|\gamma_3\|} = \frac{1}{\sqrt{14}}(3, -2, 1)$, thus the orthonormalized set is

$$S_1 = \left\{ \frac{1}{\sqrt{3}}(1, 1, -1), \frac{1}{\sqrt{42}}(1, 4, 5), \frac{1}{\sqrt{14}}(3, -2, 1)\right\}.$$

Definition 9.1.4. Let V be a euclidean vector space, and let W be a subspace. One defines the orthogonal complement of W as $W^\perp := \{\alpha \in V : \langle \alpha, \beta\rangle = 0 \text{ for every } \beta \in W\}$.

Exercise 9.1.1. Prove that $W^\perp$ is a subspace.

Example 9.1.4. Let $\langle \cdot, \cdot \rangle$ denote the usual inner product in $\mathbb{R}^n$, if $W = \{(x_1, x_2, \ldots, x_n) \mid x_1 + x_2 + \cdots + x_n = 0\}$, then $W^\perp$ has dimension 1 and is generated by $v = (1, 1, \ldots, 1)$.

Discussion. For every $i \in \{2, 3, \ldots, n\}$, let S be the set consisting of the elements E_{1i} whose entries are: 1 in the first place, -1 in the ith place for $i \neq 1$, and zero in the jth entry for every $j \notin \{1, i\}$. It is straightforward to verify that $S \subseteq W$ is linearly independent and that $\langle v, E_{1i}\rangle = 0$ for every $i \in \{2, 3, \ldots, n\}$. Moreover, $u = (u_1, u_2, \ldots, u_n) \in W^\perp$ if and only if $\langle u, E_{1i}\rangle = 0$ for every $i \in \{2, 3, \ldots, n\}$. The latter condition implies that $u_1 = u_2 = \cdots = u_n$, hence u is a multiple of v, as claimed.

Theorem 9.1.5. *If V is a finite-dimensional euclidean vector space, and W is a subspace, then $V = W \oplus W^\perp$.*

Proof. Let n denote $\dim(V)$. If $W = \{0\}$ or $W = V$, then $W^\perp = V$ or $W^\perp = \{0\}$, respectively. The result follows. Assume that W has positive dimension less than n. By the Gram–Schmidt process, we can choose an orthonormal basis of W, say $\{\alpha_1, \alpha_2, \ldots, \alpha_r\}$. This basis can be expanded to a basis of V, moreover, again by the Gram–Schmidt process, we can assume that the expanded basis $\{\alpha_1, \alpha_2, \ldots, \alpha_r, \ldots, \alpha_n\}$ is orthonormal.

Claim. The set $\{a_{r+1}, \ldots, a_n\}$ is a basis of $W^\perp$.

It is clear that $a_i \in W^\perp$ for every $i \in \{r+1, \ldots, n\}$. If $y \in W^\perp$, then it can be expressed as $y = c_1 a_1 + \cdots + c_r a_r + c_{r+1} a_{r+1} + \cdots + c_n a_n$.

For each $i \in \{1, 2, \ldots, r\}$, one has $0 = \langle a_i, y \rangle = c_i$, then y belongs to the span of $\{a_{r+1}, \ldots, a_n\}$, proving the claim. To finish the proof, we need to prove that $W \cap W^\perp = \{0\}$. If $a \in W \cap W^\perp$, then $\langle a, a \rangle = 0$, hence $a = 0$ and the theorem has been proved. $\qquad\square$

If V is a finite-dimension euclidean vector space and W is a subspace, then for every $a \in V$ one has $a = \hat{a} + \beta$, with $\hat{a} \in W$ and $\beta \in W^\perp$. The vector $\hat{a}$ is called the orthogonal projection of a along W. Figure 9.5 illustrate this situation.

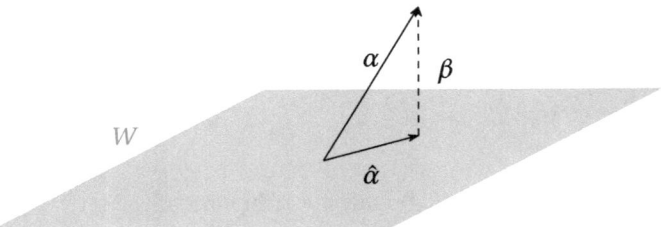

Figure 9.5: Orthogonal projection of a vector along the subspace W.

If $\{a_1, a_2, \ldots, a_r\}$ is an orthonormal basis of W, then

$$\hat{a} = \langle a, a_1 \rangle a_1 + \langle a, a_2 \rangle a_2 + \cdots + \langle a, a_r \rangle a_r.$$

In fact, if $\hat{a} = c_1 a_1 + \cdots + c_r a_r$, then, computing the inner product of $\hat{a}$ with a_i, it follows that $\langle \hat{a}, a_i \rangle = c_i$.

Theorem 9.1.6. *Let V be a finite-dimensional euclidean vector space, and let W be a subspace of V. If $a \in V$ and $\hat{a}$ is the orthogonal projection of a along W, then*

$$\|a - \hat{a}\| < \|a - y\|, \tag{9.5}$$

for every $y \in W \setminus \{\hat{a}\}$.

Proof. Let $y \in W \setminus \{\hat{a}\}$. One has $a - y = a - y + \hat{a} - \hat{a} = (a - \hat{a}) + (\hat{a} - y)$ and, by definition, $\beta = (a - \hat{a}) \in W^\perp$. Applying the Pythagorean theorem and taking into account that $y \neq \hat{a}$ yields

$$\|a - y\|^2 = \|a - \hat{a}\|^2 + \|\hat{a} - y\|^2 > \|a - \hat{a}\|^2. \tag{9.6}$$

The final conclusion is obtained by taking the square root in inequality (9.6) and using the fact that the square root function is increasing. $\qquad\square$

9.1.1 Method of least squares

Given that the points $(1, 2)$, $(2, 1)$, $(3, 3)$, and $(4, 6)$ do not lie on any line, you want to find the line that "best fits" them all. We know that the equation of any nonvertical line can be represented in the form

$$y = mx + b, \tag{9.7}$$

for some real numbers m and b. Hence, we want to find values of m and b so that all the given points are as close as possible to the line determine by m and b.

If we substitute the coordinates of the given points in (9.7), we obtain the system of linear equations

$$\begin{aligned} m + b &= 2, \\ 2m + b &= 1, \\ 3m + b &= 3, \\ 4m + b &= 6, \end{aligned} \tag{9.8}$$

which we already know has no solution, hence a natural question is: What is the "best possible solution" to (9.8)?

In what follows, we will discuss some generalities from which we can provide an answer to the posed question.

Assuming that the linear system of equations $AX = B$ has no solution, how do we find the best approximation? One possible approach is to consider finding a vector $\hat{X}$ so that $\|A\hat{X} - B\|$ is as small as possible. What does this mean in more precise terms?

Definition 9.1.5. Let A be an $m \times n$ matrix, and let B be a column vector in $\mathbb{R}^m$. A least squares solution to $AX = B$ is an $\hat{X} \in \mathbb{R}^n$ such that

$$\|A\hat{X} - B\| \leq \|AX - B\|, \tag{9.9}$$

for every X in $\mathbb{R}^n$.

Recall that the system of linear equations $AX = B$ has a solution if and only if B is in the column space of A. More precisely, if $A_1, A_2, \ldots, A_n$ represent the columns of A, then the system of equations $AX = B$ is equivalent to $x_1 A_1 + x_2 A_2 + \cdots + x_n A_n = B$, where $x_1, x_2, \ldots, x_n$ are the entries of X and the latter equation establishes that B belongs to the column space of A, exactly when the system has a solution. Then, finding $\hat{X}$ that yields the least value of $\|A\hat{X} - B\|$ is equivalent to finding a vector $\hat{B}$ in the column space of A so that $\|\hat{B} - B\|$ is minimal. Assume that W denotes the column space of A, then, applying Theorem 9.1.6, one has that the vector that satisfies the minimality condition is precisely the orthogonal projection of B along W.

How do we find $\hat{B}$? Given that $\hat{B}$ belongs to the column space of A, the system $AX = \hat{B}$ has a solution, called $\hat{X}$.

On the other hand, the vector $B - \hat{B}$ is located in the orthogonal complement of the column space of A, by Theorem 9.1.5. Then $B - \hat{B} = B - A\hat{X}$ is orthogonal to every column of A, which is equivalent to $A^t(B - A\hat{X}) = 0$ or, alternatively,

$$A^t A\hat{X} = A^t B. \tag{9.10}$$

Notice that the solutions of (9.10) correspond exactly to the solutions of the least squares problem formulated in Definition 9.1.5, which always has a solution. We have that the matrix $A^t A$ is an $n \times n$ matrix and, from what we have observed, the system (9.10) has a unique solution if and only if $A^t A$ is invertible. The following result establishes necessary and sufficient condition in order to have this requirement fulfilled.

Theorem 9.1.7. *Let A be an $m \times n$ matrix. Then the matrix $A^t A$ is invertible if and only if the columns of A are linearly independent.*

Proof. Recall that the system $AX = 0$ has only the zero solution if and only if the columns of A are linearly independent. To finish the proof, we will prove that the solutions of $A^t AX = 0$ coincide with the solutions of $AX = 0$.

Assume that C is a solution of $AX = 0$, then $AC = 0$ and, multiplying by A^t, we have $A^t AC = A^t 0 = 0$, hence C is a solution of $A^t AX = 0$. Conversely, if C is a solution of $A^t AX = 0$, then $A^t AC = 0$. Multiplying by C^t we have $C^t A^t AC = C^t 0 = 0$. Now, notice that the left-hand side of this equation is the inner product of AC with itself, and this inner product is zero if and only if $AC = 0$, that is, C is a solution of $AX = 0$, finishing the proof. □

To provide an idea of how to apply the discussed method, we solve (9.8):

$$m + b = 2,$$
$$2m + b = 1,$$
$$3m + b = 3,$$
$$4m + b = 6.$$

We have that $A = \begin{bmatrix} 1 & 1 \\ 2 & 1 \\ 3 & 1 \\ 4 & 1 \end{bmatrix}$, $B = \begin{bmatrix} 2 \\ 1 \\ 3 \\ 6 \end{bmatrix}$, and $\hat{X} = [\begin{smallmatrix} m \\ b \end{smallmatrix}]$. From these data, we obtain $A^t A = [\begin{smallmatrix} 30 & 10 \\ 10 & 4 \end{smallmatrix}]$ and $A^t B = [\begin{smallmatrix} 37 \\ 12 \end{smallmatrix}]$. Now, the system of linear equations to be solved is

$$30m + 10b = 37,$$
$$10m + 4b = 12,$$

whose solution is $\left(\frac{7}{5}, -\frac{1}{2}\right)$.

From this result, we have that the line that best fits the points $(1, 2)$, $(2, 1)$, $(3, 3)$, and $(4, 6)$ has equation $y = \frac{7}{7}x - \frac{1}{2}$, see Figure 9.6.

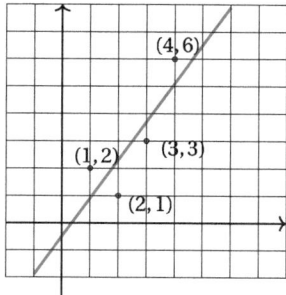

Figure 9.6: Best fitting line whose equation is $y = \frac{7}{5}x - \frac{1}{2}$.

Exercise 9.1.2. Find the equation of the plane that best fits the points $(1, 1, 0)$, $(3, 4, 3)$, $(2, 1, 3)$, $(-1, 1, 2)$, $(3, 0, 1)$, and $(-1, 1, 1)$.

Another result that will help us solve a least squares problem and which is very closely related to the Gram–Schmidt orthonormalization process is the following.

Theorem 9.1.8. *Let A be an $m \times n$ matrix whose columns, as elements of $\mathbb{R}^m$, are linearly independent. Then $A = QR$, where Q is a matrix whose columns are orthonormal elements of $\mathbb{R}^m$ and R is an upper triangular matrix with positive elements on its diagonal.*

Proof. Let $A_1, A_2, \ldots, A_n$ be the columns of A, and let V be the column space of A. We will recall how the construction of an orthonormal basis for V was carried out, see Theorem 9.1.4. The first step to construct an element of the orthonormal basis is by declaring the element $B_1 = \frac{A_1}{\|A_1\|}$. Now consider the matrix M_1 whose first column is B_1 and the others are $A_2, \ldots, A_n$, in other words,

$$M_1 = [B_1 \, A_2 \, \ldots \, A_n] = AE_1, \tag{9.11}$$

where E_1 represents the elementary column operation that multiplies the first column of A by $\frac{1}{\|A_1\|}$. According to the Gram–Schmidt orthonormalization process, to construct the second orthonormal element B_2, we define $C_2 = A_2 - \langle A_2, B_1 \rangle B_1$ and normalize it.

Translating into matrix terms, if M_2 is the matrix whose columns are $B_1, C_2, A_3, \ldots,$ A_n, then

$$M_2 = [B_1 \, C_2 \, A_3 \, \ldots \, A_n] = M_1 E_2 = AE_1 E_2, \tag{9.12}$$

where E_2 is the matrix that represents the column operation of multiplying the first column of M_1 by $-\langle A_2, B \rangle$ and subtracting the result from its second column. Now let E_3 be the matrix that represents the column operation that multiplies the second column of M_2 by $\frac{1}{\|C_2\|}$, then the resulting matrix is

$$M_3 = M_2 E_3 = [B_1 \, B_2 \, A_3 \, \ldots \, A_n] E_3 = AE_1 E_2 E_3. \tag{9.13}$$

Assuming that $i > 2$ and that $B_1, B_2, \ldots, B_{i-1}$ have been constructed, which are orthonormal, also define the matrix $M_{i+1} = [B_1 \, B_2 \, \ldots \, B_{i-1} \, A_i \, \ldots \, A_n]$.

Emulating the Gram–Schmidt process, that is, invoking equation (9.4), p. 228, we define C_i by

$$C_i = A_i - \langle A_i, B_1 \rangle B_1 - \cdots - \langle A_i, B_{i-1} \rangle B_{i-1}, \tag{9.14}$$

and one readily verifies that $\langle C_i, B_j \rangle = 0$ for every $j \in \{1, 2, \ldots, i-1\}$. Translating to matrix notation, if F_i is obtained from I_m by substituting the ith column with the transpose of $(-\langle A_i, B_1 \rangle, -\langle A_i, B_2 \rangle, \ldots, -\langle A_i, B_{i-1} \rangle, 1, 0, \ldots, 0)$, then $M_{i+1}F_i = [B_1 \, B_2 \, \ldots B_{i-1} \, C_i \, A_{i+1} \ldots A_n]$. Now, multiplying $M_{i+1}F_i$ on the right by G_i, which represents the elementary column operation that consists of multiplying by $\frac{1}{\|C_i\|}$, the resulting matrix $M_{i+2} = M_{i+1}F_iG_i = [B_1 \, B_2 \, \ldots \, B_i \, A_{i+1} \, \ldots \, A_n]$ has its first i columns orthonormal.

Summarizing, in a finite number of steps, one finishes the construction of elementary column matrices $E_1, E_2, \ldots, E_k$ such that

$$AE_1 E_2 \cdots E_k = [B_1 \, B_2 \, \ldots \, B_n] = Q, \tag{9.15}$$

were Q is an orthogonal matrix. From (9.15) we obtain $A = QR$, where $R = E_k^{-1}E_{k-1}^{-1} \cdots E_1^{-1}$ is upper triangular and its entries in the diagonal are $\|C_1\|, \|C_2\|, \ldots, \|C_n\|$.

Second proof. From equation (9.14) and $C_i = \|C_i\|B_i$, we have

$$A_i = \|C_i\|B_i + \langle A_i, B_1 \rangle B_1 + \cdots + \langle A_i, B_{i-1} \rangle B_{i-1}. \tag{9.16}$$

Define the matrix R, whose ith column R_i is

$$R_i = \begin{bmatrix} \langle A_i, B_1 \rangle \\ \vdots \\ \langle A_i, B_{i-1} \rangle \\ \|C_i\| \\ 0 \\ \vdots \\ 0 \end{bmatrix}.$$

In terms of matrix multiplication, equation (9.16) tells us that the ith column of A is obtained by multiplying the matrix Q, whose columns are $B_1, B_2, \ldots, B_n$, by R_i, that is, $A = QR$. We notice that R is a triangular matrix whose (i, i)th entry is $\|C_i\| > 0$, as required. $\qquad \square$

In the following example it is illustrated how to find the matrices Q and R such that $A = QR$. This is obtained by using the first proof method of Theorem 9.1.8.

Example 9.1.5. Let $A = \begin{bmatrix} 1 & 0 & 1 \\ 1 & 1 & -2 \\ -1 & 2 & 1 \end{bmatrix}$. Find the QR representation of A.

Discussion. Let A_1, A_2, and A_3 be the columns of A. According to the first proof of Theorem 9.1.8, the first step consists of multiplying the first column of A by the reciprocal of its norm which is $\|(1, 1, -1)\| = \sqrt{3}$. By doing this, we obtain the matrix

$$M_1 = \begin{bmatrix} 1 & 0 & 1 \\ 1 & 1 & -2 \\ -1 & 2 & 1 \end{bmatrix} \begin{bmatrix} \frac{1}{\sqrt{3}} & 0 & 0 \\ 0 & 1 & 0 \\ 0 & 0 & 1 \end{bmatrix} = \begin{bmatrix} \frac{1}{\sqrt{3}} & 0 & 1 \\ \frac{1}{\sqrt{3}} & 1 & -2 \\ \frac{-1}{\sqrt{3}} & 2 & 1 \end{bmatrix}.$$

Call B_1 the first column of M_1. The next step consists of multiplying the first column of M_1 by $-\frac{\langle B_1, A_2 \rangle}{\sqrt{3}} = \frac{1}{\sqrt{3}}$ and adding the result to its second column to obtain

$$M_2 = \begin{bmatrix} \frac{1}{\sqrt{3}} & 0 & 1 \\ \frac{1}{\sqrt{3}} & 1 & -2 \\ \frac{-1}{\sqrt{3}} & 2 & 1 \end{bmatrix} \begin{bmatrix} 1 & \frac{1}{\sqrt{3}} & 0 \\ 0 & 1 & 0 \\ 0 & 0 & 1 \end{bmatrix} = \begin{bmatrix} \frac{1}{\sqrt{3}} & \frac{1}{3} & 1 \\ \frac{1}{\sqrt{3}} & \frac{4}{3} & -2 \\ \frac{-1}{\sqrt{3}} & \frac{5}{3} & 1 \end{bmatrix}.$$

To normalize the second column of M_2, multiply it by the reciprocal of the norm of $\frac{1}{3}(1, 4, 5)$ which is $\frac{\sqrt{42}}{3}$. Then

$$M_3 = M_2 \begin{bmatrix} 1 & 0 & 0 \\ 0 & \frac{3}{\sqrt{42}} & 0 \\ 0 & 0 & 1 \end{bmatrix} = \begin{bmatrix} \frac{1}{\sqrt{3}} & \frac{1}{3} & 1 \\ \frac{1}{\sqrt{3}} & \frac{4}{3} & -2 \\ \frac{-1}{\sqrt{3}} & \frac{5}{3} & 1 \end{bmatrix} \begin{bmatrix} 1 & 0 & 0 \\ 0 & \frac{3}{\sqrt{42}} & 0 \\ 0 & 0 & 1 \end{bmatrix} = \begin{bmatrix} \frac{1}{\sqrt{3}} & \frac{1}{\sqrt{42}} & 1 \\ \frac{1}{\sqrt{3}} & \frac{4}{\sqrt{42}} & -2 \\ \frac{-1}{\sqrt{3}} & \frac{5}{\sqrt{42}} & 1 \end{bmatrix}.$$

To obtain from M_3 a matrix whose columns are orthogonal, we multiply it by a matrix as indicated below:

$$M_4 = M_3 \begin{bmatrix} 1 & 0 & \frac{2}{\sqrt{3}} \\ 0 & 1 & \frac{2}{\sqrt{42}} \\ 0 & 0 & 1 \end{bmatrix} = \begin{bmatrix} \frac{1}{\sqrt{3}} & \frac{1}{\sqrt{42}} & 1 \\ \frac{1}{\sqrt{3}} & \frac{4}{\sqrt{42}} & -2 \\ \frac{-1}{\sqrt{3}} & \frac{5}{\sqrt{42}} & 1 \end{bmatrix} \begin{bmatrix} 1 & 0 & \frac{2}{\sqrt{3}} \\ 0 & 1 & \frac{2}{\sqrt{42}} \\ 0 & 0 & 1 \end{bmatrix} = \begin{bmatrix} \frac{1}{\sqrt{3}} & \frac{1}{\sqrt{42}} & \frac{36}{21} \\ \frac{1}{\sqrt{3}} & \frac{4}{\sqrt{42}} & -\frac{24}{21} \\ \frac{-1}{\sqrt{3}} & \frac{5}{\sqrt{42}} & \frac{12}{21} \end{bmatrix}.$$

Finally, we normalize the third row of M_4, obtaining

$$Q = M_4 \begin{bmatrix} 1 & 0 & 0 \\ 0 & 1 & 0 \\ 0 & 0 & \frac{7}{4\sqrt{14}} \end{bmatrix} = \begin{bmatrix} \frac{1}{\sqrt{3}} & \frac{1}{\sqrt{42}} & \frac{3}{\sqrt{14}} \\ \frac{1}{\sqrt{3}} & \frac{4}{\sqrt{42}} & -\frac{2}{\sqrt{14}} \\ \frac{-1}{\sqrt{3}} & \frac{5}{\sqrt{42}} & \frac{1}{\sqrt{14}} \end{bmatrix} = \frac{1}{\sqrt{42}} \begin{bmatrix} \sqrt{14} & 1 & 3\sqrt{3} \\ \sqrt{14} & 4 & -2\sqrt{3} \\ -\sqrt{14} & 5 & \sqrt{3} \end{bmatrix}.$$

From the previous calculations we can obtain R, since $AE_1E_2E_3E_4E_5 = Q$, where E_1, E_2, E_3, E_4, and E_5 are the matrices that appear above to obtain Q. Then, since R and Q satisfy $A = QR$, from the following equation:

$$A\begin{bmatrix} \frac{1}{\sqrt{3}} & 0 & 0 \\ 0 & 1 & 0 \\ 0 & 0 & 1 \end{bmatrix}\begin{bmatrix} 1 & \frac{1}{\sqrt{3}} & 0 \\ 0 & 1 & 0 \\ 0 & 0 & 1 \end{bmatrix}\begin{bmatrix} 1 & 0 & 0 \\ 0 & \frac{3}{\sqrt{42}} & 0 \\ 0 & 0 & 1 \end{bmatrix}\begin{bmatrix} 1 & 0 & \frac{2}{\sqrt{3}} \\ 0 & 1 & \frac{2}{\sqrt{42}} \\ 0 & 0 & 1 \end{bmatrix}\begin{bmatrix} 1 & 0 & 0 \\ 0 & 1 & 0 \\ 0 & 0 & \frac{7}{4\sqrt{14}} \end{bmatrix} = Q,$$

we obtain R, that is, $R = E_5^{-1}E_4^{-1}E_3^{-1}E_2^{-1}E_1^{-1}$. In explicit form,

$$R = \begin{bmatrix} 1 & 0 & 0 \\ 0 & 1 & 0 \\ 0 & 0 & \frac{4\sqrt{14}}{7} \end{bmatrix}\begin{bmatrix} 1 & 0 & -\frac{2}{\sqrt{3}} \\ 0 & 1 & -\frac{2}{\sqrt{42}} \\ 0 & 0 & 1 \end{bmatrix}\begin{bmatrix} 1 & 0 & 0 \\ 0 & \frac{\sqrt{42}}{3} & 0 \\ 0 & 0 & 1 \end{bmatrix}\begin{bmatrix} 1 & -\frac{1}{\sqrt{3}} & 0 \\ 0 & 1 & 0 \\ 0 & 0 & 1 \end{bmatrix}\begin{bmatrix} \sqrt{3} & 0 & 0 \\ 0 & 1 & 0 \\ 0 & 0 & 1 \end{bmatrix}$$

$$= \begin{bmatrix} \sqrt{3} & -\frac{1}{\sqrt{3}} & -\frac{2}{\sqrt{3}} \\ 0 & \frac{\sqrt{42}}{3} & -\frac{2}{\sqrt{42}} \\ 0 & 0 & \frac{4\sqrt{14}}{7} \end{bmatrix}.$$

After the latter example, we explore a little bit further the way in which a QR representation of a matrix A can be used to solve a least squares problem.

Assume that the columns of a matrix A are linearly independent and let $A = QR$ be the QR decomposition given in Theorem 9.1.8, then the system

$$A^t A \hat{X} = A^t B \tag{9.17}$$

is equivalent to

$$R^t Q^t QR\hat{X} = R^t Q^t B. \tag{9.18}$$

Since the entries on the diagonal of R are positive, R and R^t are invertible. The condition on Q implies that $Q^t Q = I$, hence equation (9.18) reduces to $R\hat{X} = Q^t B$, or

$$\hat{X} = R^{-1} Q^t B. \tag{9.19}$$

From the computational point of view, equation

$$R\hat{X} = Q^t B \tag{9.20}$$

has several advantages, since it is triangular and solving it is reduced to back substitution.

9.2 Complex vector spaces

When dealing with problems that come from science or engineering, in several cases it is needed to introduce complex vector spaces. From the linear algebra point of view, the need to use complex numbers arises right away. For instance, the minimal polynomial of the nice matrix $A = \begin{bmatrix} 0 & -1 \\ 1 & 0 \end{bmatrix}$ has no real roots. Actually, the minimal polynomial of A is $m_A(x) = x^2 + 1$ and its roots are $\pm i$, thus it is necessary to amplify the context of the real numbers to approach those cases.

Before going any further into the discussion of this section, we encourage the reader to review Section A.3 in Appendix A to get an account of basic properties of the complex numbers.

So far we have been discussing vector spaces where the scalars were real numbers. Also, the entries of the matrices that we have studied were real numbers. In this section the vector spaces to be considered are over the complex numbers, which means that the scalars will be complex numbers such as $1 + i$.

The definition of a complex vector space is the same as that of a real vector space, Definition 4.4.1, p. 110, except that the scalars are complex numbers instead of real numbers.

One prototype of a real vector space is $\mathbb{R}^n$, so it is natural to expect the same behavior for complex vector spaces, that is, the typical example of a complex vector space is $\mathbb{C}^n = \{(x_1, x_2, \ldots, x_n) \mid x_j \in \mathbb{C}, j = 1, 2, \ldots, n\}$. The sum and multiplication by scalars in $\mathbb{C}^n$ are defined as follows.

Let $\alpha = (x_1, x_2, \ldots, x_n)$ and $\beta = (y_1, y_2, \ldots, y_n)$ be elements of $\mathbb{C}^n$, then the sum of α and β is given by

$$\alpha + \beta = (x_1, x_2, \ldots, x_n) + (y_1, y_2, \ldots, y_n) = (x_1 + y_1, x_2 + y_2, \ldots, x_n + y_n). \tag{9.21}$$

Also, if $r \in \mathbb{C}$, then

$$r\alpha = (rx_1, rx_2, \ldots, rx_n). \tag{9.22}$$

From equations (9.21) and (9.22), the following properties are obtained readily:

1. Properties of the sum in $\mathbb{C}^n$:
   (a) (commutativity) For every α and $\beta \in \mathbb{C}^n$, $\alpha + \beta = \beta + \alpha$.
   (b) (associativity) For every α, β and γ in $\mathbb{C}^n$, $(\alpha + \beta) + \gamma = \alpha + (\beta + \gamma)$.
   (c) (existence of an additive identity) There is an element in $\mathbb{C}^n$, called zero and denoted by 0, such that $0 + \alpha = \alpha$, for every $\alpha \in \mathbb{C}^n$.
   (d) (existence of an additive inverse) For every $\alpha \in \mathbb{C}^n$, there is $\alpha' \in \mathbb{C}^n$ such that $\alpha + \alpha' = 0$.
2. Properties of the product with a scalar in $\mathbb{C}^n$:
   (a) For every $\alpha \in \mathbb{C}^n$, $1\alpha = \alpha$, with $1 \in \mathbb{C}$.
   (b) For every $\alpha \in \mathbb{C}^n$ and $\lambda, \mu \in \mathbb{C}$, one has $\lambda(\mu\alpha) = (\lambda\mu)\alpha$.
   (c) The product with a scalar is distributive, that is, $(\lambda + \mu)\alpha = \lambda\alpha + \mu\alpha$; $\lambda(\alpha + \beta) = \lambda\alpha + \lambda\beta$, for every $\lambda, \mu \in \mathbb{C}$ and $\alpha, \beta \in \mathbb{C}^n$.

The results that hold in real vector spaces also hold in complex vector spaces. Moreover, it is important to notice that a complex vector space can be considered as a real vector space, but not every real vector space can be consider a complex vector space. For instance, $\mathbb{R}^n$ is not a complex vector space, since the product of elements of $\mathbb{R}^n$, by complex numbers are not necessarily in $\mathbb{R}^n$. Another important aspect to point out is that when V is a complex vector space of dimension n, then, considering it as a real vector space, its dimension is $2n$, in particular, $\mathbb{C}$ has dimension 1 as a complex vector space and dimension 2 as a real vector space. For the case of $\mathbb{C}^n$, the set

$$A = \{(1, 0, 0, \ldots, 0), (0, 1, 0, \ldots, 0), \ldots, (0, 0, 0, \ldots, 0, 1)\}$$

is a basis of $\mathbb{C}^n$ over the complex numbers, however, this set is not a basis when the scalars are restricted to the real numbers. To have a basis of $\mathbb{C}^n$ over the real numbers, the previous set has to be enlarged, for instance, by considering the set

$$B = \{(i, 0, 0, \ldots, 0), (0, i, 0, \ldots, 0), \ldots, (0, 0, 0, \ldots, 0, i)\}.$$

We invite the reader to prove that $A \cup B$ is a basis of $\mathbb{C}^n$ over the reals, hence the dimension of $\mathbb{C}^n$, as a real vector space, is $2n$.

If A is a matrix with complex entries, we define the conjugate of A, denoted by $\bar{A}$, as the matrix whose entries are the conjugates of the entries of A. For instance, if

$$A = \begin{bmatrix} 1+i & 3-i \\ 2-i & 1+i \end{bmatrix}, \quad \text{then } \bar{A} = \begin{bmatrix} 1-i & 3+i \\ 2+i & 1-i \end{bmatrix}.$$

In particular, the conjugate of a column vector Y in $\mathbb{C}^n$ is defined as the vector whose entries are the conjugates of the entries of Y.

Another important aspect that should be pointed out is the definition of an inner product in a complex vector space. The geometric meaning of the inner product could be difficult to visualize, however, the algebraic aspect is similar to that of a real vector space.

Definition 9.2.1. A complex vector space V is called euclidean if there is a function $\langle \cdot, \cdot \rangle : V \times V \to \mathbb{C}$ that satisfies:
1. for every $\alpha, \beta \in V$, $\langle \alpha, \beta \rangle = \overline{\langle \beta, \alpha \rangle}$,
2. for every $a, b \in \mathbb{C}$ and $\alpha, \beta, \gamma \in V$, $\langle a\alpha + b\beta, \gamma \rangle = a\langle \alpha, \gamma \rangle + b\langle \beta, \gamma \rangle$,
3. for every $\alpha \in V$, $\langle \alpha, \alpha \rangle \geq 0$, and $\langle \alpha, \alpha \rangle = 0$ only if $\alpha = 0$.

The function $\langle \cdot, \cdot \rangle$ is called an inner product.

Remark 9.2.1. Notice that the first two properties in Definition 9.2.1 imply $\langle \alpha, \lambda\beta \rangle = \bar{\lambda}\langle \alpha, \beta \rangle$, for every $\alpha, \beta \in V$ and $\lambda \in \mathbb{C}$. These properties coincide with those in the real case, since the conjugate of a real number is itself. There is no general agreement on property 2; several authors call it linearity in the second entry, instead of the first.

Example 9.2.1. For any $X, Y \in \mathbb{C}^n$, considered as column vectors, the function $\langle \cdot, \cdot \rangle$: $\mathbb{C}^n \times \mathbb{C}^n \to \mathbb{C}$ given by $\langle X, Y \rangle := \overline{Y}^t X = \sum_{j=1}^{n} \overline{y_j} x_j$, defines an inner product in $\mathbb{C}^n$, where x_j and $y_j, j \in \{1, 2, \ldots, n\}$, are the coordinates of X and Y, respectively.

Discussion. To verify property 1, notice that $\langle X, Y \rangle = \overline{Y}^t X$ is a complex number which can be considered as a 1×1 matrix, thus its transpose is itself. From this and the properties of conjugation of complex numbers, we have

$$\overline{\langle Y, X \rangle} = \overline{\overline{X}^t Y} = \overline{\overline{X}^t} \, \overline{Y} = X^t \overline{Y} = (X^t \overline{Y})^t = \overline{Y}^t X = \langle X, Y \rangle,$$

proving property 1. If a and b are complex numbers and X, Y, and Z are elements of $\mathbb{C}^n$, then

$$\langle aX + bY, Z \rangle = \overline{Z}^t (aX + bY)$$
$$= a\overline{Z}^t X + b\overline{Z}^t Y$$
$$= a\langle X, Z \rangle + b\langle Y, Z \rangle,$$

proving property 2.

Finally, $\langle X, X \rangle = \overline{X}^t X = \overline{x_1} x_1 + \cdots + \overline{x_n} x_n$ is the sum of the squares of the absolute values of all entries of X, and the sum is zero if and only if $X = 0$, proving property 3.

Concepts such as orthogonal vectors and norm are established as in the real case. More precisely, if V is a euclidean complex vector space and α, β belong to V, it is said that α and β are orthogonal if $\langle \alpha, \beta \rangle = 0$. The norm of α is defined as $\|\alpha\| = \sqrt{\langle \alpha, \alpha \rangle}$.

Another important property derived from the inner product in $\mathbb{C}^n$ is as follows: given an $n \times n$ matrix A with complex entries, one verifies that $\langle AX, Y \rangle = \overline{Y}^t AX$. Using properties of the transpose and conjugation of complex numbers, we have

$$\langle AX, Y \rangle = \langle X, \overline{A}^t Y \rangle. \tag{9.23}$$

Remark 9.2.2. One of the most important differences between real and complex vector spaces is related to the existence of eigenvalues. This issue has to do with the existence of roots of polynomials. According to Theorem A.5.3, p. 279, every polynomial of positive degree with complex coefficients has all its roots in the complex numbers. Hence, every matrix with complex entries has a Jordan canonical form Corollary 8.3.1, p. 203.

9.3 Quadratic forms and bilinear functions

From Exercise 13, p. 140, we know that a linear transformation T from $\mathbb{R}^n$ to the reals has the form $T(x_1, x_2, \ldots, x_n) = a_1 x_1 + a_2 x_2 + \cdots + a_n x_n$, where $a_1, a_2, \ldots, a_n$ are fixed real numbers depending only on T. Notice that T can be represented in terms of the inner product in $\mathbb{R}^n$, that is, setting $X = (x_1, x_2, \ldots, x_n)$ and $A = (a_1, a_2, \ldots, a_n)$, then $T(x_1, x_2, \ldots, x_n) = T(X) = \langle A, X \rangle = a_1 x_1 + a_2 x_2 + \cdots + a_n x_n$. This representation of T shows the relationship between linear transformations from $\mathbb{R}^n$ to $\mathbb{R}$ and the inner product in $\mathbb{R}^n$. More generally, if V is a euclidean vector space and $\beta \in V$ is fixed, one can define a linear transformation $T_\beta : V \to \mathbb{R}$ as follows:

$$T_\beta(\alpha) = \langle \beta, \alpha \rangle. \tag{9.24}$$

In this section we will be dealing with functions that generalize the idea of an inner product, that is, by dropping some of its properties, we will arrive at the idea of a bilinear function.

Definition 9.3.1. A bilinear function on a real vector space V is a function $B : V \times V \to \mathbb{R}$ that satisfies:

1. For every $a, b \in \mathbb{R}$ and α, β, and γ elements of V, $B(a\alpha + b\beta, \gamma) = aB(\alpha, \gamma) + bB(\beta, \gamma)$.
2. For every $a, b \in \mathbb{R}$ and α, β, and γ elements of V, $B(\alpha, a\beta + b\gamma) = aB(\alpha, \beta) + bB(\alpha, \gamma)$.

An alternative way to express that B is bilinear is by considering it as a function of two variables, which is linear in each one.

When B satisfies $B(\alpha, \beta) = B(\beta, \alpha)$ for every α and β in V, we say that B is symmetric.

Remark 9.3.1. Notice that an inner product is a symmetric bilinear function, which additionally is positive definite.

Example 9.3.1. If V is the vector space of the Riemann integrable functions in $[a, b]$, then the function $B : V \times V \to \mathbb{R}$ given by $B(f, g) := \int_a^b fg$ is symmetric and bilinear.

Discussion. Consider $f, g, h \in V$ and $c, d \in \mathbb{R}$. From the properties of the integral, one has that $B(af + bg, h) := \int_a^b (cf + dg)h = c\int_a^b fh + d\int_a^b gh$, proving that the function is linear in the first entry. Linearity in the second entry is obtained likewise. It is important to notice that B is not an inner product, since there are functions f so that $B(f, f) = 0$ and f is not zero. As an exercise, the reader is invited to provide an example to show that the bilinear function above is not an inner product.

Let V be a finite-dimensional vector space with basis $\{\alpha_1, \alpha_2, \ldots, \alpha_n\}$, and let B be a bilinear function. Given $\alpha = x_1\alpha_1 + \cdots + x_n\alpha_n$ and $\beta = y_1\alpha_1 + \cdots + y_n\alpha_n$ in V, using the properties of B, one has

$$
\begin{aligned}
B(\alpha, \beta) &= B(x_1\alpha_1 + \cdots + x_n\alpha_n, y_1\alpha_1 + \cdots + y_n\alpha_n) \\
&= x_1 B(\alpha_1, y_1\alpha_1 + \cdots + y_n\alpha_n) + \cdots + x_n B(\alpha_n, y_1\alpha_1 + \cdots + y_n\alpha_n) \\
&= x_1 y_1 B(\alpha_1, \alpha_1) + \cdots + x_1 y_n B(\alpha_1, \alpha_n) + \cdots + \\
&\quad + x_n y_1 B(\alpha_n, \alpha_1) + \cdots + x_n y_n B(\alpha_n, \alpha_n) \\
&= [x_1\ x_2\ \ldots\ x_n]
\begin{bmatrix}
B(\alpha_1, \alpha_1) & \cdots & B(\alpha_1, \alpha_n) \\
B(\alpha_2, \alpha_1) & \cdots & B(\alpha_2, \alpha_n) \\
\vdots & \ddots & \vdots \\
B(\alpha_n, \alpha_1) & \cdots & B(\alpha_n, \alpha_n)
\end{bmatrix}
\begin{bmatrix}
y_1 \\
\vdots \\
y_n
\end{bmatrix}.
\end{aligned}
$$

Defining the matrix A whose entries are the elements $B(a_i, a_j)$, and recalling that X and Y are matrices with one column and n rows whose entries are $x_1, x_2, \ldots, x_n$ and $y_1, y_2, \ldots, y_n$, respectively, the representation of $B(\alpha, \beta)$, using the matrix A, is

$$B(\alpha, \beta) = X^t A Y. \tag{9.25}$$

Notice that the right-hand side of equation (9.25) depends on the chosen basis of V and it is called the bilinear form of B with respect to the basis $\{a_1, a_2, \ldots, a_n\}$.

Let $\{a_1', a_2', \ldots, a_n'\}$ be another basis of V, with P the change of basis matrix, and let X' and Y' be the coordinate vectors of α and β with respect to this new basis. Then, according to equation (5.27), p. 135, one has $X = PX'$ and $Y = PY'$. If C denotes the matrix associated to B with respect to the basis $\{a_1', a_2', \ldots, a_n'\}$, then

$$B(\alpha, \beta) = X'^t C Y' = X^t (P^{-1})^t C P^{-1} Y = X^t A Y, \tag{9.26}$$

hence, $A = (P^{-1})^t C P^{-1}$, which is equivalent to

$$C = P^t A P. \tag{9.27}$$

Summarizing, we have proved:

Theorem 9.3.1. *Let V be a finite-dimensional vector space, B a bilinear function on V. If A and C are matrices associated to B with respect to the corresponding bases $\{a_1, a_2, \ldots, a_n\}$ and $\{a_1', a_2' \ldots, a_n'\}$, and P is the change of bases matrix, then $C = P^t A P$.*

Definition 9.3.2. Let A and C be $n \times n$ matrices. It is said that C is congruent to A if there is a nonsingular matrix P such that $C = P^t A P$.

Remark 9.3.2. Two $n \times n$ matrices are congruent if and only if they represent a bilinear function with respect to some bases.

Exercise 9.3.1. Prove that being congruent is an equivalence relation in the vector space of square matrices.

Example 9.3.2. Let $B : \mathbb{R}^3 \times \mathbb{R}^3 \to \mathbb{R}$ be given by

$$B((x_1, x_2, x_3), (y_1, y_2, y_3)) = 2x_1 y_1 + x_1 y_2 + x_2 y_1 + x_2 y_3 + x_3 y_1 + x_3 y_3.$$

Find the expression for B with respect to the basis $S = \{(1, 1, 0), (1, 1, 1), (1, 0, 0)\}$.

Discussion. Let A denote the matrix of B with respect to the canonical basis, then its entries are found by using equation (9.25) and noticing that the (i, j)th entry is the coefficient of the term $x_i y_j$. Thus, $A = \begin{bmatrix} 2 & 1 & 0 \\ 1 & 0 & 1 \\ 1 & 0 & 1 \end{bmatrix}$. On the other hand, the change of basis matrix, from the canonical to S, is $P = \begin{bmatrix} 1 & 1 & 1 \\ 1 & 1 & 0 \\ 0 & 1 & 0 \end{bmatrix}$. Applying Theorem 9.3.1, we have that the matrix associated to B with respect to S is

$$P^t AP = \begin{bmatrix} 1 & 1 & 0 \\ 1 & 1 & 1 \\ 1 & 0 & 0 \end{bmatrix} \begin{bmatrix} 2 & 1 & 0 \\ 1 & 0 & 1 \\ 1 & 0 & 1 \end{bmatrix} \begin{bmatrix} 1 & 1 & 1 \\ 1 & 1 & 0 \\ 0 & 1 & 0 \end{bmatrix} = \begin{bmatrix} 4 & 5 & 3 \\ 5 & 7 & 4 \\ 3 & 3 & 2 \end{bmatrix}.$$

If the coordinates of a vector with respect to the basis $S = \{(1, 1, 0), (1, 1, 1), (1, 0, 0)\}$ are denoted by (x_1', x_2', x_3'), then

$$B((x_1', x_2', x_3'), (y_1', y_2', y_3')) = \begin{bmatrix} x_1' & x_2' & x_3' \end{bmatrix} \begin{bmatrix} 4 & 5 & 3 \\ 5 & 7 & 4 \\ 3 & 3 & 2 \end{bmatrix} \begin{bmatrix} y_1' \\ y_2' \\ y_3' \end{bmatrix}$$

$$= 4x_1'y_1' + 5x_1'y_2' + 3x_1'y_3' + 5x_2'y_1' + 7x_2'y_2' + 4x_2'y_3'$$
$$+ 3x_3'y_1' + 3x_3'y_2' + 2x_3'y_3'.$$

Example 9.3.2 shows that the representation of a bilinear function by a matrix could be simpler in some basis. Then it seems natural to ask if there is a basis where the representation of B is as simple as possible. Notice that a positive answer is equivalent to saying that there is a basis $\{a_1, a_2, \ldots, a_n\}$ such that

$$B(a_i, a_j) = \begin{cases} a_i & \text{if } i = j, \\ 0 & \text{if } i \neq j. \end{cases} \tag{9.28}$$

If such a basis exists, we will have some sort of Gram–Schmidt process to find an "orthogonal" basis with respect to B. In addition, equation (9.28) implies that B is symmetric, since if the matrix C is associated to B with respect to another basis, then

$$\text{diag}\{a_1, a_2, \ldots, a_n\} = P^t C P,$$

with P being the change of basis matrix. Transposing in this equation and using that a diagonal matrix is symmetric, we obtain $C^t = C$.

The arguments from the previous paragraph show that, for the existence of a basis to guarantee conditions of equation (9.28), it is necessary that B be symmetric. Is this condition also sufficient for the existence of such a basis?

First of all, if B is the zero function, that is, if $B(a, a) = 0$ for every $a \in V$, then B is represented by a diagonal matrix.

From the previous consideration, we may assume that there is $a_1 \in V \setminus \{0\}$ such that $B(a_1, a_1) \neq 0$. Since we are looking for "orthogonal" elements, it is natural to consider elements $\beta \in V$ such that $B(a, \beta) = 0$ and from those choose elements to construct the basis with the required properties.

We notice that the set of $\beta \in V$ which satisfy $B(a, \beta) = 0$ is the kernel of the linear function $f_{a_1} : V \to \mathbb{R}$ defined by $f_{a_1}(\beta) = B(a_1, \beta)$. Since f_{a_1} is not the zero function, it is surjective, therefore its image has dimension 1. Applying Theorem 5.3.2, p. 129, one concludes that the dimension of the kernel of f_{a_1} is $\dim(V) - 1$.

Let W be the kernel of f_{α_1}. It is clear that $\mathcal{B}$ can be consider as a bilinear function on W. An inductive argument on the dimension of V, where $\mathcal{B}$ is defined, finishes the proof. More precisely, if V has dimension 1, there is nothing to be proved. Assume that V has dimension > 1 and, when $\mathcal{B}$ acts on subspaces of lower dimension, the required basis exists.

From what we have proved, there is a basis of W, $\{\alpha_2, \alpha_3, \dots, \alpha_n\}$, such that $\mathcal{B}(\alpha_i, \alpha_j) = \mathcal{B}(\alpha_j, \alpha_i) = 0$, for every $i, j \in \{2, 3, \dots, n\}$, if $i \neq j$.

We have that α_1 satisfies $\mathcal{B}(\alpha_1, \alpha_1) \neq 0$, and hence $\alpha_1 \notin W$. From this we conclude that $\{\alpha_1, \alpha_2, \alpha_3, \dots, \alpha_n\}$ is a basis of V. From the definition of W, we have that $\mathcal{B}(\alpha_1, \alpha_j) = 0$ for every $j \in \{2, 3, \dots, n\}$. Since $\mathcal{B}$ is symmetric, $\mathcal{B}(\alpha_j, \alpha_1) = 0$ for every $j \in \{2, 3, \dots, n\}$, proving that $\{\alpha_1, \alpha_2, \alpha_3, \dots, \alpha_n\}$ is the needed basis.

Summarizing, we have proved:

Theorem 9.3.2. *Let V be a finite-dimensional vector space and $\mathcal{B}$ a bilinear function. Then $\mathcal{B}$ is represented by a diagonal matrix if and only if $\mathcal{B}$ is symmetric.*

Remark 9.3.3. If V is a finite-dimension vector space and $\mathcal{B}$ is a symmetric bilinear function, then there exists a basis $\{\alpha_1, \alpha_2, \alpha_3, \dots, \alpha_n\}$ of V such that $\mathcal{B}(\alpha, \beta) = a_1 x_1 y_1 + \cdots + a_n x_n y_n$, where $\alpha = x_1 \alpha_1 + x_2 \alpha_2 + \cdots + x_n \alpha_n$ and $\beta = y_1 \alpha_1 + y_2 \alpha_2 + \cdots + y_n \alpha_n$.

Corollary 9.3.1. *If A is a real square matrix, then A is symmetric if and only if there is a nonsingular matrix P such that $P^t A P = \mathrm{diag}\{d_1, d_2, \dots, d_n\}$.*

Exercise 9.3.2. Assume that matrices A and B are congruent. Prove that $\mathrm{Rank}(A) = \mathrm{Rank}(B)$.

We will improve Corollary 9.3.1 much more.

Theorem 9.3.3. *Let A be a real symmetric matrix of rank r. Then there is a nonsingular matrix P such that*

$$P^t A P = \mathrm{diag}\{1, 1, \dots, 1, -1, -1, \dots, -1, 0, 0, \dots, 0\}. \tag{9.29}$$

The number of 1s is p and the number of -1s is $r - p$. This representation of $P^t A P$ is unique in the sense that p, $r - p$, and r only depend on A.

Proof. If A is the zero matrix, there is nothing to be done. Hence we will assume that A has rank $r > 0$, thus, according to Corollary 9.3.1, r is the number of nonzero elements in $\mathrm{diag}\{d_1, d_2, \dots, d_n\}$.

We notice that the elementary row matrix obtained by swapping the ith and jth rows in the identity matrix is the same as the elementary column matrix obtained by swapping the ith and jth columns in the identity matrix, hence there is a nonsingular matrix P_1 such that

$$P_1^t A P_1 = \mathrm{diag}\{d_1, d_2, \dots, d_p, d_{p+1}, \dots, d_r, 0 \dots, 0\}, \tag{9.30}$$

where $d_i > 0$ for $i \in \{1, 2, \dots, p\}$ and $d_j < 0$ for $p < j \leq r$.

Now defining $P_2 = \text{diag}\{\frac{1}{\sqrt{d_1}}, \frac{1}{\sqrt{d_2}}, \ldots, \frac{1}{\sqrt{d_p}}, \frac{1}{\sqrt{-d_{p+1}}}, \ldots, \frac{1}{\sqrt{-d_r}}, 1 \ldots, 1\}$, one obtains

$$P^t A P = \text{diag}\{1, 1, \ldots, 1, -1, -1, \ldots, -1, 0, 0, \ldots, 0\}, \tag{9.31}$$

where $P = P_1 P_2$.

To finish the proof, let us assume that there is a nonsingular matrix Q such that $Q^t A Q$ is a diagonal matrix with q and $r - q$ of 1s and -1s, respectively. We will prove that $p = q$. We will argue by contradiction, that is, we may assume that $p > q$ and that there are nonsingular matrices P and Q such that $D_1 = P^t A P$ and $D_2 = Q^t A Q$ have p and q 1s, respectively. From this equation, one shows that there is a nonsingular matrix R such that $D_1 = R^t D_2 R$. Let $W_p = \{(x_1, x_2, \ldots, x_p, 0, \ldots, 0) \in \mathbb{R}^n \mid x_i \in \mathbb{R}\}$. It is verified directly that $X^t D_1 X = x_1^2 + x_2^2 + \cdots + x_p^2 \geq 0$ for every $X \in W_p$. Since R is nonsingular, $U_p = \{RX : X \in W_p\}$ has dimension p as does W_p. Let $W_q = \{(0, 0, \ldots, 0, y_{q+1}, \ldots, y_n) \in \mathbb{R}^n \mid y_j \in \mathbb{R}\}$. It is clear that $Y^t D_2 Y = -y_{q+1}^2 - \cdots - y_r^2 \leq 0$, for every $Y \in W_q$. Hence $U_p \cap W_q = \{0\}$. Also $U_p + W_q$ is a subspace of $\mathbb{R}^n$, thus $\dim(U_p + W_q) \leq n$. On the other hand, $\dim(U_p + W_q) = \dim(U_p) + \dim(W_q) = p + (n - q) = n + (p - q) > n$, a contradiction. If we assume that $q > p$, an analogous procedure again produces a contradiction, thus $p = q$. $\qquad\square$

Definition 9.3.3. Given a real symmetric matrix A, the number $p - (r - p) = 2p - r$ is called the signature of A.

Corollary 9.3.2. *Two $n \times n$ matrices, A and B, are congruent if and only if they have the same rank and signature.*

9.3.1 Quadratic forms

If $\mathcal{B}$ is a bilinear function in V, represented by a matrix A, then $\mathcal{B}(\alpha, \alpha) = X^t A X$, where X is the coordinate vector of α with respect to a given basis. The expression $X^t A X$ is called a *quadratic form*.

There is an alternative way to define a quadratic form.

Definition 9.3.4. Let V be a vector space. A function $Q : V \to \mathbb{R}$ is called a quadratic form if it satisfies:
1. For every $\alpha \in V$ and $r \in \mathbb{R}$, one has $Q(r\alpha) = r^2 Q(\alpha)$.
2. The function $\mathcal{B}(\alpha, \beta) = \frac{Q(\alpha+\beta) - Q(\alpha-\beta)}{4}$ is bilinear.

Remark 9.3.4. Notice that from the second part of Definition 9.3.4 and from the discussion above, one obtains that the bilinear function in the definition can be represented by a symmetric matrix, since this bilinear function is symmetric.

As a consequence of Theorem 9.3.2 and the previous remark, one has:

Theorem 9.3.4. *Let V be vector space of dimension n. If Q is a quadratic form in V, then there is a basis of V with respect to which $Q(a) = d_1 x_1^2 + d_2 x_2^2 + \cdots + d_n x_n^2$, with $x_1, x_2, \ldots, x_n$ being the coordinates of a with respect to the given basis.*

When a quadratic form Q has been represented by a matrix A with respect to a given basis, and a vector a has a coordinate vector X with respect to the same basis, the expression $Q(X) = X^t A X$ is usually called a *quadratic form*.

We notice that the sign of the values of Q, represented by $Q(a) = d_1 x_1^2 + d_2 x_2^2 + \cdots + d_n x_n^2$, can be obtained by examining the signs of the coefficients $d_1, d_2, \ldots, d_n$. For instance, if $d_i > 0$ for every $i \in \{1, 2, \ldots, n\}$, then $Q(a) > 0$ for every $a \neq 0$; if $d_i < 0$ for every $i \in \{1, 2, \ldots, n\}$, then $Q(a) < 0$ for every $a \neq 0$.

This remark is formalized in:

Definition 9.3.5. A quadratic form $Q(X)$ is positive (negative) definite if $Q(X) > 0$ ($Q(X) < 0$) for every $X \neq 0$.

We notice that the definition of a quadratic form, being positive or negative definite, can be approached in terms of the matrix representing it. First of all, we notice that the matrix is symmetric. In the following sections we will discuss positive definite matrices which are closely related to positive definite quadratic forms.

9.3.2 Principal-axis theorem

In this section we will prove that any symmetric matrix over the reals is orthogonal diagonalizable, that is, if A is a symmetric matrix with real entries, then there exists an orthogonal matrix P such that $P^t A P = P^{-1} A P = D$, where D is diagonal. This result will have important geometrical implications to represent conics in a "canonical way".

We start by proving the next lemma.

Lemma 9.3.1. *If A is a symmetric matrix with real entries, then the eigenvalues of A are real.*

Proof. Let λ be an eigenvalue of A associated to the eigenvector X. Since A has real entries and is symmetric, applying equation (9.23), p. 239, one has $\langle AX, X \rangle = \langle X, AX \rangle$. From this equation and one of the properties of the inner product in a complex vector space, one has $\lambda \langle X, X \rangle = \bar{\lambda} \langle X, X \rangle$; equivalently, $(\lambda - \bar{\lambda}) \langle X, X \rangle = 0$. Since X is not zero, $\langle X, X \rangle \neq 0$, hence $(\lambda - \bar{\lambda}) = 0$, that is, λ is real. □

Lemma 9.3.2. *Let A be a real symmetric matrix, and let μ and λ be different eigenvalues of A. If X_1 and X_2 are eigenvectors associated to μ and λ, respectively, then X_1 is orthogonal to X_2.*

Proof. From equation (9.23) and the assumption on A, one has that $\langle AX_1, X_2 \rangle = \langle X_1, AX_2 \rangle$. Using the definition of an eigenvalue, the linearity of the inner product, and grouping, we have $(\mu - \lambda)\langle X_1, X_2 \rangle = 0$. The conclusion follows from the assumption on μ and λ. □

Lemma 9.3.3. *Let B be a real symmetric matrix. Assume that $B^m X = 0$ for some positive integer m, then $BX = 0$.*

Proof. Choose m to be the minimum integer that satisfies $B^m X = 0$. If $m > 1$ then $B^{m-1}X = Y \neq 0$ and $BY = 0$. From this, one has $0 = \langle BY, B^{m-2}X \rangle = \langle Y, B^{m-1}X \rangle = \langle Y, Y \rangle$, contradicting the condition of Y. □

Definition 9.3.6. Let P be a nonsingular real matrix. We say that P is orthogonal if $P^{-1} = P^t$.

Remark 9.3.5. A matrix P is orthogonal if and only if its columns are orthonormal.

Proof. Let P be a matrix whose columns are denoted by $P_1, P_2, \ldots, P_n$, then the entries of $P^t P$, c_{ij}, are given by $c_{ij} = \langle P_i, P_j \rangle = \begin{cases} 1 & \text{if } i = j, \\ 0 & \text{otherwise.} \end{cases}$ From this we have that P is orthogonal if and only if the columns of P are orthonormal. □

Theorem 9.3.5 (Principal-axis theorem). *Let A be a real symmetric matrix. Then there is an orthogonal matrix P such that $P^{-1}AP = P^t AP = \text{diag}\{\lambda_1, \ldots, \lambda_n\}$, with $\lambda_1, \ldots, \lambda_n$ being the eigenvalues of A.*

Proof. First of all, we will show that A is diagonalizable, which is equivalent to showing that the minimal polynomial has simple roots, see Theorem 8.1.1 p. 189. Let $m_A(x) = p(x)^l q(x)$ be the minimal polynomial of A, where $p(x)$ is irreducible and relatively prime to $q(x)$. From Lemma 9.3.1, $p(x)$ must be linear. To finish this part of the proof, we will show that $l = 1$. Let $X \in \mathbb{R}^n$, then $m_A(A)X = p(A)^l q(A)X = 0$. It is straightforward to show that $p(A)$ is a symmetric matrix. Applying Lemma 9.3.3 to $p(A)$, one has $p(A)q(A)X = 0$, that is, $p(A)q(A)$ annihilates any $X \in \mathbb{R}^n$, which implies that $p(A)q(A)$ is the zero matrix. Then $m_A(x)$ divides $p(x)q(x)$, concluding that $l = 1$.

Let $m_A(x) = (x - \lambda_1) \cdots (x - \lambda_r)$ be the factorization of the minimal polynomial of A, and let $W_1, W_2, \ldots, W_r$ be the kernels of $A - \lambda_1 I, A - \lambda_2 I, \ldots, A - \lambda_r I$, respectively. Applying Theorem 7.2.1, p. 181, one has

$$\mathbb{R}^n = W_1 \oplus W_2 \oplus \cdots \oplus W_r.$$

Let $\mathcal{B}_i$ be an orthonormal basis of W_i, for $i \in \{1, 2, \ldots, r\}$. From Lemma 9.3.2, the union of $\mathcal{B}_1, \mathcal{B}_2, \ldots, \mathcal{B}_r$ is an orthonormal basis of $\mathbb{R}^n$. Define P as the matrix whose columns are the elements of the basis just defined. From Exercise 18, p. 221, we have that $P^{-1}AP = P^t AP$ is diagonal, with entries on the diagonal being the eigenvalues of A. It should be noticed that the elements on the diagonal could be repeated, actually λ_i appears as many times as $\dim(W_i)$. □

As an interesting corollary of the principal axes' theorem, we have:

Corollary 9.3.3. *The only nilpotent and symmetric matrix is the zero matrix.*

The following exercise provides a connection between the QR decomposition of a matrix A and the condition of having real eigenvalues.

Exercise 9.3.3. Let A be an $n \times n$ real matrix. Prove that all eigenvalues of A are real if and only if there are two matrices, an orthogonal Q and an upper triangular R, such that $A = QRQ^t$.

(*Hint.*) If $A = QRQ^t$, then $Q^tAQ = R$, that is, A is similar to R and the eigenvalues of R are the elements on its diagonal. Conversely, assume that the eigenvalues of A are real. Let U_1 be a vector of norm 1 associated to λ_1, that is, $AU_1 = \lambda_1 U_1$ and $\|U_1\| = 1$. Expanding U_1 in an orthonormal basis of $\mathbb{R}^n$, we can construct an orthogonal matrix Q_1 whose columns are the elements of this basis, then Q_1^tAQ is a matrix whose first entry in the first column is λ_1 and the rest are zero. The eigenvalues of the submatrix obtained by dropping the first row and first column of Q_1^tAQ are $\lambda_2, \ldots, \lambda_n$. Perform appropriate constructions and use induction to finish the proof.

Remark 9.3.6. The Principal-axis theorem is a corollary of Exercise 9.3.3.

9.3.3 Positive definite matrices

One important problem in calculus of one variable is to classify maxima or minima of functions. One method to approach the problem is by using the second derivative criterion. The corresponding problem for functions $f : \mathbb{R}^n \to \mathbb{R}$ is dealt with using a similar criterion. For this case, the second derivative criterion is given in terms of the hessian matrix, which is symmetric. Taking these ideas into account, we will define when a symmetric matrix is "positive".

Definition 9.3.7. Let A be a symmetric matrix. We say that A is positive definite, if $X^tAX > 0$ for every $X \neq 0$. If $X^tAX < 0$ for every $X \neq 0$, we say that A is negative definite.

Theorem 9.3.6. *Let A be a symmetric matrix, then A is positive definite if and only if its eigenvalues are positive.*

Proof. Assume that the eigenvalues of A, $\lambda_1, \lambda_2, \ldots, \lambda_n$ are positive. From Theorem 9.3.5, there is a orthogonal matrix P such that $P^tAP = \text{diag}\{\lambda_1, \lambda_2, \ldots, \lambda_n\} = D$.

Since P is invertible, given $X \neq 0$, there is $Y \neq 0$ such that $X = PY$. From this we have

$$X^tAX = (PY)^tAPY$$
$$= Y^tP^tAPY = Y^tDY$$
$$= \lambda_1 y_1^2 + \lambda_2 y_2^2 + \cdots + \lambda_n y_n^2 > 0,$$

proving that A is positive definite.

For the converse, we argue by contradiction. Assume that $\lambda_i < 0$ for some i, then, taking E_i, the ith canonical vector of $\mathbb{R}^n$, and using that P is invertible, one has $PE_i = Y_i \neq 0$. From this we have $Y_i^t A Y_i = E_i^t P^t A P E_i = E_i^t D E_i = \lambda_i < 0$, contradicting that A is positive definite, hence all the eigenvalues of A are positive. $\qquad\square$

In many cases, it could be difficult to apply Theorem 9.3.6, since in general it is not easy to find the eigenvalues of A. The next theorem might help to deal with this situation, or to construct many examples of positive definite matrices.

Theorem 9.3.7. *Let A be an $n \times n$ matrix, then A is positive definite if and only if there is a nonsingular matrix B such that $A = B^t B$.*

Proof. Assume that A is positive definite. From Theorem 9.3.6, the eigenvalues of A are positive. Also, Theorem 9.3.5 implies that there is P such that $P^t A P = \text{diag}\{\lambda_1, \lambda_2, \ldots, \lambda_n\}$, with $\lambda_i > 0$ for every $i \in \{1, 2, \ldots, n\}$. Set $D = \text{diag}\{\sqrt{\lambda_1}, \sqrt{\lambda_2}, \ldots, \sqrt{\lambda_n}\}$, then $P^t A P = D^2$ or $A = PDDP^t = (DP^t)^t(DP^t)$. Defining $B = DP^t$, it has the required properties, that is, B is nonsingular and $A = B^t B$.

Conversely, if $A = B^t B$, with B nonsingular, then $BX \neq 0$ for every $X \neq 0$. From this we have $X^t A X = X^t B^t B X = \langle BX, BX \rangle > 0$, proving that A is positive definite. $\qquad\square$

One more result that can help to deal with positive definite matrices is the forthcoming result, Theorem 9.3.8. In the proof of this result, we will use Descartes' rule of signs, Theorem A.5.5, p. 281. We think that Theorem 9.3.8 could be useful, since we have algorithms to compute the minimal polynomial of a matrix, Algorithm 7.1.1, p. 176, and Algorithm 8.2.1, p. 200.

Theorem 9.3.8. *Let A be a symmetric matrix with minimal polynomial*

$$m_A(x) = x^k + a_{k-1}x^{k-1} + \cdots + a_1 x_1 + a_0.$$

Then the following conditions are equivalent:
1. *The matrix A is positive definite.*
2. *For every $i \in \{1, 2, \ldots, k-1\}$, $a_i a_{i-1} < 0$.*
3. *For every $i \in \{1, 2, \ldots, k-1\}$, $(-1)^i a_{k-i} > 0$.*

Proof. Since the number of changes of sign in the coefficients of $m_A(x)$ will play and important role, let v denote that number.

$(1 \Rightarrow 2)$ Since A is positive definite, $m_A(x)$ has positive simple roots, thus

$$m_A(x) = (x - \lambda_1)(x - \lambda_2) \cdots (x - \lambda_k)$$
$$= x^k - (\lambda_1 + \lambda_2 + \cdots + \lambda_k)x^{k-1} + a_{k-2}x^{k-2} + \cdots + a_1 x + a_0. \tag{9.32}$$

From equation (9.32), we have $v \leq k$. On the other hand, from Theorem A.5.5, p. 281, we have that $v = k + 2l$, with $l \geq 0$, hence, $v = k$. Therefore $a_i a_{i-1} < 0$, for every $i \in \{1, 2, \ldots, k\}$.

$(2 \Rightarrow 3)$ Since $m_A(x)$ is monic and $a_i a_{i-1} < 0$, one has $a_{k-1} < 0$ and $a_{k-2} > 0$. In general, $a_{k-j} > 0$ if and only if j is even, hence $(-1)^j a_{k-j} > 0$, for every $j \in \{1, 2, \ldots, k\}$.

$(3 \Rightarrow 1)$ Since A is symmetric, all the roots of $m_A(x)$ are real. We will prove that all the roots of $m_A(x)$ are positive, which is equivalent to proving that that $m_A(-x)$ has no positive or zero roots.

The condition $a_i(-1)^i > 0$ for every $i \in \{1, 2, \ldots, k\}$ implies that the coefficients of $m_A(-x)$, which are $(-1)^i a_i$, keep sign, thus from Theorem A.5.5, p. 281, $m_A(-x)$ has no positive or zero roots, finishing the proof of the theorem. $\qquad\square$

Remark 9.3.7. In Theorem 9.3.8, the minimal polynomial can be replaced by the characteristic polynomial and the conclusions still hold. One possible advantage of the theorem as it is stated is that the leading coefficient is one.

Example 9.3.3. Let $A = \begin{pmatrix} a & b \\ b & c \end{pmatrix}$ be a real matrix. Then the characteristic polynomial of A is $f_A(x) = x^2 - (a + c)x + ac - b^2$. Applying the previous theorem, $f_A(x)$ has positive roots if and only if $a + c > 0$ and $ac - b^2 > 0$, if and only if $a > 0$ and $ac - b^2 > 0$. This criterion is useful when a 2×2 real matrix represents the second derivative of a function from $\mathbb{R}^2$ to $\mathbb{R}$, and it is needed to characterize maxima or minima.

9.3.4 Singular value decomposition (SVD)

EVERYWHERE YOU GO, ALWAYS TAKE THE SVD WITH YOU. The SVD is a fascinating, fundamental object of study that has provided a great deal of insight into a diverse array of problems, which range from social network analysis and quantum information theory to applications in geology. The matrix SVD has also served as the foundation from which to conduct data analysis of multi-way data by using its higher-dimensional tensor versions. The abundance of workshops, conference talks, and journal papers in the past decade on multilinear algebra and tensors also demonstrates the explosive growth of applications for tensors and tensor SVDs. The SVD is an omnipresent factorization in a plethora of application areas. We recommend it highly. [18]

We rephrase the cited quotation by saying that the singular value decomposition of an $m \times n$ matrix A, as the product $A = Q\Sigma P^t$, where P and Q are orthogonal and Σ is diagonal whose nonzero entries are positive, is a useful representation of A which has several implications to approach a wide range of problems. Also, the representation $A = Q\Sigma P^t$ has a nice geometric meaning, as we will see.

If $B = A^t A$, then B is symmetric and, from Theorem 9.3.5, there are orthonormal vectors $v_1, v_2, \ldots, v_n$ such that

$$Bv_i = A^t A v_i = \lambda_i v_i, \tag{9.33}$$

for every $i \in \{1, 2, \ldots, n\}$.

Assume that $\lambda_i \neq 0$ for every $i \in \{1, 2, \ldots, r\}$, where r is the rank of A. From the choice of r, we have $r \leq n, m$.

Using equation (9.33), we have

$$AA^t A v_i = A \lambda_i v_i = \lambda_i A v_i, \tag{9.34}$$

for every $i \in \{1, 2, \ldots, n\}$. That is, if $A v_i \neq 0$, then it is an eigenvector of AA^t. Also, $\|A v_i\|^2 = v_i^t A^t A v_i = v_i^t \lambda_i v_i = \lambda_i$. For $i \in \{1, 2, \ldots, r\}$, define $u_i = \frac{A v_i}{\|A v_i\|}$. It is clear that u_i has norm 1. We also have

$$
\begin{aligned}
u_i^t u_j &= \left(\frac{A v_i}{\|A v_i\|} \right)^t \frac{A v_j}{\|A v_j\|} \\
&= \frac{v_i^t A^t}{\|A v_i\|} \frac{A v_j}{\|A v_j\|} \\
&= \frac{v_i^t}{\|A v_i\|} \frac{\lambda_j v_j}{\|A v_j\|} \\
&= \frac{\lambda_j v_i^t v_j}{\|A v_i\| \|A v_j\|}.
\end{aligned}
\tag{9.35}
$$

Hence, the set $\{u_1, u_2, \ldots, u_r\} \subseteq \mathbb{R}^m$ is orthonormal. From Theorems 4.2.4 (see p. 103) and 9.1.4 (see p. 227), we may assume that this set has been extended to

$$\{u_1, u_2, \ldots, u_r, u_{r+1}, \ldots, u_m\},$$

an orthonormal basis of $\mathbb{R}^m$.

Let Q and P be the matrices whose columns are $u_1, u_2 \ldots, u_m$ and $v_1, v_2, \ldots, v_n$, respectively. Also, let Σ be the $m \times n$ matrix whose entries c_{ij} are defined by

$$
c_{ij} = \begin{cases} \|A v_i\|, & \text{if } i, j \in \{1, 2, \ldots, r\} \text{ and } i = j, \\ 0 & \text{otherwise.} \end{cases}
\tag{9.36}
$$

From the definition of P, Q, and Σ, it is straightforward to verify that

$$Q^t A P = \Sigma. \tag{9.37}$$

Using the orthogonality condition on the columns of P and Q, we have that equation (9.37) is equivalent to

$$A = Q \Sigma P^t. \tag{9.38}$$

From the discussion above, the following theorem has been proved.

Theorem 9.3.9 (Singular value decomposition). *If A is an $m \times n$ matrix, with real entries, then there are orthogonal matrices P and Q and a matrix Σ, defined by equation (9.36), such that*

$$A = Q\Sigma P^t. \tag{9.39}$$

Definition 9.3.8. Given an $m \times n$ matrix A, the representation of A given in Theorem 9.3.9, equation (9.39), is called the singular value decomposition of A. The nonzero elements of Σ are called the singular values of A and the matrix Σ is called the singular matrix of A.

Example 9.3.4. Let $A = \left(\begin{smallmatrix} 1 & 3 \\ 2 & 1 \end{smallmatrix}\right)$. Find the singular value decomposition of A, i. e., Q and P such that $A = Q\Sigma P^t$.

Discussion. Start by computing the eigenvalues and eigenvectors of $B = A^t A = \left(\begin{smallmatrix} 5 & 5 \\ 5 & 10 \end{smallmatrix}\right)$. The minimal and characteristic polynomials of B are the same and are equal to $m_B(x) = x^2 - 15x + 25$, whose roots are $\lambda_1 = \frac{5}{2}(3 + \sqrt{5})$ and $\lambda_2 = \frac{5}{2}(3 - \sqrt{5})$, with corresponding eigenvectors $a_1 = \left(1, \frac{1+\sqrt{5}}{2}\right)$ and $a_2 = \left(1, \frac{1-\sqrt{5}}{2}\right)$, respectively.

According to the notation in Theorem 9.3.9, $v_1 = \frac{a_1}{\|a_1\|}$ and $v_2 = \frac{a_2}{\|a_2\|}$ are the columns of P. Performing calculations and simplifying, we obtain

$$v_1 = \frac{(2, 1 + \sqrt{5})}{\sqrt{10 + 2\sqrt{5}}}, \quad v_2 = \frac{(2, 1 - \sqrt{5})}{\sqrt{10 - 2\sqrt{5}}}. \tag{9.40}$$

Also, according to the cited theorem, the columns of matrix Q are the vectors $u_1 = \frac{Av_1}{\|Av_1\|}$ and $u_2 = \frac{Av_2}{\|Av_2\|}$, which are given in the form

$$u_1 = \frac{(3 + \sqrt{5}, 1 + \sqrt{5})}{2\sqrt{5 + 2\sqrt{5}}}, \quad u_2 = \frac{(\sqrt{5} - 3, \sqrt{5} - 1)}{2\sqrt{5 - 2\sqrt{5}}}. \tag{9.41}$$

Since the numbers $a_1 = \|a_1\| = \sqrt{10 + 2\sqrt{5}}$ and $a_2 = \|a_2\| = \sqrt{10 - 2\sqrt{5}}$ will play an important role in what follows, we explicitly state two properties that they satisfy and invite the reader to justify them, Exercise 22 (this chapter):

$$a_1 a_2 = 4\sqrt{5},$$
$$2a_2 + (1 - \sqrt{5})a_1 = 0. \tag{9.42}$$

Using equation (9.42), we find that the matrix $P^t = \left(\begin{smallmatrix} \frac{2}{a_1} & \frac{1+\sqrt{5}}{a_1} \\ \frac{2}{a_2} & \frac{1-\sqrt{5}}{a_2} \end{smallmatrix}\right)$ has determinant -1 and minimal polynomial $m(x) = x^2 - 1$, thus the eigenvalues of P^t are ± 1. This justifies that P^t is a rotation around the line passing through the origin and defined by one eigenvector associated to the eigenvalue 1, which is $w = \left(1, \frac{a_1 - 2}{1 + \sqrt{5}}\right)$.

We also have the matrix $\Sigma = \left(\begin{smallmatrix} \|Av_1\| & 0 \\ 0 & \|Av_2\| \end{smallmatrix} \right)$, where $\|Av_1\| = \sqrt{\frac{5}{2}(3 + \sqrt{5})}$ and $\|Av_2\| = \sqrt{\frac{5}{2}(3 - \sqrt{5})}$. With these matrices, we are in a position to verify that $A = Q\Sigma P^t$.

To illustrate the geometric meaning of the decomposition $A = Q\Sigma P^t$, we consider the circle with equation $x^2 + y^2 = 4$ and take a point u on it. The action of P^t on u produces point E, which is obtained by reflecting u across the line L, with equation $y = \frac{a_1 - 2}{1 + \sqrt{5}}x$. Applying Σ to E produces point F, and applying Q to F produces point G.

The factorization of A can be interpreted as first applying a reflection, then stretching, and finally, a rotation. It can be shown, see Exercise 25, that A transforms the circle with equation $x^2 + y^2 = 4$ into an ellipse with equation $x^2 + 2y^2 - 2xy = 20$. □

The geometric meaning of Figure 9.7 can be decomposed in a diagram as shown in Figure 9.8.

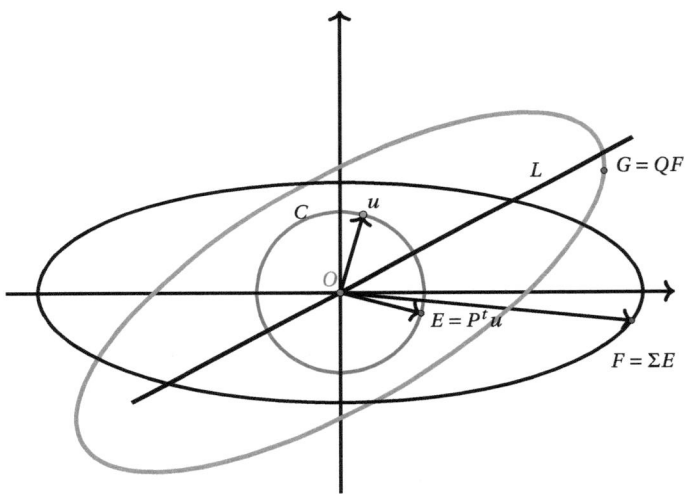

Figure 9.7: Geometric representation of $A = Q\Sigma P^t$.

Having shown an example, using pencil and paper, it would be a good idea to show one more using GeoGebra.

Example 9.3.5. Let $A = \left(\begin{smallmatrix} 1 & 2 \\ 3 & -1 \end{smallmatrix} \right)$. Find the singular value decomposition of A and, using GeoGebra, illustrate the geometric representation of the equation $A = Q\Sigma P^t$.

Discussion. Start by computing the eigenvalues and eigenvectors of $C = A^t A = \left(\begin{smallmatrix} 10 & -1 \\ -1 & 5 \end{smallmatrix} \right)$. To this end, we use GeoGebra click here. The eigenvalues of C are $\left\{ \frac{\sqrt{29}+15}{2}, \frac{-\sqrt{29}+15}{2} \right\}$ and the corresponding eigenvectors, written in matrix form are $\left(\begin{smallmatrix} \sqrt{29}+5 & -\sqrt{29}+5 \\ -2 & -2 \end{smallmatrix} \right)$. □

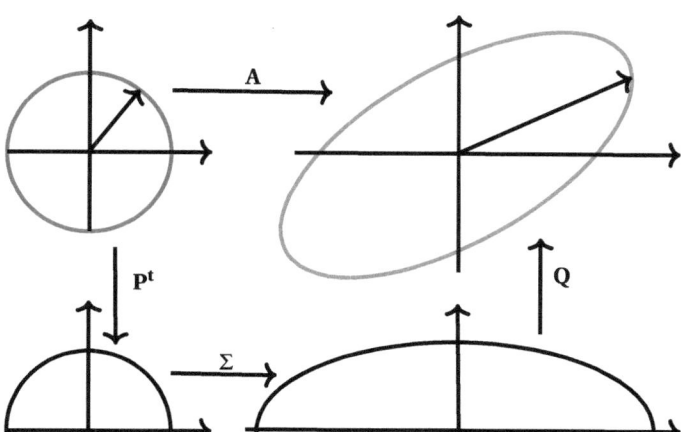

Figure 9.8: Diagram illustrating the geometric interpretation of A as the composition of P^t, Σ, and Q.

With the discussion up to this point and Definition 9.3.8, we are ready to estate a theorem that links several equivalent conditions characterizing matrices that have inverses. Since this result is one of the most important in linear algebra, we strongly recommend the reader to give a discussion of the proof.

Theorem 9.3.10 (Invertible matrix theorem). *Let A be an $n \times n$ matrix. Then the following statements are equivalent:*

1. *The matrix A has an inverse.*
2. *The matrix A is row equivalent to I_n.*
3. *The rows of A are linearly independent.*
4. *The matrix A is a product of elementary matrices.*
5. *The system $AX = 0$ has a unique solution.*
6. *The columns of A are linearly independent.*
7. *For every $B = \begin{bmatrix} b_1 \\ b_2 \\ \vdots \\ b_n \end{bmatrix}$, the system $AX = B$ has a unique solution.*
8. *The columns of A span $\mathbb{R}^n$.*
9. *The rank of A is n.*
10. $\det(A) \neq 0$.
11. *Zero is not an eigenvalue of A.*
12. *The set of singular values of A has cardinality n.*
13. *The eigenvalues of $A^t A$ are positive.*

9.3.5 Classification of conics

One interesting application of the principal axes' theorem occurs in geometry, which is quite natural, since it resembles geometric aspects. From the algebraic point of view, the main problem in analytic geometry is to classify the conics, starting from the general quadratic equation in two variables. More precisely, the problem is to classify the conics starting from the equation

$$ax^2 + bxy + cy^2 + dx + ey + f = 0, \tag{9.43}$$

where a, b, c, d, e, and f are fixed real numbers.

We notice that equation (9.43) has a quadratic part, which can be explored from the quadratic forms perspective, thus we use a symmetric matrix to represent that part of the equation:

$$\begin{bmatrix} x & y \end{bmatrix} \begin{bmatrix} a & \frac{b}{2} \\ \frac{b}{2} & c \end{bmatrix} \begin{bmatrix} x \\ y \end{bmatrix} + dx + ey + f = 0. \tag{9.44}$$

Assuming that $A = \begin{bmatrix} a & \frac{b}{2} \\ \frac{b}{2} & c \end{bmatrix}$ and then applying the principal axes' theorem, Theorem 9.3.5, to A, we have that there is an orthogonal matrix P such that

$$P^t A P = \text{diag}\{\lambda_1, \lambda_2\}, \tag{9.45}$$

where λ_1 and λ_2 are the eigenvalues of A.

Introducing new coordinates, say (u, v), and using the matrix $P = \begin{bmatrix} p_{11} & p_{12} \\ p_{21} & p_{22} \end{bmatrix}$ to describe the relation between coordinates, we have

$$\begin{bmatrix} x \\ y \end{bmatrix} = P \begin{bmatrix} u \\ v \end{bmatrix} = \begin{bmatrix} p_{11}u + p_{12}v \\ p_{21}u + p_{22}v \end{bmatrix}. \tag{9.46}$$

Now using equations (9.44), (9.45), and (9.46), one has

$$ax^2 + bxy + cy^2 + dx + ey + f$$
$$= \begin{bmatrix} x & y \end{bmatrix} \begin{bmatrix} a & \frac{b}{2} \\ \frac{b}{2} & c \end{bmatrix} \begin{bmatrix} x \\ y \end{bmatrix} + dx + ey + f$$
$$= \lambda_1 u^2 + \lambda_2 v^2 + d(p_{11}u + p_{12}v) + e(p_{21}u + p_{22}v) + f$$
$$= \lambda_1 u^2 + (dp_{11} + ep_{21})u + \lambda_2 v^2 + (dp_{12} + ep_{22})v + f. \tag{9.47}$$

From (9.47), we have that equation (9.43) has been transformed to the equivalent equation

$$\lambda_1 u^2 + (dp_{11} + ep_{21})u + \lambda_2 v^2 + (dp_{12} + ep_{22})v + f = 0, \tag{9.48}$$

which can be used to identify the conic represented by (9.43).

Exercise 9.3.4. Prove that equation (9.48) represents:
1. a parabola if exactly one of the eigenvalues is zero;
2. a circle if $\lambda_1 = \lambda_2$;
3. an ellipse if $\lambda_1\lambda_2 > 0$;
4. a hyperbola if $\lambda_1\lambda_2 < 0$.

To determine each of the cases in Exercise 9.3.4, we need to compute λ_1 and λ_2, which are the roots of the characteristic polynomial of A.

To this end we have,

$$f_A(x) = \begin{vmatrix} a - x & \frac{b}{2} \\ \frac{b}{2} & c - x \end{vmatrix} = x^2 - (a + c)x + ac - \frac{b^2}{4},$$

whose roots are given by

$$\lambda_1 = \frac{a + c + \sqrt{(a + c)^2 - 4ac + b^2}}{2} \quad \text{and} \quad \lambda_2 = \frac{a + c - \sqrt{(a - c)^2 + b^2}}{2}. \tag{9.49}$$

Notice that if the eigenvalues of A are zero, then from equation (9.45) one has that $A = 0$, and we do not have a quadratic equation at all. Therefore, we may assume that $A \neq 0$.

From this assumption, we have that exactly one eigenvalue is zero if and only if one of the equations $a + c + \sqrt{(a - c)^2 + b^2} = 0$ and $a + c - \sqrt{(a - c)^2 + b^2} = 0$ is true, and the latter occurs if and only if $4ac - b^2 = 0$. That is, equation (9.45) represents a parabola if $4ac - b^2 = 0$.

Equation (9.45) represents a circle if and only if $a + c - \sqrt{(a - c)^2 + b^2} = a + c + \sqrt{(a - c)^2 + b^2}$, and this occurs if and only if $(a - c)^2 + b^2 = 0$, the latter occurs if and only if $a = c$ and $b = 0$.

Equation (9.45) represents an ellipse if and only if $\lambda_1\lambda_2 > 0$, that is, if and only if $(a + c - \sqrt{(a - c)^2 + b^2})(a + c + \sqrt{(a - c)^2 + b^2}) > 0$. Performing calculations, the latter inequality holds if and only if $(a + c)^2 - (a - c)^2 - b^2 > 0$, or, in equivalent form, if and only if $4ac - b^2 > 0$.

Equation (9.45) represents a hyperbola if and only if $\lambda_1\lambda_2 < 0$, equivalently, if and only if $(a + c - \sqrt{(a - c)^2 + b^2})(a + c + \sqrt{(a - c)^2 + b^2}) < 0$. Again, performing calculations, the latter inequality is true if and only if $4ac - b^2 < 0$.

The number $\Delta = -(b^2 - 4ac) = 4 \begin{vmatrix} a & \frac{b}{2} \\ \frac{b}{2} & c \end{vmatrix}$ plays an important role in the classification of conics.

The above results are stated as:

Theorem 9.3.11. *The equation*

$$ax^2 + bxy + cy^2 + dx + ey + f = 0, \tag{9.50}$$

represents:
1. *an ellipse if $b^2 - 4ac < 0$; since a circle is a special case of an ellipse, the equation represents a circle when $a = c \neq 0$, $b = 0$, and $d^2 + e^2 - 4af > 0$;*
2. *a parabola if $b^2 - 4ac = 0$;*
3. *a hyperbola if $b^2 - 4ac > 0$; a special case is when $a + c = 0$, in this case the hyperbola is rectangular, that is, its asymptotes are orthogonal.*

The axes with respect to which one has the new coordinates (u, v) are determined by the eigenvectors of $A = \begin{bmatrix} a & \frac{b}{2} \\ \frac{b}{2} & c \end{bmatrix}$.

Example 9.3.6. Decide which type of conic is represented by the equation

$$x^2 + xy + y^2 + x + y = 10 \tag{9.51}$$

and sketch its graph.

Discussion. Using the number $b^2 - 4ac = -3$, we recognize immediately that the equation represents an ellipse. To sketch its graph, we need to find the new axes with respect to which the cross term disappears. Recall that these axes are parallel to the eigenvectors of the matrix associated to the quadratic part of (9.51), which is $A = \begin{bmatrix} 1 & \frac{1}{2} \\ \frac{1}{2} & 1 \end{bmatrix}$. To find the new axes, we might need a translation as well.

Computing the minimal polynomial of A, we have that $m_A(x) = (1 - x)^2 - \frac{1}{4}$. From this, its eigenvalues are $\frac{1}{2}$ and $\frac{3}{2}$. To compute the eigenvectors, we solve the systems of linear equations, $\left(A - \frac{1}{2}I_2\right)X = 0$ and $(A - \frac{3}{2}I_2)X = 0$. Once we have solved them, the obtained vectors need to be normalized. After some calculations, the normal eigenvectors associated to the eigenvalues are as follows: $\frac{1}{\sqrt{2}}(-1, 1)$ is associated to $\frac{1}{2}$, while $\frac{1}{\sqrt{2}}(1, 1)$ is associated to $\frac{3}{2}$. From this we have that the matrix P that diagonalizes A is $P = \frac{1}{\sqrt{2}}\begin{bmatrix} 1 & 1 \\ -1 & 1 \end{bmatrix}$. On the other hand, from equation (5.27), p. 135, we also know that the new coordinates u, v and the x, y coordinates are related by $\begin{bmatrix} x \\ y \end{bmatrix} = \frac{1}{\sqrt{2}}\begin{bmatrix} 1 & 1 \\ -1 & 1 \end{bmatrix}\begin{bmatrix} u \\ v \end{bmatrix}$, equivalently, $x = \frac{1}{\sqrt{2}}(u + v)$ and $y = \frac{1}{\sqrt{2}}(u - v)$.

Directly substituting these equations into (9.51) and simplifying leads to

$$u^2 + 3v^2 + 2\sqrt{2}v = 20, \tag{9.52}$$

from where we will arrive at an equation of the form

$$\frac{(u - u_0)^2}{a^2} + \frac{(v - v_0)^2}{b^2} = 1, \tag{9.53}$$

which is very helpful to sketch the graph of the ellipse. We encourage the reader to compare its graph with that in Figure 9.9.

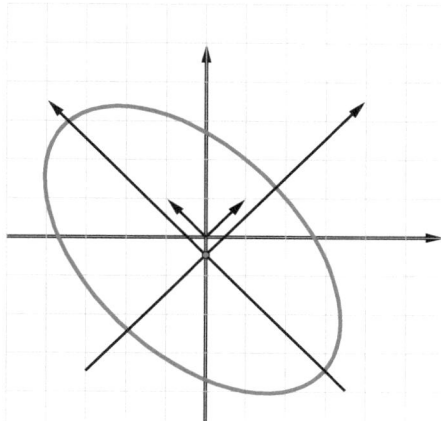

Figure 9.9: Graph of the equation $x^2 + xy + y^2 + x + y = 10$.

Exercise 9.3.5. With respect to the new axes, represent equation (9.52) in the form given by (9.53). Find the coordinates of the foci and the lengths of ellipse axes.

9.4 Adjoint and normal operators

In this part of the discussion we present a couple of fundamental concepts from operator theory. We start by proving that, given an operator T on a finite-dimensional euclidean vector space, there is another operator called the *adjoint of* T whose properties are given in the following theorem.

Theorem 9.4.1. *Let V be a finite-dimensional inner product vector space and T an operator on V. Then there is a unique operator T^* which satisfies*

$$\langle T(\alpha), \beta \rangle = \langle \alpha, T^*(\beta) \rangle \quad \text{for every } \alpha, \beta \in V. \tag{9.54}$$

Proof. We will prove the existence of T^*, the uniqueness will follow from its definition. The proof for the real case is the one that we present here, the complex case is obtained by conjugation of coefficients in the definition of T^*. From the Gram–Schmidt process, Theorem 9.1.4, we may assume that $\{\alpha_1, \alpha_2, \ldots, \alpha_n\}$ is an orthonormal basis of V. Then, given $\beta \in V$, define

$$T^*(\beta) = \langle T(\alpha_1), \beta \rangle \alpha_1 + \langle T(\alpha_2), \beta \rangle \alpha_2 + \cdots + \langle T(\alpha_n), \beta \rangle \alpha_n. \tag{9.55}$$

Using properties of an inner product, one verifies directly that T^* is linear.

If $a, \beta \in V$ and $\alpha = a_1\alpha_1 + a_2\alpha_2 + \cdots + a_n\alpha_n$, then $\langle T(\alpha), \beta \rangle = a_1\langle T(\alpha_1), \beta \rangle + a_2\langle T(\alpha_2), \beta \rangle + \cdots + a_n\langle T(\alpha_n), \beta \rangle$. On the other hand, using the bilinearity property of the inner product, the definition of T^*, as well as the condition that the basis $\{\alpha_1, \alpha_2, \ldots, \alpha_n\}$ is orthonormal, we have

$$\begin{aligned}
\langle \alpha, T^*(\beta) \rangle &= \langle a_1\alpha_1 + a_2\alpha_2 + \cdots + a_n\alpha_n, T^*(\beta) \rangle \\
&= \langle a_1\alpha_1 + a_2\alpha_2 + \cdots + a_n\alpha_n, \langle T(\alpha_1), \beta \rangle\alpha_1 + \langle T(\alpha_2), \beta \rangle\alpha_2 \\
&\quad + \cdots + \langle T(\alpha_n), \beta \rangle\alpha_n \rangle \\
&= a_1\langle T(\alpha_1), \beta \rangle + a_2\langle T(\alpha_2), \beta \rangle + \cdots + a_n\langle T(\alpha_n), \beta \rangle \\
&= \langle T(\alpha), \beta \rangle.
\end{aligned}$$

The uniqueness follows directly from the condition that T^* satisfies, finishing the proof. □

For matrices, given a matrix A whose entries are complex numbers, the adjoint of A is declared as the transposed conjugate of A. In symbols,

Definition 9.4.1. Let A be a matrix with complex entries. The adjoint of A is defined by $A^* = \overline{A^t}$.

Theorem 9.4.2. *Let V be a finite-dimensional inner product vector space, and let T be an operator in V. If A is the matrix associated to T with respect to an orthonormal basis, then A^* is the matrix associated to T^* with respect to the same basis.*

Proof. Let $\{\alpha_1, \alpha_2, \ldots, \alpha_n\}$ be an orthonormal basis. If a_{ij} denotes the (i,j)th entry of A, then for each $i \in \{1, 2, \ldots, n\}$ and $j \in \{1, 2, \ldots, n\}$, one has

$$\begin{aligned}
\langle T(\alpha_j), \alpha_i \rangle &= \langle a_{1j}\alpha_1 + a_{2j}\alpha_2 + \cdots + a_{nj}\alpha_n, \alpha_i \rangle \\
&= a_{1j}\langle \alpha_1, \alpha_i \rangle + a_{2j}\langle \alpha_2, \alpha_i \rangle + \cdots + a_{nj}\langle \alpha_n, \alpha_i \rangle \\
&= a_{ij}.
\end{aligned}$$

On the other hand, denoting by B the matrix associated to T^* with respect to the same basis, we have

$$\begin{aligned}
\langle \alpha_j, T^*(\alpha_i) \rangle &= \langle \alpha_j, b_{1i}\alpha_1 + b_{2i}\alpha_2 + \cdots + b_{ni}\alpha_n \rangle \\
&= \overline{b_{1i}}\langle \alpha_j, \alpha_1 \rangle + \overline{b_{2i}}\langle \alpha_j, \alpha_2 \rangle + \cdots + \overline{b_{ni}}\langle \alpha_j, \alpha_n \rangle \\
&= \overline{b_{ji}}.
\end{aligned}$$

Using the condition that defines the adjoint operator and the previous calculations, one has $a_{ij} = \langle T(\alpha_j), \alpha_i \rangle = \langle \alpha_j, T^*(\alpha_i) \rangle = \overline{b_{ji}}$, proving what was claimed. □

Definition 9.4.2. Let V be a inner product vector space, and let T be an operator on V that admits an adjoint.

1. It is said that T is normal if $TT^* = T^*T$.

2. When $T = T^*$, T is called self-adjoint or Hermitian.
3. It is said that T is unitary if $TT^* = T^*T = I$.

Theorem 9.4.3. *Let V be an operator over an inner product finite-dimensional complex vector space, and let T be an operator on V. Then T is normal if and only if V has an orthonormal basis of eigenvectors of T.*

Proof. If T is normal, the proof follows the same lines as the proof of the principal axes' theorem, Theorem 9.3.5, p. 246, changing a symmetric matrix with a normal operator.

Conversely, if V has an orthonormal basis of eigenvectors of T, then T is represented by a diagonal matrix D. From Theorem 9.4.2, the associated matrix of T^* is $\overline{D^t}$. It is clear that D and $\overline{D^t}$ commute, then so do T and T^*, proving that T is normal. □

Definition 9.4.3. Let C be a complex matrix.
1. If $C^{-1} = C^*$, then C is called unitary.
2. If $CC^* = C^*C$, then C is called normal.
3. If $C = C^*$, then C is called self-adjoint or Hermitian.

Remark 9.4.1. A complex matrix C is unitary if and only if its columns form an orthonormal basis of $\mathbb{C}^n$.

We end the discussion by presenting the version of Theorem 9.4.3 for the matrix case. The proof is obtained by translating what must be translated, from an operator to a matrix.

Corollary 9.4.1. *If A is a complex matrix, then A is normal if and only if there is a unitary matrix C such that $C^*AC = \mathrm{diag}\{a_1, a_2, \ldots, a_n\}$, where $a_1, a_2, \ldots, a_n$ are the eigenvalues of A.*

9.5 Exercises

1. Let A be a positive definite matrix and k a positive integer. Prove that there is B such that $A = B^k$.
2. Let A be an $n \times n$ real matrix. Prove that A is positive definite if and only if the function $\langle A\cdot, \cdot \rangle : \mathbb{R}^n \times \mathbb{R}^n \to \mathbb{R}$ defined by $\langle A\cdot, \cdot \rangle(X, Y) := \langle AX, Y \rangle$ is an inner product.
3. Let $z_1, z_2, \ldots, z_n$ be complex numbers. Prove that

$$|z_1 + z_2 + \cdots + z_n|^2 \leq n(z_1\overline{z_1} + z_2\overline{z_2} + \cdots + z_n\overline{z_n})^2.$$

4. Let $X = (x_1, x_2, \ldots, x_n)$ and $Y = (y_1, y_2, \ldots, y_n)$ be elements of $\mathbb{C}^n$. Define $\langle X, Y \rangle_1 = \overline{x_1}y_1 + \overline{x_2}y_2 + \cdots + \overline{x_n}y_n$. Is $\langle \cdot, \cdot \rangle_1$ an inner product? See Remark 9.2.1 and Example 9.2.1.
5. Let C be an $n \times n$ complex matrix. Prove that there exist Hermitian matrices A and B such that $C = A + iB$.

6. Let $X_1, X_2, \ldots, X_m$ be elements of $\mathbb{R}^n$, linearly independent with integral coordinates. Prove that there are elements $Y_1, Y_2, \ldots, Y_m$ in $\mathbb{R}^n$ such that:
    (a) $\langle Y_i, Y_j \rangle = 0$, if $i \neq j$.
    (b) Each Y_i has integral entries and $\{Y_1, Y_2, \ldots, Y_m\}$ spans the same subspace as $\{X_1, X_2, \ldots, X_m\}$.

7. (Polarization identity) Let V be an inner product vector space. Prove that $\langle a, \beta \rangle = \frac{1}{4}(\|a + \beta\|^2 - \|a - \beta\|^2)$.

8. Let V be an inner product vector space, and let $a_1, a_2, \ldots, a_r$ be elements of V. Define the matrix A whose entries are given by $\langle a_i, a_j \rangle$, $i, j \in \{1, 2, \ldots, n\}$. Prove that $a_1, a_2, \ldots, a_r$ are linearly independent if and only if A is nonsingular.

9. Let A be a symmetric matrix with real entries, and let $k \geq 1$ be an integer such that $A^k = I$. Prove that $A^2 = I$.

10. Let V be an inner product vector space, and let T be an operator on V. It is said that T is orthogonal if $\langle T(a), T(\beta) \rangle = \langle a, \beta \rangle$ for every $a, \beta \in V$. Prove that
    (a) Operator T is orthogonal if and only if $\langle T(a), T(a) \rangle = \langle a, a \rangle$, for every $a \in V$.
    (b) Operator T is orthogonal if and only if T transforms an orthonormal basis into an orthonormal basis.
    (c) Let $\mathcal{B} = \{a_1, a_2, \ldots, a_n\}$ be an orthonormal basis of V, T an orthogonal operator on V, and A the associated matrix of T with respect to basis $\mathcal{B}$. Prove that the rows and columns of A are orthogonal in the appropriate space. What is the determinant of A?

11. Let V be an inner product vector space and let $u \in V \setminus \{0\}$. Show that the function $T_u : V \to V$ given by $T_u(v) = \frac{\langle u, v \rangle}{\|u\|^2} u$ is a linear transformation. If V has dimension $n > 1$, what is the dimension of the kernel of T? What is the geometric meaning of the image of T?

12. Let V be an inner product vector space and let $W \neq \{0\}$ be a subspace different from V. Assume that $\{a_1, a_2, \ldots, a_r\}$ is an orthonormal basis of W. Let $T_W : V \to W$ be given by $T_W(a) = \langle a, a_1 \rangle a_1 + \langle a, a_2 \rangle a_2 + \cdots + \langle a, a_r \rangle a_r$. Prove that T_W is a linear transformation whose kernel is $W^\perp$ and its image is W.

13. Let V be an inner product vector space, and let T be an operator on V.
    (a) Prove that $TT^* = T^*T$ if and only if $\|T(a)\| = \|T^*(a)\|$ for every $a \in V$.
    (b) Prove that $TT^* = I$ if and only if $\|T(a)\| = \|a\|$ for every $a \in V$.

14. Classify the orthogonal operators in $\mathbb{R}^2$.

15. Let V be an inner product vector space, and let $T : V \to V$ be a function that satisfies $\langle T(a), T(\beta) \rangle = \langle a, \beta \rangle$ for every $a, \beta \in V$. Prove that T is an operator, that is, prove that T is a linear transformation. Notice that if the condition $\langle T(a), T(\beta) \rangle = \langle a, \beta \rangle$ is changed to $\langle T(a), T(a) \rangle = \langle a, a \rangle$, T is not linear as the example shows: $T(a) = \sqrt{\langle a, a \rangle} \beta$, where $\beta \in V$ with norm 1.

16. Let V be the space of continuous functions on the interval $[0, 2\pi]$, with the inner product defined by $\langle f, g \rangle = \int_0^{2\pi} fg$. Prove that the subset of V,

$$\mathcal{B}_n = \{1, \sin(x), \sin(2x), \ldots, \sin(nx), \cos(x), \cos(2x), \ldots, \cos(nx)\},$$

is orthogonal. If W_n is the subspace spanned by $\mathcal{B}_n$ and $f \in V$, determine the orthogonal projection f along W_n. Take $n = 5$ and $f(x) = x^2$ to illustrate the process.

17. Let V be the vector space of continuous functions in $[a, b]$ and $S = \{f_1, f_2, \ldots, f_n\} \subseteq V$. For every $i, j \in \{1, 2, \ldots, n\}$, define $a_{ij} = \int_a^b f_i f_j$. Let A be the matrix that has entries a_{ij}. Prove that $f_1, f_2, \ldots, f_n$ are linearly independent if and only if A is nonsingular.

18. Classify the conics given by the following equations, sketch their graphs, and find the axes of $\mathbb{R}^2$ with respect to which there is no cross term xy:
    (a) $4x^2 - 24xy + 11y^2 + 56x - 58y + 95 = 0$,
    (b) $12x^2 + 8xy - y^2 + x + y - 2 = 0$,
    (c) $8x^2 + 4y^2 - xy + x - y + 3 = 0$,
    (d) $2x^2 + 4xy + 2y^2 + x + y - 8 = 0$,
    (e) $4x^2 - 4xy + 7y^2 = 24$.

19. Find necessary and sufficient conditions in order that equation (9.43) represents two lines, or equivalently, $ax^2 + bxy + cy^2 + dx + ey + f = (Ax + By + C)(A_1x + B_1y + C_1) = 0$.

20. For each of the following matrices, find a QR representation according to Theorem 9.1.8:
    (a) $A = \begin{bmatrix} 1 & 3 \\ -1 & 5 \end{bmatrix}$,
    (b) $B = \begin{bmatrix} 1 & 1 & 1 \\ 1 & 1 & 0 \\ 0 & 1 & 0 \end{bmatrix}$,
    (c) $C = \begin{bmatrix} 1 & 1 & 1 \\ 1 & 0 & -2 \\ 0 & 1 & 1 \end{bmatrix}$.

21. Let A be a real symmetric matrix. Prove that $\det(A) > 0$ if and only if A is nonsingular and the number of negative eigenvalues is even.

22. Justify that equations (9.42), p. 251, are correct.

23. Given the matrix $A = \begin{bmatrix} 1 & 0 & 1 \\ 0 & -1 & 1 \end{bmatrix}$, find the singular value decomposition of A.

24. Consider the matrices P, Q, and Σ as defined in Example 9.3.4 and show that

$$\begin{bmatrix} 1 & 3 \\ 2 & 1 \end{bmatrix} = Q\Sigma P^t.$$

25. Show that the matrix A in Example 9.3.4 transforms the circle $x^2 + y^2 = 4$ into the ellipse $x^2 + 2y^2 - 2xy = 20$.

26. Let A be a 2×2 real matrix. Use the singular value decomposition of A to find necessary and sufficient conditions in order that A maps circles into circles. What are the conditions if A is $n \times n$, with $n > 2$?

27. Let A be an $m \times n$ matrix. Prove that the eigenvalues of $A^t A$ are real and nonnegative. More precisely, if $\{v_1, v_2, \ldots, v_n\}$ is an orthonormal basis of $\mathbb{R}^n$ for $A^t A$, then $\lambda_i = \|Av_i\|^2$ is an eigenvalue of $A^t A$ for every $i \in \{1, 2, \ldots, n\}$.

28. Assume that A is a square invertible matrix with a singular value decomposition $A = Q\Sigma P^t$. What is a singular value decomposition for A^{-1}?

29. If $A = Q\Sigma P^t$ is a singular value decomposition for A. What is a singular value decomposition for A^t?

30. Is there a way to say that a singular value decomposition of an $m \times n$ matrix A is unique?

31. (Sylvester's criterion of positive definiteness, [14]) Let A be a real $n \times n$ symmetric matrix, and let A_k be the $k \times k$ submatrix of A defined by deleting the last $(n - k)$ columns and rows of A. That is, if the entries of A are a_{ij}, with $i, j \in \{1, 2, \ldots, n\}$, then the entries of A_k are a_{ij} where $i, j \in \{1, 2, \ldots, k\}$. Prove that A is positive definite, if and only if $\det(A_k) > 0$ for every $k \in \{1, 2, \ldots, n\}$.

    (*Hint.*) The map $\mathbb{R}^k \to \mathbb{R}^n$, given by $X = (x_1, x_2, \ldots, x_k) \to \overline{X} = (x_1, x_2, \ldots, x_k, 0, \ldots, 0)$ allows us to write $X^t A_k X = \overline{X}^t A \overline{X} > 0$, thus if A is positive definite, then so is A_k, concluding that $\det(A_k) > 0$. To prove the converse, we use induction on n and the following two facts:

    Fact 1. If $W \subseteq V$ is a subspace of dimension k, $v_1, v_2, \ldots, v_n$ is a basis of V, $m < k$, and W_1 is the span of $\{v_{m+1}, v_{m+2}, \ldots, v_n\}$ then $W \cap W_1 \neq \{0\}$.

    The proof of Fact 1 follows by noticing that $n \geq \dim(W + W_1) = \dim(W) + \dim(W_1) - \dim(W \cap W_1) = k + (n - m) - \dim(W \cap W_1)$.

    Fact 2. If A is a symmetric matrix and $X^t A X > 0$ for every $X \neq 0$ in a subspace W of dimension k, then A has at least k positive eigenvalues.

    Proof of Fact 2. Since A is symmetric, $\mathbb{R}^n$ has an orthonormal basis of eigenvectors of A, say $X_1, X_2, \ldots, X_n$ so that $AX_i = c_i X_i$. Now, if A has $m < k$ positive eigenvalues, from Fact 1, there is $X \in W \setminus \{0\}$ which is a linear combination $X = b_{m+1} X_{m+1} + \cdots + c_n X_n$. From this, one has $0 < X^t A X = c_{m+1} b_{m+1}^2 + \cdots + c_n b_n^2 \leq 0$, a contradiction.

    To finish the proof, notice that A_{n-1} is positive definite in the subspace of $\mathbb{R}^n$ consisting of the elements whose last entry is zero. Then $X^t A X > 0$ in a subspace of dimension $n - 1$, hence A has at least $n - 1$ positive eigenvalues. The proof is completed by observing that $\det(A) > 0$ is the product of the eigenvalues of A, hence all the eigenvalues are positive.

A Integers, complex numbers, and polynomials

In this appendix we present several mathematical results and terms that are used throughout the discussions in this book. Usually, these results and terms are not presented in a linear algebra course. That is why we consider them to be appropriate to be included in this appendix. The results included here are related to properties of the positive integers, permutations, the complex numbers, and polynomials in one variable. The eager reader who wants to explore these topics in a deeper and structured way can consult the abundant literature available on the websites or elsewhere.

A.1 Well-ordering and mathematical induction

Several mathematical statements have to do with properties of the positive integers. That might explain why there are several methods of proof based on specific properties of these numbers. In this section we present four properties that are very useful to carry out proofs that involve the positive integers. However, our attention will be centered on only two of them: the well-ordering principle and the principle of mathematical induction.

However, we think that it is important to point out that these four principles are logically equivalent. For a proof of their equivalence, the interested reader should consult [25].

Principle A.1.1 (Dirichlet's box principle or pigeonhole principle). Assume that m pigeons are to be distributed into n pigeonholes. If $m > n$, then at least one pigeonhole contains more than one pigeon.

Principle A.1.2 (Multiplication principle). If there exist n ways of performing one operation and m ways of performing another operation (independent of the first), then there exist mn ways of performing both operations, one followed by the other.

Principle A.1.3 (Well-ordering principle). If S is a nonempty subset of the positive integers, then there exists $n_0 \in S$ so that $n_0 \leq n$ for every $n \in S$.

We illustrate how the well-ordering principle can be used to prove the following result.

Fact. Every positive integer greater than 1 can be expressed as the product of prime numbers.

Proof. Let S denote the set of integers greater than 1 which cannot be expressed as the product of primes. We will prove that S is empty.

We will argue by contradiction. Suppose that S is not empty, then, by Principle A.1.3, there is $n_0 \in S$ such that $n_0 \leq n$ for every $n \in S$. From the choice of n_0, it is not prime, hence $n_0 = mq$ with $1 < m < n_0$, and $1 < q < n_0$. By the choice of n_0, we have that m and

https://doi.org/10.1515/9783111135915-010

q are not in S. Therefore each of m and q is a product of primes, thus $n_0 = mq$ is also a product a primes. This contradicts the choice of n_0, therefore S is empty, as claimed. □

Principle A.1.4 (Mathematical induction principle). If S is a subset of the positive integers that satisfies:
1. $1 \in S$, and
2. whenever $k \in S$, one has $k + 1 \in S$,

then S is the set of all positive integers.

There are plenty of examples where the mathematical induction principle is used to prove a proposition that involves the positive integers. For instance, assume that r is a real number different from 1, then the nth sum of the geometric series, $S_n = 1 + r + r^2 + \cdots + r^n$, is given by

$$S_n = \frac{1 - r^{n+1}}{1 - r}. \tag{A.1}$$

Proof. We prove equation (A.1) by mathematical induction.

Base case. We need to prove that (A.1) holds for $n = 1$. For this case, $S_1 = 1 + r$. On the other hand,

$$\frac{1 - r^{n+1}}{1 - r} = \frac{1 - r^{1+1}}{1 - r} = \frac{1 - r^2}{1 - r} = \frac{(1 - r)(1 + r)}{1 - r} = 1 + r = S_1,$$

that is, our claim in the case $n = 1$ is true.

Induction hypothesis. Assume that $n > 1$ and that (A.1) is true for this value of n. We need to prove, based on the validity of the nth case, that $S_{n+1} = \frac{1 - r^{n+2}}{1 - r}$.
  We notice that $S_{n+1} = S_n + r^{n+1}$. From this we can proceed as follows:

$$\begin{aligned} S_{n+1} &= S_n + r^{n+1} \\ &= \frac{1 - r^{n+1}}{1 - r} + r^{n+1} \quad \text{(induction hypothesis)} \\ &= \frac{1 - r^{n+1} + r^{n+1}(1 - r)}{1 - r} \\ &= \frac{1 - r^{n+2}}{1 - r}, \end{aligned}$$

finishing the proof. □

There are other ways to formulate the mathematical induction principle, all of them equivalent, for instance, the strong mathematical induction principle stated below. A good exercise for the intrepid reader would be to prove the equivalence of those and other statements concerning the mathematical induction principle.

Principle A.1.5 (Strong mathematical induction principle). If S is a subset of the positive integers that satisfies:

1. $1 \in S$, and
2. whenever $\{1, 2, \ldots, k\} \subseteq S$, one has $k + 1 \in S$,

then S is the set of positive integers.

A.2 Permutations

Permutations play an important role in several branches of mathematics, and linear algebra is not an exception. In this section we present basic results concerning permutations.

Definition A.2.1. A permutation on the set $\{1, 2, \ldots, n\}$ is a bijective function $\sigma : \{1, 2, \ldots, n\} \to \{1, 2, \ldots, n\}$. The set of all permutations on $\{1, 2, \ldots, n\}$ is denoted by S_n.

One way to give meaning to the definition above is to think of a permutation as an action that interchanges the positions among a set of n objects. Thinking in this way of a permutation, it should be clear that there is a way to revert this action and restore the objects to their original position. This restoration is again a permutation and it called the *inverse* of the original permutation. If σ denotes a permutation, we use σ^{-1} to denote its inverse.

There are special permutations that play an important role when describing their properties. These permutations are associated to subsets of $\{1, 2, \ldots, n\}$ containing two or more elements. Let us formulate this in a more precise way.

Definition A.2.2. Let $k \geq 2$ be an integer and $\tau \in S_n$. We say that τ is a k-cycle if there is a subset $\{i_1, i_2, \ldots, i_k\} \subseteq \{1, 2, \ldots, n\}$ so that $\tau(i_j) = i_{j+1}$, for every $j \in \{1, 2, \ldots, k - 1\}$, $\tau(i_k) = i_1$, and $\tau(x) = x$ for every $x \notin \{i_1, i_2, \ldots, i_k\}$. If $k = 2$, τ is called a transposition.

It should be noted that the inverse of a transposition is the transposition itself. If τ is a cycle determined by $\{i_1, i_2, \ldots, i_k\}$, we denote it as $\tau = (i_1 i_2 \cdots i_k)$ and the meaning is $\tau(i_j) = i_{j+1}$ for $j \in \{1, 2, \ldots, k - 1\}$ and $\tau(i_k) = i_1$.

In general, the composition of two permutations does not commute, that is, $\sigma \circ \tau \neq \tau \circ \sigma$. However, there is a special case under which $\sigma \circ \tau = \tau \circ \sigma$.

Definition A.2.3. The permutations σ and τ are said to be disjoint, if the following two conditions hold:

1. If $\sigma(x) \neq x$, then $\tau(x) = x$, and
2. if $\tau(x) \neq x$, then $\sigma(x) = x$.

Lemma A.2.1. *If σ and τ are disjoint permutations, then $\sigma \circ \tau = \tau \circ \sigma$.*

Proof. We need to prove that $\sigma \circ \tau(x) = \tau \circ \sigma(x)$ for every x. We proceed to show this by proving two statements. First, we will prove that $\sigma \circ \tau(x) = \tau \circ \sigma(x)$ for every x such that $\tau(x) \neq x$. The second part of the proof will show that $\sigma \circ \tau(x) = \tau \circ \sigma(x)$ for every x such that $\tau(x) = x$

If $\tau(x) \neq x$, then $\sigma(x) = x$, thus $\tau \circ \sigma(x) = \tau(x)$. We will prove that $\sigma \circ \tau(x) = \tau(x)$. We argue by contradiction and assume that $\sigma \circ \tau(x) \neq \tau(x)$; this implies that σ moves $\tau(x)$. From Definition A.2.3(1), we have that $\tau \circ \tau(x) = \tau(x)$. Applying τ^{-1} to this equation, we obtain $\tau(x) = x$, contradicting that $\tau(x) \neq x$. From above, we have that $\tau \circ \sigma(x) = \sigma \circ \tau(x)$, for the case $\tau(x) \neq x$.

For the second part, assume that $\tau(x) = x$, thus $\sigma(x) \neq x$. Now switching the roles of τ and σ in the previous argument, we conclude that $\tau \circ \sigma(x) = \sigma \circ \tau(x)$. This finishes the proof of the lemma. $\square$

The next result is analogous to the fundamental theorem of arithmetic applied to permutations, where cycles play the role of prime numbers.

Theorem A.2.1. *Every permutation different from the identity can be represented as a product of disjoint cycles. Moreover, this cycle representation of a permutation is unique.*

Proof. Let σ be a permutation and define $X_\sigma = \{x \in \{1, 2, \ldots, n\} \mid \sigma(x) \neq x\}$. We will apply mathematical induction on $|X_\sigma| \geq 2$. If $|X_\sigma| = 2$, then σ is a transposition and we have finished. Hence we may assume that $|X_\sigma| > 2$ and the result to be true for every τ such that $|X_\tau| < |X_\sigma|$. Assume that $y \in X_\sigma$ and set $Y = \{\sigma^k(y) \mid k \geq 1\}$. If $Y = X_\sigma$, then σ is a cycle and we have finished. Hence we may assume that $X_1 = X_\sigma \setminus Y$ is not empty. Define σ_1 and σ_2 as follows:

$$\sigma_1(x) = \begin{cases} \sigma(x) & \text{if } x \in Y, \\ x & \text{otherwise,} \end{cases} \qquad \sigma_2(x) = \begin{cases} \sigma(x) & \text{if } x \in X_1, \\ x & \text{otherwise.} \end{cases} \tag{A.2}$$

From (A.2) one sees directly that σ_1 and σ_2 are disjoint and $\sigma = \sigma_1 \sigma_2$. Also, $|X_{\sigma_2}| < |X_\sigma|$, hence, by the induction assumption, σ_2 is a product of disjoint cycles, hence so is σ.

The proof of uniqueness of the representation is left to the reader as an exercise. $\square$

The next result establishes that a permutation can be represented as a product of transpositions. The representation is not unique, since any transposition and its inverse can be added to a given representation without changing it. What is true is that the number of transpositions in the representation of a permutation is always even or always odd.

Theorem A.2.2. *Every permutation different from the identity can be represented as a product of transpositions.*

Proof. From Theorem A.2.1, it is enough to prove that any cycle is the product of transpositions. Without loss of generality, we may assume that a k-cycle is $\tau = (1\,2\,3\,\ldots\,k)$. A direct computation shows that $\tau = (1\,k)(1\,k-2)\cdots(1\,3)(1\,2)$. $\square$

Remark A.2.1. Note that the representation from Theorem A.2.2 is not unique, since a transposition τ can be written as, say $\tau = \tau \circ \tau \circ \tau$.

The following theorem is presented without proof. The adventurous reader can consult [5, Theorem 2.1.7] for a proof.

Theorem A.2.3. *Let m and k be the numbers of cycles in any two representations of a permutation as a product of transpositions. Then m and k have the same parity, that is, either m and k are both even or m and k are both odd.*

Definition A.2.4. Let σ be a permutation, and let k be the number of transpositions in a representation of σ as a product of transpositions. The sign of σ is defined as $\operatorname{sign}(\sigma) = (-1)^k$.

Corollary A.2.1. *If σ and τ are permutations, then $\operatorname{sign}(\sigma \circ \tau) = \operatorname{sign}(\sigma)\operatorname{sign}(\tau)$.*

Proof. If σ is written as the product of m transpositions and τ is written as the product of k transpositions, then $\sigma\tau$ is written as the product of $m + k$ transpositions. Thus $\operatorname{sign}(\sigma \circ \tau) = (-1)^{m+k} = (-1)^m(-1)^k = \operatorname{sign}(\sigma)\operatorname{sign}(\tau)$. $\qquad\square$

A.3 Complex numbers

An introduction to the study of complex numbers may start at high school, when solving quadratic equations as $x^2 + 1 = 0$. Maybe the way the complex numbers were presented was by defining the symbol $i = \sqrt{-1}$ and establishing that $i^2 = -1$. From this starting point, a complex number is an expression of the form $a + bi$, with a and b real numbers. Arithmetically speaking, there is nothing wrong with this; however, a more formal presentation of this important number system is needed to uncover the mysteries of the number i. In this section we present a formal description, basic results, and terminology concerning the complex numbers.

We start by declaring that the set of complex numbers is denoted by $\mathbb{C}$ and defined by

$$\mathbb{C} = \{(a, b) \mid a, b \in \mathbb{R}\}. \tag{A.3}$$

It should be noted that $\mathbb{C}$, as a set, is the usual cartesian plane. What will be added to the plane is a couple of operations, with which it will resemble properties analogous to those of the real numbers.

Sum and product
Given $(a, b), (c, d) \in \mathbb{C}$:
–   The sum of (a, b) and (c, d) is defined by

$$(a, b) + (c, d) = (a + c, b + d). \tag{A.4}$$

–   The product of (a, b) and (c, d) is by definition

$$(a, b)(c, d) = (ac - bd, ad + bc). \tag{A.5}$$

The product definition is not intuitive at all, however, if we think of (a, b) as "$a + bi$", with $i = \sqrt{-1}$, and consider that the rules for the sum and product should hold as for the real numbers, then definition (A.5) seems "reasonable".

It can be verified that with definitions (A.4) and (A.5), the complex numbers satisfy the same arithmetic properties as the real numbers.

According to (A.5), we have that $(a, b)(1, 0) = (a(1) - b(0), a(0) + b(1)) = (a, b)$. Thus $(1, 0)$ is the identity for the multiplication of complex numbers. It can be verified directly that $(0, 0)$ is the identity for the addition in $\mathbb{C}$. Also, the subset $\{(a, 0) \mid a \in \mathbb{R}\}$ of $\mathbb{C}$ can be identified with $\mathbb{R}$, by the map $(a, 0) \mapsto a$, hence, under this identification, we can think of the real numbers as being a subset of the complex numbers. Furthermore, setting $i = (0, 1)$ the identification can be extended to $(a, b) \mapsto a + bi$, which is the usual way to represent a complex number. We point out that $(1, 0) \mapsto 1$ and $(0, 0) \mapsto 0$.

It is readily noticed that $(0, 1)(0, 1) = (-1, 0)$, hence $(0, 1)^2 = (-1, 0)$ corresponds to the equation $i^2 = -1$.

If $a + bi$ is not zero, then its multiplicative inverse is $(a + bi)^{-1} = \frac{a}{a^2 + b^2} - \frac{b}{a^2 + b^2} i$, as can be verified.

Definition A.3.1. If $z = a + bi \in \mathbb{C}$, then we define:
1.   The norm or absolute value of z, denoted by $|z|$, as $|z| = \sqrt{a^2 + b^2}$.
2.   The conjugate of z, denoted by $\bar{z}$, as $\bar{z} = a - bi$.
3.   The real part of z, namely $\text{Re}(z) = a$, and the imaginary part of z, that is, $\text{Im}(z) = b$.

In the following theorem, we present some of the basic properties of the complex numbers.

Theorem A.3.1. *If z and w are complex numbers, then the following conditions hold:*
1. $|zw| = |z|\,|w|$,
2. $|z + w| \leq |z| + |w|$ *(triangle inequality),*
3. $\overline{z + w} = \bar{z} + \bar{w}$,
4. $\overline{zw} = \bar{z}\,\bar{w}$,
5.   *if $z \neq 0$, then $\frac{1}{z} = \frac{\bar{z}}{z\bar{z}}$.*

Proof. We prove properties 1 and 2, the others are left to the reader as an exercise.

Let us begin by proving property 1. Write $z = a + bi$ and $w = c + di$, then $zw = (ac - bd) + (ad + bc)i$. Now, from the definition of the absolute value of a complex number, proving $|zw| = |z|\,|w|$ reduces to proving

$$\sqrt{(ac - bd)^2 + (ad + bc)^2} = \sqrt{a^2 + b^2}\sqrt{c^2 + d^2}. \tag{A.6}$$

By squaring both sides in (A.6), we need to prove that $(ac - bd)^2 + (ad + bc)^2 = (a^2 + b^2)(c^2 + d^2)$, which is attained by expanding the products.

To prove property 2, we notice that $|z + w| \leq |z| + |w|$ is equivalent to proving that

$$\sqrt{(a + c)^2 + (b + d)^2} \leq \sqrt{a^2 + b^2} + \sqrt{c^2 + d^2}. \qquad \text{(A.7)}$$

Before continuing, we want to point out that the function $f(x) = x^2$ is bijective on the nonnegative real numbers, hence all the following inequalities are equivalent.

Squaring equation (A.7), it remains to prove

$$(a + c)^2 + (b + d)^2 \leq \left(\sqrt{a^2 + b^2} + \sqrt{c^2 + d^2}\right)^2. \qquad \text{(A.8)}$$

Expanding the products and simplifying, one is left to show that

$$2(ac + bd) \leq 2\left(\sqrt{a^2 + b^2}\sqrt{c^2 + d^2}\right) \qquad \text{(A.9)}$$

is true.

Again, squaring (A.9) and simplifying leads to showing

$$2acbd \leq a^2 d^2 + b^2 c^2, \qquad \text{(A.10)}$$

which is true, since it is equivalent to $0 \leq (ad - bc)^2$, finishing the proof of property 2. $\quad\square$

Polar or trigonometric representation of a complex number

According to (A.3), the complex numbers are pairs of real numbers, hence a geometric representation of the complex numbers can be done by means of a coordinate plane, sometimes called the complex plane. From this, and taking into account the representation of a complex number in the form $z = a + ib$, we can think of z as the point (a, b) in a coordinate plane, see Figure A.1.

Using this representation of the complex numbers, the horizontal axis is called the real axis, while the vertical axis is called the imaginary axis. If $z = a + ib \neq 0$, then $|z| \neq 0$ and from this we have

$$a = |z| \cos(\theta),$$
$$b = |z| \sin(\theta). \qquad \text{(A.11)}$$

From the equations in (A.11), we have

$$z = |z|(\cos(\theta) + i\sin(\theta)), \qquad \text{(A.12)}$$

which is called the polar or trigonometric representation of z. The representation of z in polar form has an important consequence, the so-called De Moivre's formula or De Moivre's theorem.

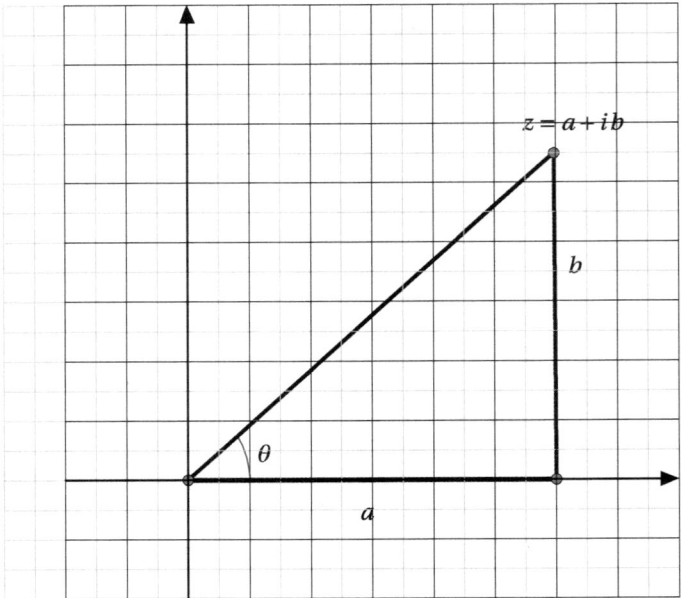

Figure A.1: Representation of the complex number $z = a + ib$ as the point (a, b).

Theorem A.3.2 (Moivre's formula). *If n is an integer number and $z = |z|(\cos(\theta) + i\sin(\theta))$, then*

$$z^n = |z|^n(\cos(n\theta) + i\sin(n\theta)). \tag{A.13}$$

Proof. We use the principle of mathematical induction to prove the theorem. Let $z = |z|(\cos(\theta) + i\sin(\theta))$ and $n = 1$. For this case the conclusion follows directly, hence we may assume that $n > 1$ and the result to be true for $n - 1$ (induction hypothesis), that is, we may assume that

$$z^{n-1} = |z|^{n-1}(\cos((n-1)\theta) + i\sin((n-1)\theta)). \tag{A.14}$$

In the following part of the proof, we will use the addition formulas for sine and cosine which are:

$$\sin(\alpha + \theta) = \sin(\theta)\cos(\alpha) + \sin(\alpha)\cos(\theta), \quad \cos(\alpha + \theta) = \cos(\alpha)\cos(\theta) - \sin(\alpha)\sin(\theta).$$
$$\tag{A.15}$$

For $n > 1$, we have

$$z^n = zz^{n-1}$$
$$= |z|(\cos(\theta) + i\sin(\theta))|z|^{n-1}(\cos((n-1)\theta) + i\sin((n-1)\theta)$$
$$= |z|^n[(\cos(\theta)\cos((n-1)\theta) - \sin(\theta)\sin((n-1)\theta))$$
$$+ i(\cos(\theta)\sin((n-1)\theta) + \sin(\theta)\cos((n-1)\theta))]$$
$$= |z|^n(\cos(n\theta) + i\sin(n\theta)), \tag{A.16}$$

proving the theorem for positive integers. If $n < 0$, then $-n = k > 0$, and from this we have

$$z^{-n} = z^k$$
$$= |z|^k(\cos(k\theta) + i\sin(k\theta))$$
$$= |z|^{-n}(\cos(-n\theta) + i\sin(-n\theta))$$
$$= |z|^{-n}(\cos(n\theta) - i\sin(n\theta)). \tag{A.17}$$

Now, from equation (A.17), we obtain $z^n = |z|^n(\cos(n\theta) - i\sin(n\theta))$ and, multiplying the latter equation by $(\cos(n\theta) + i\sin(n\theta))$, the result follows, since

$$(\cos(n\theta) - i\sin(n\theta))(\cos(n\theta) + i\sin(n\theta)) = 1. \qquad \square$$

A.4 Polynomials in one variable

Polynomials or polynomial functions play a crucial role in the study of several mathematical problems. For instance, when solving a homogeneous second-order differential equation, the solutions are obtained by finding the roots of a second-degree polynomial. There are many more important examples where polynomials play a crucial role, its discussion will lead us to another direction, not that intended to provide basic results concerning polynomials.

Here we present basic results and terminology to be used in the discussion of the theory of eigenvalues; see Chapter 7. Hence, we do not present the formal definition of a polynomial, and the polynomials that we consider will have coefficients that are either real or complex numbers.

A polynomial $p(x)$, with real or complex coefficients, in the variable x is an expression of the form

$$p(x) = a_0 + a_1x + \cdots + a_nx^n, \tag{A.18}$$

where $a_0, a_1, \ldots, a_n$ are fixed real or complex numbers, called the coefficients of $p(x)$. If $a_n \neq 0$, n is called the degree of $p(x)$, denoted by $\deg(p(x)) = n$, and a_n is called the leading coefficient. If $a_n = 1$, the polynomial is called *monic*. A few more terms are needed. Two polynomials $p(x) = a_0 + a_1 x + \cdots + a_n x^n$ and $q(x) = b_0 + b_1 x + \cdots + b_m x^m$ are equal, if $m = n$ and $a_j = b_j$ for every $j \in \{1, 2, \ldots, n\}$. The zero polynomial has all its coefficients equal to zero. The degree of the zero polynomial is usually not defined. However, in case we need to define the degree of the zero polynomial, it will be taken as $-\infty$.

Important concepts, when dealing with polynomials, are the divisibility of polynomials and irreducible polynomials. The next definition points toward that direction.

Definition A.4.1. Let $f(x)$ and $p(x)$ be polynomials.
1. We say that $p(x)$ divides $f(x)$, or that $p(x)$ is a divisor of $f(x)$, if there is another polynomial $q(x)$ such that $f(x) = p(x)q(x)$.
2. A polynomial $p(x)$ of positive degree is called irreducible, if it has no divisors of positive degree.

One useful consequence of Definition A.4.1 concerning the degree of a polynomials is

Remark A.4.1. If $f(x)$ and $g(x)$ are polynomials, then
1. $\deg(f(x) + g(x)) \leq \max\{\deg(f(x)), \deg(g(x))\}$,
2. $\deg(f(x)g(x)) \leq \deg(f(x)) + \deg(g(x))$, with equality if both are nonzero.

One of the most important results in arithmetic (number theory) is the division algorithm, which also holds for polynomials.

Theorem A.4.1 (Division algorithm). *Let $f(x)$ and $g(x)$ be polynomials with coefficients in the real or complex numbers. Furthermore, assume that $g(x) \neq 0$. Then there exist unique polynomials $q(x)$ and $r(x)$ such that*

$$f(x) = q(x)g(x) + r(x), \quad \text{with } \deg(r(x)) < \deg(g(x)) \text{ or } r(x) = 0. \quad \text{(A.19)}$$

Proof. The existence of the polynomials $q(x)$ and $r(x)$ will be carried out by mathematical induction on the degree of $f(x)$. It is clear that when $f(x)$ is constant, the polynomials $q(x)$ and $r(x)$ are obtained readily, hence we assume that $f(x)$ has degree at least one.

Let n be the degree of $f(x)$. If $n = 1$ and $1 < \deg(g(x))$, then take $q(x) = 0$ and $r(x) = f(x)$, which clearly satisfy $f(x) = 0g(x) + f(x)$ and the required condition on $\deg(r(x)) < \deg(g(x))$. If $\deg(g(x)) \leq 1$, then $g(x) = cx + d$ with at least one of c or d not zero, since $g(x) \neq 0$. The assumption on $\deg(f(x)) = 1$ implies $f(x) = ax + b$. If $c = 0$, then $g(x) = d \neq 0$ is a constant which clearly divides $f(x)$ with quotient $q(x) = \frac{1}{d}(ax + b)$ and remainder $r(x) = 0$.

If $c \neq 0$, then

$$ax + b = \frac{a}{c}(cx + d) + b - \frac{ad}{c}, \quad \text{(A.20)}$$

and we take $q(x) = \frac{a}{c}$ and $r(x) = b - \frac{ad}{c}$.

Now, assume that $n > 1$ and the result to be true for every polynomial of degree less than n. As before, if $n < \deg(g(x))$, then take $q(x) = 0$ and $r(x) = f(x)$. Thus, we may assume that $m = \deg(g(x)) \le \deg(f(x)) = n$. Write $f(x) = a_n x^n + a_{n-1} x^{n-1} + \cdots + a_1 x + a_0$, $g(x) = b_m x^m + b_{m-1} x^{m-1} + \cdots + b_1 x + b_0$ and set

$$f_1(x) = f(x) - \frac{a_n}{b_m} x^{n-m} g(x) = \left(a_{n-1} - \frac{a_n}{b_m} b_{m-1} \right) x^{n-1} + h(x), \tag{A.21}$$

where

$$h(x) = \left(a_{n-2} - \frac{a_n}{b_m} b_{m-2} \right) x^{n-2} + \left(a_{n-3} - \frac{a_n}{b_m} b_{m-3} \right) x^{n-3} + \left(a_{n-4} - \frac{a_n}{b_m} b_{m-4} \right) x^{n-4} + \cdots. \tag{A.22}$$

From equation (A.22), we have that $h(x)$ has degree at most $n - 2$, hence $\deg(f_1(x)) \le n - 1 < n = \deg(f(x))$. Applying the induction hypothesis to $f_1(x)$, we have that there exist polynomials $q_1(x)$ and $r_1(x)$ such that $f_1(x) = q_1(x)g(x) + r_1(x)$, with $r_1(x) = 0$ or $\deg(r_1(x)) < \deg(g(x))$.

Substituting this equation into (A.21) and regrouping, we obtain

$$f(x) = \left(q_1(x) + \frac{a_n}{b_m} x^{n-m} \right) g(x) + r_1(x). \tag{A.23}$$

Take $q(x) = q_1(x) + \frac{a_n}{b_m} x^{n-m}$ and $r(x) = r_1(x)$. With this we have shown the existence of $q(x)$ and $r(x)$.

To prove the uniqueness of $q(x)$ and $r(x)$, let us assume that there are $q(x)$, $q_1(x)$, $r(x)$, and $r_1(x)$, satisfying (A.19) and

$$f(x) = q(x)g(x) + r(x) = q_1(x)g(x) + r_1(x). \tag{A.24}$$

Equation (A.24) implies

$$(q(x) - q_1(x))g(x) = r_1(x) - r(x). \tag{A.25}$$

Also, from (A.19) we have that $r(x)$ and $r_1(x)$ have degree $< \deg(g(x))$. If $r(x) - r_1(x) \neq 0$, applying Remark A.4.1 to (A.25) will lead to a contradiction, hence $r(x) = r_1(x)$. From this, equation (A.25) reduces to $(q(x) - q_1(x))g(x) = 0$. By assumption, $g(x) \neq 0$, hence $q(x) - q_1(x) = 0$, or $q(x) = q_1(x)$, finishing the proof of the theorem. □

One more term that is needed is the greatest common divisor of two polynomials. This concept is important to decide if a polynomial has multiple roots. The way it is used is by considering the greatest common divisor of the polynomial and its derivative, see Corollary A.5.2. If the greatest common divisor has a linear factor, then the polynomial has multiple roots. For example, the polynomial $f(x) = x^3 - 1$ and its derivative, which is $f'(x) = 3x^2$, have no common factors. This tells us that $f(x)$ has no repeated roots. Intuitively speaking, the greatest common divisor of two polynomials $f(x)$ and $g(x)$ is the polynomial of highest degree that divides both $f(x)$ and $g(x)$. We formalize this as follows.

Definition A.4.2. Given polynomials $f(x)$ and $g(x)$, we say that the polynomial $d(x)$ is the greatest common divisor of $f(x)$ and $g(x)$ if:
1. $d(x)$ divides both $f(x)$ and $g(x)$,
2. if $d_1(x)$ is another polynomial that divides $f(x)$ and $g(x)$, then $d_1(x)$ divides $d(x)$.

If $d(x)$ is the greatest common divisor of $f(x)$ and $g(x)$, then it is denoted by $d(x) = \gcd(f(x), g(x))$. When $d(x) = 1$, the polynomials $f(x)$ and $g(x)$ are called *relatively prime*.

The existence of the greatest common divisor of two polynomials will be established in Theorem A.4.2.

Theorem A.4.2. *If $f(x)$ and $g(x)$ are polynomials with at least one of them not zero, then the greatest common divisor $\gcd(f(x), g(x))$ exists and can be represented in the form*

$$\gcd(f(x), g(x)) = h_0(x)f(x) + k_0(x)g(x), \tag{A.26}$$

for some polynomials $h_0(x)$ and $k_0(x)$.

Proof. The existence and representation of $\gcd(f(x), g(x))$ is obtained in the same step.

Let $T = \{f(x)h(x) + g(x)k(x) \mid h(x) \text{ and } k(x) \text{ are polynomials}\}$. By taking appropriate polynomials $h(x)$ and $k(x)$, we notice that $f(x), g(x) \in T$. Thus T has nonzero elements. By the well-ordering principle, Principle A.1.3, applied to the set of degrees of polynomials in T, we can choose one with minimum degree. Let us denote it by $d(x)$. Since $d(x) \in T$, then $d(x) = f(x)h_0(x) + k_0(x)g(x)$, for some polynomials $h_0(x)$ and $k_0(x)$.

We claim that $d(x) = \gcd(f(x), g(x))$. First, we will prove that $d(x)$ divides $f(x)$ and $g(x)$; in fact, we will prove more, namely that $d(x)$ divides every element of T.

If $f_1(x) = f(x)h(x) + k(x)g(x) \in T$, then, applying Theorem A.4.1 to $f_1(x)$ and $d(x)$, one has

$$f_1(x) = q(x)d(x) + r(x), \quad \text{with } \deg(r(x)) < \deg(s(x)) \text{ or } r(x) = 0. \tag{A.27}$$

Now using the representation of $f_1(x)$ and $d(x)$, we can see that $r(x) \in T$. Since $d(x)$ is the polynomial of minimum degree, we must have $r(x) = 0$.

If $d_1(x)$ divides $f(x)$ and $g(x)$, then it divides $f(x)h(x) + k(x)g(x)$ for any choice of $h(x)$ and $k(x)$, in particular, $d_1(x)$ divides $d(x) = f(x)h_0(x) + k_0(x)g(x)$, finishing the proof of the theorem. ☐

Corollary A.4.1. *Let $f(x)$ and $g(x)$ be two polynomials. Then $f(x)$ and $g(x)$ are relatively prime if and only if there are polynomials $f_0(x)$ and $g_0(x)$ such that*

$$f(x)f_0(x) + g_0(x)g(x) = 1. \tag{A.28}$$

Proof. The proof of the corollary follows directly from the representation of

$$\gcd(f(x), g(x)) = f_0(x)f(x) + g_0(x)g(x).$$ □

Theorem A.4.2 and its corollary can be generalized to any finite number of polynomials $f_1(x), f_2(x), \ldots, f_n(x)$. Their proofs follow the same lines as above.

Theorem A.4.3. *If $f_1(x), f_2(x), \ldots, f_n(x)$ are polynomials with at least one of them not zero, then the greatest common divisor, $\gcd(f_1(x), f_2(x), \ldots, f_n(x))$, can be represented in the form*

$$\gcd(f_1(x), f_2(x), \ldots, f_n(x)) = h_1(x)f_1(x) + h_2(x)f_2(x) + \cdots + h_n(x)f_n(x), \quad (A.29)$$

for some polynomials $h_1(x), h_2(x), \ldots, h_n(x)$.

Corollary A.4.2. *Let $f_1(x), f_2(x), \ldots, f_n(x)$ be polynomials. Then $f_1(x), f_2(x), \ldots, f_n(x)$ are relatively prime if and only if there are polynomials $g_1(x), g_2(x), \ldots, g_n(x)$ such that*

$$g_1(x)f_1(x) + g_2(x)f_2(x) + \cdots + g_n(x)f_n(x) = 1. \quad (A.30)$$

Corollary A.4.3. *Let $f_1(x), f_2(x), \ldots, f_n(x)$ be pairwise relatively prime polynomials. Assume that for every $j \in \{1, 2, \ldots, n\}$, $f_j(x)$ divides $f(x)$. Then $f_1(x)f_2(x) \cdots f_n(x)$ divides $f(x)$.*

Sketch. From the cited corollary, there are polynomials $g_1(x), g_2(x)$ such that

$$g_1(x)f_1(x) + g_2(x)f_2(x) = 1. \quad (A.31)$$

Multiplying equation (A.31) by $f(x)$, we obtain

$$f(x)g_1(x)f_1(x) + f(x)g_2(x)f_2(x) = f(x). \quad (A.32)$$

Now, using the assumption that $f_j(x)$ divides $f(x)$ for $j = 1, 2$, we have that $f(x) = f_j(x)h_j(x)$. Hence, equation (A.32) becomes

$$f_2(x)h_2(x)g_1(x)f_1(x) + f_1(x)h_1(x)g_2(x)f_2(x) = f(x). \quad (A.33)$$

Factorizing $f_1(x)f_2(x)$ in the left-hand side of (A.33), the conclusion follows. □

Another important term that is used when considering divisibility of polynomials is the *least common multiple* of two or more nonzero polynomials.

Definition A.4.3. Given nonzero polynomials $f(x)$ and $g(x)$, we say that the polynomial $m(x)$ is the least common multiple of $f(x)$ and $g(x)$ if:
1. $f(x)$ and $g(x)$ divide $m(x)$,
2. if $f(x)$ and $g(x)$ divide $m_1(x)$ then $m(x)$ divides $m_1(x)$.

The least common multiple of $f(x)$ and $g(x)$ will be denoted by LCM$\{f(x), g(x)\}$.

Theorem A.4.4. *Let $f(x)$ and $g(x)$ be nonzero polynomials. Then*

$$\text{LCM}\{f(x), g(x)\} = \frac{f(x)g(x)}{\gcd(f(x), g(x))}. \tag{A.34}$$

Proof. Let $m(x) = \text{LCM}\{f(x), g(x)\}$. Clearly, $f(x)$ and $g(x)$ divide $\frac{f(x)g(x)}{\gcd(f(x), g(x))}$, thus Definition A.4.3(2) implies that $m(x)$ divides $\frac{f(x)g(x)}{\gcd(f(x), g(x))}$. On the other hand, $f(x)$ and $\frac{g(x)}{\gcd(f(x), g(x))}$ are relatively prime and both divide $m(x)$. Applying Corollary A.4.3, we obtain that $\frac{f(x)g(x)}{\gcd(f(x), g(x))}$ divides $m(x)$, finishing the proof. □

A recursive procedure to compute the least common multiple of several polynomials is given in the next result.

Theorem A.4.5. *Let $f_1(x), f_2(x), \ldots, f_n(x)$ be nonzero polynomials. Then*

$$\text{LCM}\{f_1(x), f_2(x), \ldots, f_n(x)\} = \text{LCM}\{\text{LCM}\{f_1(x), f_2(x), \ldots, f_{n-1}(x)\}, f_n(x)\}. \tag{A.35}$$

Proof. Define the polynomials $m(x)$ and $m_1(x)$ as follows:

$$m(x) = \text{LCM}\{f_1(x), f_2(x), \ldots, f_n(x)\},$$
$$m_1(x) = \text{LCM}\{\text{LCM}\{f_1(x), f_2(x), \ldots, f_{n-1}(x)\}, f_n(x)\}.$$

Using Definition A.4.3(2), one verifies that $m(x)$ divides $m_1(x)$ and $m(x)$ divides $m_1(x)$; with this the proof is finished. □

An important result concerning the arithmetic of polynomials is the analogue to the fundamental theorem of arithmetic. In the proof of this result, we will use a version of the well-known theorem, credited to Euclid, concerning the divisibility by primes.

Recall that a polynomial $p(x)$ of degree $n \geq 1$ is called irreducible if it cannot be divided by a polynomial $q(x)$ whose degree is positive and less than n.

Theorem A.4.6. *Let $f(x)$, $g(x)$, and $p(x)$ be polynomials, with $p(x)$ irreducible. If $p(x)$ divides $f(x)g(x)$, then $p(x)$ divides at least one of $f(x)$ or $g(x)$.*

Proof. From the assumption, we have that $f(x)g(x) = p(x)q(x)$ for some $q(x)$. If $p(x)$ does not divide $f(x)$, then $\gcd(f(x), p(x)) = 1$. Hence, from Corollary A.4.1, there are polynomials $h(x)$ and $k(x)$ so that

$$f(x)h(x) + p(x)k(x) = 1. \tag{A.36}$$

Multiplying equation (A.36) by $g(x)$, using $f(x)g(x) = p(x)q(x)$, and performing a few calculations, we obtain

$$p(x)(q(x)h(x) + g(x)k(x)) = g(x), \tag{A.37}$$

from where we see that $p(x)$ divides $g(x)$, finishing the proof. □

Theorem A.4.7 (Unique factorization of polynomials). *Let $f(x)$ be a polynomial of positive degree. Then $f(x)$ can be represented in the form*

$$f(x) = p_1^{e_1}(x)p_2^{e_2}(x)\cdots p_k^{e_k}(x), \tag{A.38}$$

where $p_i(x)$ is irreducible for every $i \in \{1, 2, \ldots, k\}$, $p_i(x) \neq p_j(x)$ for every $i \neq j$. Furthermore, the representation in (A.38) is unique up to order and nonzero constants.

Proof. Existence of the factorization. We proceed by induction on the degree of $f(x)$. If $\deg(f(x)) = 1$, then $f(x)$ is irreducible and the factorization in (A.38) has only one factor.

We may assume that $\deg(f(x)) = n > 1$ and the result to be true for every polynomial whose degree is $< n$. If $f(x)$ is irreducible, then there is nothing to prove, since (A.38) is satisfied. Hence we may assume that $f(x) = g(x)h(x)$, with $g(x)$ and $h(x)$ polynomials satisfying $1 \leq \deg(g(x)), \deg(h(x)) < n$. From the induction assumption, each of $g(x)$ and $h(x)$ can be factorized as the product of irreducible polynomials. Hence so can $f(x)$, proving the existence of the factorization.

Uniqueness of the factorization. Let us assume that

$$f(x) = p_1^{e_1}(x)p_2^{e_2}(x)\cdots p_k^{e_k}(x) = q_1^{a_1}(x)q_2^{a_2}(x)\cdots q_m^{a_m}(x), \tag{A.39}$$

with $p_i(x)$ and $q_j(x)$ irreducible polynomials for all $i \in \{1, 2, \ldots, k\}, j \in \{1, 2, \ldots, m\}$, and e_i, a_j positive integers.

Let l be the minimum of k and m. By the existence part, already proved, we have that $l \geq 1$.

We will apply induction on l. Assume that $l = 1$ and, without loss of generality, $k = l$. Then equation (A.39) becomes

$$f(x) = p_1^{e_1}(x) = q_1^{a_1}(x)q_2^{a_2}(x)\cdots q_m^{a_m}(x). \tag{A.40}$$

If $e_1 = 1$, then the irreducibility assumption on $p_1(x)$ implies that $m = 1$ and $a_1 = 1$.

If $e_1 > 1$, using that $p_1(x)$ is irreducible and applying Theorem A.4.6 to equation (A.40), we conclude that $p_1(x)$ divides some $q_j(x)$. We may assume that $j = 1$. Since $p_1(x)$ and $q_1(x)$ are irreducible, necessarily $p_1(x) = q_1(x)$. Also $e_1 = a_1$, otherwise, we may assume that $e_1 < a_1$ and, canceling out $p_1^{e_1}$ in equation (A.40), we have

$$1 = q_1^{a_1-e_1}q_2^{a_2}(x)\cdots q_m^{a_m}(x), \tag{A.41}$$

which is not possible, unless $a_1 = e_1$ and $m = 1$. Hence, we have finished the proof for the case $l = 1$.

We may assume that $l > 1$ and the result to be true for all integers $s < l$, where l is the minimum number of factors in (A.39).

Since $p_1(x)$ is irreducible, applying Theorem A.4.6 in equation (A.39), we have that $p_1(x)$ divides $q_j(x)$ for some $j \in \{1, 2, \ldots, m\}$. We may assume that $j = 1$. Arguing as before, we obtain that $p_1(x) = q_1(x)$ and $e_1 = a_1$.

From this and equation (A.39), we obtain

$$p_2^{e_2}(x) \cdots p_k^{e_k}(x) = q_2^{a_2}(x) \cdots q_m^{a_m}(x). \tag{A.42}$$

Now, applying the induction hypothesis to (A.42), we conclude that $k - 1 = m - 1$, $e_i = a_j$, and $p_i(x) = q_j(x)$ for every $i, j \in \{2, 3, \ldots, l\}$. With this we have finished the proof of the theorem. □

Remark A.4.2. It is interesting to note that Theorems A.4.6 and A.4.7 are equivalent.

A.5 Fundamental theorem of algebra

Solving polynomial equations is an old problem. Historians date it at least to Babylonian times, some 3800 years ago.

Here we state the main result concerning the existence of solutions to polynomial equations in one variable – the fundamental theorem of algebra.

If $p(x)$ is a polynomial and c is a real or complex number so that $p(c) = a_0 + a_1c + \cdots + a_nc^n = 0$, c is called a *root* or *zero* of $p(x)$. For instance, if $p(x) = x^3 - 1$ and $c = \frac{-1+\sqrt{3}i}{2}$, one verifies that c is a root of $p(x)$.

The next result provides necessary and sufficient conditions under which an element c is a root.

Theorem A.5.1 (Factor theorem). *Let $p(x)$ be a polynomial and let c be a real or complex number. Then c is a root of $p(x)$, if and only if $x - c$ divides $p(x)$.*

Proof. Applying Theorem A.4.1 to $p(x)$ and $x - c$, we have that there are polynomials $q(x)$ and $r(x)$ so that

$$p(x) = q(x)(x - c) + r(x), \tag{A.43}$$

with $\deg(r(x)) < \deg(x - c) = 1$. From equation (A.43), we have that $r(x) = p(c)$, hence $p(c) = 0$ if and only if $f(x) = (x - c)q(x)$, that is, c is a root of $p(x)$ if and only if $x - c$ divides $p(x)$. □

From Theorem A.5.1 we have that if c is a root of $p(x)$, then $p(x) = (x - c)q(x)$, for some polynomial $q(x)$. It might be the case that c is also a root of $q(x)$. Hence, continuing this way, we have that $p(x) = (x - c)^k q(x)$, with $k \geq 1$ and $q(c) \neq 0$. In this case we say that c is a root with multiplicity k. When $k = 1$, c is called a simple root.

Corollary A.5.1. *If $p(x)$ is a polynomial of degree n, then $p(x)$ has at most n roots, counting multiplicities.*

Proof. We argue by induction on $\deg(p(x)) = n$. If $n = 1$, say $p(x) = ax + b$, then the only root of $p(x)$ is $c = -\frac{b}{a}$ and the result is true.

Assume that $n > 1$ and the result to be true for polynomials of degree at most $n - 1$. If c is a root of $p(x)$, then Theorem A.5.1 implies that $p(x) = (x - c)q(x)$, with $q(x)$ a polynomial of degree $n - 1$. Notice that the remaining roots of $p(x)$ are exactly the roots of $q(x)$. By the induction assumption, $q(x)$ has at most $n - 1$ roots, hence $p(x)$ has at most n roots. □

The next result is useful when trying to find out if a polynomial has multiple roots. We need to recall what the derivative of a polynomial is. If $p(x) = a_n x^n + a_{n-1} x^{n-1} + \cdots + a_1 x + a_0$, the derivative of $p(x)$ is $p'(x) = n a_n x^{n-1} + (n-1) a_{n-1} x^{n-2} + \cdots + 2 a_2 x + a_1$.

Theorem A.5.2. *Let $p(x)$ be a polynomial and let $p'(x)$ be its derivative. Assume that c is a root of $p(x)$. Then c has multiplicity $k > 1$, if and only if $p'(c) = 0$.*

Proof. Since c is a root of $p(x)$,

$$p(x) = (x - c)^k q(x), \tag{A.44}$$

with $k \geq 1$ and $q(c) \neq 0$. From the product rule to compute derivatives and equation (A.44), we obtain

$$p'(x) = k(x - c)^{k-1} q(x) + (x - c)^k q'(x). \tag{A.45}$$

We also have that $x - c$ is not a factor of $q(x)$, since $q(c) \neq 0$. Hence, from equation A.45 we have that $x - c$ is a factor of $p'(x)$, if and only if $k > 1$, that is, $p'(c) = 0$ if and only if $k > 1$, finishing the proof of the theorem. □

Corollary A.5.2. *Let $p(x)$ be a polynomial and let $p'(x)$ be its derivative. The polynomial $p(x)$ has multiple roots if and only if $d(x) = \gcd(p(x), p'(x))$ has a factor of degree one.*

Proof. Assume that c is a multiple root of $p(x)$. From Theorem A.5.2, c is also a root of $p'(x)$. Now from Theorem A.5.1, $x - c$ is a factor of $p(x)$ and a factor of $p'(x)$. From Definition A.4.2, we have that $x - c$ is a factor of $d(x) = \gcd(p(x), p'(x))$.

Conversely, if $x - c$ is a linear factor of $d(x)$, then $x - c$ is a factor of $p(x)$ and of $p'(x)$. From this we have that c is a root of $p(x)$ as well as of $p'(x)$. Hence, c is a multiple root of $p(x)$. □

The next result is one of the most important theorems in mathematics. Its proof is out of the scope of this appendix. There are many proofs of this result, none of which can be presented in a first approach to the theory of polynomials. The interested reader can find several proofs by searching the web. For a short and nice proof, I recommend to consult [2].

Theorem A.5.3 (Fundamental theorem of algebra). *If $p(x)$ is a polynomial with complex coefficients of degree $n \geq 1$, then $p(x)$ has n roots in $\mathbb{C}$.*

An important result for polynomials with real coefficients is the following.

Theorem A.5.4. *If $p(x)$ is a polynomial with real coefficients and z is a nonreal complex root of $f(x)$, then the conjugate of z is also a root of $p(x)$.*

Proof. Let us write $p(x) = a_0 + a_1 x + \cdots + a_n x^n$. Since z is a root of $p(x,)$, we have $p(z) = a_0 + a_1 z + \cdots + a_n z^n = 0$. Now applying Theorem A.3.1 to this equation, we have

$$\overline{a_0 + a_1 z + \cdots + a_n z^n} = \overline{a_0} + \overline{a_1}\,\overline{z} + \cdots + \overline{a_n}\,\overline{z}^n = 0.$$

On the other hand, since each a_j is real, $a_j = \overline{a_j}$. From this we have that $p(\overline{z}) = 0$, that is, $\overline{z}$ is a root of $p(x)$. $\qquad\square$

Corollary A.5.3. *The only irreducible polynomials whose coefficients are real numbers are linear or quadratic.*

Proof. Let $p(x)$ be an irreducible polynomial with real coefficients. If $p(x)$ has a real root, say c, then, by Theorem A.5.1, $x - c$ divides $p(x)$, that is, $p(x) = (x - c)q(x)$. The irreducibility condition on $p(x)$ implies that $q(x)$ must be a constant, so $p(x)$ has degree one.

If $z = a + bi$ is a nonreal root of $p(x)$, then, by Theorem A.5.4, $\overline{z} = a - bi$ is also a root of $p(x)$. Applying Theorem A.5.1, we have that $(x - (a + bi))(x - (a - bi))$ divides $p(x)$, but $(x - (a + bi))(x - (a - bi)) = x^2 - 2ax + a^2 + b^2$ is a polynomial with real coefficients. The irreducibility condition on $p(x)$ implies $p(x) = x^2 - 2ax + a^2 + b^2$, finishing the proof. $\quad\square$

Remark A.5.1. Corollary A.5.3 is the foundation for the partial fractions decomposition method when computing primitives of real rational functions.

Descartes' rule of signs

One important result that can provide a way to give an estimate of the number of real roots of a polynomial with real coefficients is Descartes' rule of signs. This is especially useful because its applications rely only on counting the number of changes of sign in the coefficients of the polynomial. Here we present a short proof of Descartes' rule of signs, due to X. Wang [27], who also refers to a proof of the same result in [6].

Lemma A.5.1. *Let $p(x) = (x - a)^l q(x)$ be a polynomial with real coefficients and $q(a) \neq 0$. Then $p(x)$ changes sign in a neighborhood of a if and only if l is odd.*

Proof. Since $q(a) \neq 0$ and $q(x)$ is continuous, there is a neighborhood B of a where we have $q(x)q(a) > 0$ for every $x \in B$. From this we can see that $p(x)$ changes sign in B if and only if l is odd. $\qquad\square$

Lemma A.5.2. *Let $p(x) = a_0 x^{b_0} + a_1 x^{b_1} + \cdots + a_n x^{b_n}$ be a polynomial with real coefficients. Additionally, assume that $a_0 a_n \neq 0$ and $0 \leq b_0 < b_1 < b_2 < \cdots < b_n$. Then the number of positive roots of $p(x)$ is even if and only if $a_0 a_n > 0$.*

Proof. We discuss only the case $a_0 > 0$ and $a_n > 0$, since the other is handled by considering $-p(x)$.

Let $z(p)$ denote the number of positive roots of $p(x)$ counting multiplicities. The assumptions $a_0 > 0$ and $a_n > 0$ imply $a_0 = p(0) > 0$ and $\lim_{x \to \infty} p(x) = +\infty$. Hence the number of times that the graph of $p(x)$ crosses the horizontal axis is even. Hence, from Lemma A.5.1, $z(p)$ is even. $\qquad\square$

Theorem A.5.5 (Descartes' rule of signs). *Let $p(x) = a_0 x^{b_0} + a_1 x^{b_1} + \cdots + a_n x^{b_n}$ be a polynomial as in Lemma A.5.2. If $v(p)$ denotes the number of sign changes in the coefficients of $p(x)$ and $z(p)$ denotes the number of positive roots of $p(x)$, then $v(p) = v(p) + 2l$, where $l \geq 0$ is an integer.*

Proof. Without loss of generality, we may suppose that $b_0 = 0$ and proceed by induction on n. If $n = 1$, then $p(x) = a_0 + a_1 x^{b_1}$ and it is straightforward to verify that $z(p) = v(p)$. Assume that $n > 1$ and the result to be true for every integer $m < n$. Two cases need to be considered.

Case I. $a_0 a_1 > 0$. Let $p'(x)$ denote the derivative of $p(x)$, then $v(p) = v(p')$ and, by the induction assumption, $v(p') = z(p') + 2k$, for some integer $k \geq 0$. Also, by Lemma A.5.2, both $z(p)$ and $z(p')$ are even or odd. The same holds for $z(p)$ and $v(p)$. Invoking Rolle's theorem, we have $z(p') \geq z(p) - 1$. It follows that

$$v(p) = v(p') = z(p') + 2k \geq z(p) + 2k - 1 > z(p) + 2k - 2. \tag{A.46}$$

Inequality (A.46) and the condition that $z(p)$ and $v(p)$ have the same parity show that $v(p) = z(p) + 2l$, for some integer $l \geq 0$.

Case II. $a_0 a_1 < 0$. In this case $v(p) = v(p') + 1$ and, by Lemma A.5.2, $z(p)$ and $z(p')$ have different parity. Again, by the induction assumption, $v(p') = z(p') + 2k$. From this condition, we have that $v(p)$ and $z(p)$ have the same parity. Applying Rolle's theorem, one has $z(p') \geq z(p) - 1$. Thus $v(p) = v(p') + 1 \geq z(p') + 1 \geq z(p)$, from which, together with the fact that $v(p)$ and $z(p)$ have the same parity, we conclude that $v(p) = z(p) + 2l$, for some integer $l \geq 0$. $\qquad\square$

B SageMath code to compute the minimal polynomial

B.1 SageMath source code for the minimal polynomial algorithm

```
def anihilator( v, T ):
    if not T.is_square():
        raise TypeException("Not a square matrix.")
    n = T.nrows()
    powers = [identity_matrix(T.base_ring(),n)]
    applications = [v]
    V=VectorSpace(T.base_ring(),n)
    for i in range(n):
        powers.append( T*powers[-1] )
        applications.append( powers[-1]*v )
        coeffs = V.linear_dependence( applications )
        if len(coeffs)>0:
            # the list applications is linearly dependent
            break
    return PolynomialRing( T.base_ring(), 'x')(coeffs[0].list())


def minpoly( A, basis ):
    R = PolynomialRing( A.base_ring(), 'x')
    min_poly = R(1)
    for v in basis:
        if min_poly(A).is_zero():
            break
        min_poly = min_poly.lcm(anihilator(v,A))
    return min_poly.monic()
```

https://doi.org/10.1515/9783111135915-011

Bibliography

[1] R. P. Agarwal and S. K. Sen. *Creators of Mathematical and Computational Sciences*, 1st edition. Springer, Cham–Heidelberg–New York–Dordrecht–London, 2014.

[2] S. Anindya. Fundamental theorem of algebra – yet another proof. *The American Mathematical Monthly*, **107**:842–843 (2000).

[3] S. Axler. Down with determinants. *The American Mathematical Monthly*, **102**:139–154 (1995).

[4] S. Axler. *Linear Algebra Done Right*, 2nd edition. Springer, New York, 1997.

[5] F. Barrera-Mora. *Introducción a la teoría de grupos*, 1st edition. Mexican Mathematical Association, Mexico City, 2004.

[6] J. Bochnak, M. Coste, and M.-F. Roy. *Real Algebraic Geometry*, expanded English version edition. Springer, Berlin–Heidelberg–New York, 1998.

[7] R. Borrelli and C. S. Colleman. *Ecuaciones Diferenciales: Una perspectiva de modelación*. Oxford University Press, México, 2002.

[8] R. Bru, J.-J. Climent, J. Mas, and A. Urbano. *Álgebra Lineal*. Alfaomega, Universidad Politécnica de Valencia, México, 2004.

[9] A. Cayley. A memoir on the theory of matrices. *Philosophical Transactions of the Royal Society of London*, **148**:17–37 (1858).

[10] H. Cohen. *A Course in Computational Algebraic Number Theory*, 1st edition. Springer, Berlin–Heidelberg–New York, 1993.

[11] C. W. Curtis. *Linear Algebra: An Introductory Approach*, 3rd edition. Springer, New York, 1991.

[12] J.-L. Dorier (Editor). *On the Teaching of Linear Algebra*. Kluwer Academic Publisher, Dordrecht, Boston–London, 2000.

[13] L. Edelstein-Keshet. *Mathematical Models in Biology*, republication edition. SIAM, New York, 2005.

[14] G. T. Gilbert. Definite matrices and Sylvester's criterion. *The American Mathematical Monthly*, **98**:44–46 (1991).

[15] L. Hogben. *Canonical forms*. In: Handbook of Linear Algebra, 1st edition. Edited by L. Hogben. CRC Press, Boca Raton, FL, 2006.

[16] D. C. Lay. *Álgebra lineal y sus aplicaciones*. Pearson, México, 1999.

[17] N. Loehr. A direct proof that row rank equals column rank. *The College Mathematics Journal*, **38**:300–301 (2007).

[18] C. D. Martin and M. A. Porter. The extraordinary SVD. *The American Mathematical Monthly*, **119**:838–851 (2012).

[19] A. Mas-Colell, M. Whinston, and J. Green. *Microeconomic Theory*, 1st edition. Oxford University Press, New York, 1995.

[20] J. T. Moore. *Elements of Linear Algebra and Matrix Theory*, 1st edition. McGraw-Hill, New York, 1968.

[21] M. Newman. *Integral Matrices: Pure and Applied Mathematics*, 1st edition. Academic Press, Orlando, Florida, 1972.

[22] S. M. Robinson. A short proof of Cramer's rule. *Mathematics Magazine*, **43**:94–95 (1970).

[23] I. R. Shafarevich and A. O. Remizov. *Linear Algebra and Geometry*, 1st English edition. Springer, Berlin–Heidelberg, 2013.

[24] M. Spivak. *Calculus*, 4th edition. Publish or Perish, Inc., Houston, Texas, USA, 2008.

[25] L. G. Swanson and R. T. Hansen. The equivalence of the multiplication, pigeonhole, induction, and well-ordering principles. *International Journal of Mathematical Education in Science and Technology*, **19**(1):129–131 (1988).

[26] A. Tucker. *A Unified Introduction to Linear Algebra: Models, Methods, and Theory*, 1st edition. Macmillan Publishing Company, New York, 1988.

[27] X. Wang. A simple proof of Descartes' rule of signs. *The American Mathematical Monthly*, **6**:525–526 (2004).

[28] D. E. Whitford and M. S. Klamkin. On an elementary derivation of Cramer's rule. *The American Mathematical Monthly*, **60**:186–187 (1953).

https://doi.org/10.1515/9783111135915-012

Symbols list

$\dim(\mathbb{R}^n)$	dimension of $\mathbb{R}^n$ 102
$\max\{a_1, a_2, \ldots, a_n\}$	maximum of the elements $a_1, a_2, \ldots, a_n$ 66
$a < b$	a is less than b 23
$a \le b$	a is less or equal to b 23
$u \times v$	cross product of u and v 89
$U \oplus W$	direct sum of U and W 113
$:=$	equality by definition 24
$\text{Proj}_v u$	orthogonal projection of u on v 83
$\gcd(f(x), g(x))$	greatest common divisor of $f(x)$ and $g(x)$ 274
$\ge$	greater or equal to 3
$\in$	is an element of or belongs to 26
$\mathbb{Q}$	field of rational numbers 105
$\mathbb{R}$	field of real numbers 26
$\mathbb{R}^2$	cartesian plane 68
$\mathbb{R}^n$	$\mathbb{R}^n$ as a vector space 96
$\mathbb{Z}$	set of integers 105
$\mathcal{M}_{m\times n}(\mathbb{R})$	$m \times n$ matrices with entries in $\mathbb{R}$ 43
$\ne$	is not equal 44
$\notin$	is not an element of 105
$\bar{z}$	conjugate of z 268
$\sim$	row equivalent matrices 13
$\sqrt{a}$	square root of a 70
$\text{diag}\{d_1, d_2, \ldots, d_r\}$	diagonal matrix 63
$\text{Im}(z)$	imaginary part of z 268
$\text{LCM}\{f(x), g(x)\}$	least common multiple of $f(x)$ and $g(x)$ 276
$\text{Re}(z)$	real part of z 268
$\text{tr}(A)$	trace of A 52
$\underbrace{\mathbb{R}^n \times \cdots \times \mathbb{R}^n}_{n \text{ times}}$	cartesian product of $\mathbb{R}^n$ n times 152
$\sum_{i=1}^{n} a_i$	sum of elements $a_1, a_2, \ldots, a_n$ 52
$\|a\|$	norm of a 80
$|A|$	determinant of matrix A 148
$|x|$	absolute value of x 83
$A \subseteq B$	A subset of B 105
$A \cap B$	intersection of A and B 105
$A \cup B$	union of A and B 105
A^*	adjoint of A 258
A^t	transpose of the matrix A 52
A^{-1}	multiplicative inverse of A 45
$A_a(x)$	T-annihilator of a 173
$C(\beta, T)$	T-cyclic subspace spanned by β 173
$f_A(x)$	characteristic polynomial of A 166
$m_T(x)$	minimal polynomial of T 168
N_T	kernel of T 127
R_T	range or image of T 127
T^*	adjoint of T 257
T^n	composition of T with itself n times 140

https://doi.org/10.1515/9783111135915-013

$T_1 \circ T$	composition of T_1 and T 131
$U \setminus W$	difference of sets U and W 109
$V \cong W$	isomorphism between V and W 128
V^*	dual space of V 141
$W(g_1, g_2, \ldots, g_n)(x)$	wronskian of the functions $g_1, g_2, \ldots, g_n$ 158
$W^\perp$	orthogonal complement of W 229
$\mathcal{L}(S)$	subspace spanned by S 105
$\mathcal{L}(V; W)$	linear transformations from V to W 138
Locus	A set of points that share a property 71

Index

https://doi.org/10.1515/9783111135915-014